ENCYCLOPEDIA

OF

ELECTROCHEMISTRY

OF THE ELEMENTS

VOLUME VI

ORGANIZATIONAL CHART

	SYMBOL	ELEMENT NO.	VOLUME	CHAPTER
Actinium	Ac	24		
Aluminum	Al	21	VI	3
Americium	Am	24		
Antimony	Sb	11	IV	1
Argon	Ar	1	I	8
Arsenic	As	9	II	2
Astatine	At	6	I	3
Barium	Ba	18	I	7
Berkelium	Bk	24		
Beryllium	Be	16	V	4
Bismuth	Bi	12		
Boron	B	14	II	1
Bromine	Br	5	I	2
Cadmium	Cd	20	I	4
Calcium	Ca	18	I	7
Californium	Cf	24		
Carbon	C			
Cerium	Ce	22	VI	2
Cesium	Cs	15		
Chlorine	Cl	5	I	1
Chromium	Cr	31		
Cobalt	Co	36	III	2
Copper	Cu	38	II	6
Curium	Cm	24		
Dysprosium	Dy	22	VI	2
Einsteinium	Es	24		
Erbium	Er	22	VI	2
Europium	Eu	22	VI	2
Fermium	Fm	24		
Fluorine	F	4	IV	2
Francium	Fr	15		
Gadolinium	Gd	22	VI	2
Gallium	Ga	14		
Germanium	Ge	13	V	5
Gold	Au	40	IV	3
Hafnium	Hf	26		
Helium	He	1	I	8
Holmium	Ho	22	VI	2
Hydrogen	H	2		
Indium	In	14	VI	1
Iodine	I	6	I	3
Iridium	Ir	41	VI	5
Iron	Fe	37		
Krypton	Kr	1	I	8
Lanthanum	La	22	VI	2
Lawrencium	Lw	24		
Lead	Pb	28	I	5
Lithium	Li	15		
Lutetium	Lu	22	VI	2
Magnesium	Mg	17		
Manganese	Mn	34	I	6
Mendelevium	Md	24		

ORGANIZATIONAL CHART

	Symbol	Element No.	Volume	Chapter
Mercury	Hg	10		
Molybdenum	Mo	32	V	3
Neodymium	Nd	22	VI	2
Neon	Ne	1	I	8
Neptunium	Np	24		
Nickel	Ni	35	III	3
Niobium	Nb	30	II	3
Nitrogen	N	8		
Nobelium	No	24		
Osmium	Os	41	VI	6
Oxygen	O	3	II	5
Palladium	Pd	41	VI	7
Phosphorus	P	9	III	1
Platinum	Pt	41	VI	4
Plutonium	Pu	24		
Polonium	Po	7	IV	9
Potassium	K	15		
Praseodymium	Pr	22	VI	2
Promethium	Pm	22	VI	2
Protactinium	Pa	24		
Radium	Ra	18	I	7
Radon	Rn	1	I	8
Rhenium	Re	42	II	4
Rhodium	Rh	41	VI	9
Rubidium	Rb	15		
Ruthenium	Ru	41	VI	8
Samarium	Sm	22	VI	2
Scandium	Sc	22	VI	2
Selenium	Se	7	IV	7
Silicon	Si	13	V	6
Silver	Ag	39		
Sodium	Na	15		
Strontium	Sr	18	I	7
Sulfur	S	7	IV	6
Tantalum	Ta	30	II	3
Technetium	Tc	42	II	4
Tellurium	Te	7	IV	8
Terbium	Tb	22	VI	2
Thallium	Tl	14	IV	4
Thorium	Th	24		
Thulium	Tm	22	VI	2
Tin	Sn	27	IV	5
Titanium	Ti	25	V	7
Tungsten	W	33	V	2
Uranium	U	24		
Vanadium	V	29		
Xenon	Xe	1	I	8
Ytterbium	Yb	22	VI	2
Yttrium	Y	22	VI	2
Zinc	Zn	19	V	1
Zirconium	Zr	26		

ENCYCLOPEDIA OF ELECTROCHEMISTRY OF THE ELEMENTS

EDITOR

Allen J. Bard

Department of Chemistry
University of Texas
Austin, Texas

VOLUME VI

Al Pt
In Rh
Ir Ru
Os Sc
Pd Y
Lanthanides

MARCEL DEKKER, INC. New York and Basel

MARCEL DEKKER, INC.
270 Madison Avenue, New York, New York 10016

LIBRARY OF CONGRESS CATALOG CARD NUMBER 73-88796

ISBN 0-8247-2506-9

PRINTED IN THE UNITED STATES OF AMERICA

This volume is dedicated to the memory of

Klaus J. Vetter

whose many contributions in the field of electrochemistry and service on the advisory board of this series will be remembered with deep appreciation.

CONTENTS

LIST OF CONTRIBUTORS

F. COLOM, Department of Surface Chemistry and Electrochemistry, Insitute of Physical Chemistry, C. S. I. C., Madrid, Spain

HARVEY B. HERMAN, Department of Chemistry, University of North Carolina at Greensboro, Greensboro, North Carolina

J. F. LLOPIS,* Department of Surface Chemistry and Electrochemistry, Insitute of Physical Chemistry, C. S. I. C., Madrid, Spain

V. V. LOSEV, Karpov Institute of Physical Chemistry, Moscow, USSR

PAUL A. MALACHESKY,† Tyco Laboratories, Inc., Chemistry Department, Waltham, Massachusetts

A. I. MOLODOV, Karpov Institute of Physical Chemistry, Moscow, USSR

J. R. RAIRDEN, General Electric Research and Development Center, General Electric Company, Schenectady, New York

I. M. TORDESILLAS, Department of Surface Chemistry and Electrochemistry, Institute of Physical Chemistry, C. S. I. C., Madrid, Spain

* Deceased

† Present Address: Exxon Research and Engineering Co., Linden, New Jersey

INTRODUCTION

The aim of this series is to provide a critical, systematic, and comprehensive review of the electrochemical behavior of the elements and their compounds. Part I deals with inorganic electrochemistry and Part II with organic electrochemistry. The field of electrochemistry has undergone extensive growth in recent years and electrochemical techniques and concepts have been applied to many areas of basic and applied research and technology. While the fundamentals of electrochemistry and the recent advances in this field have been described in a number of textbooks, monographs, and review series, the vast literature concerning the descriptive electrochemistry of inorganic and organic compounds has been relatively neglected. This series is designed to provide the natural starting point for new electrochemical investigations and to suggest areas where further research is needed. Each chapter, written by experts on that subject, contains the best available information on the electrochemical behavior and applications of the element.

The classification of elements generally follows the scheme used in Gmelin, with some grouping and rearrangement. Each element is assigned a number (n). This number determines in what chapter a compound or alloy will be considered; a chapter on element n will discuss compounds of element n with elements of number lower than n. For example, zinc-amalgam is treated in the zinc chapter (element number 19), rather than in the mercury chapter (element number 10). A listing of the elements and the element numbers is given in the organizational chart at the front of the book. To minimize delays in publication of the chapters, these are published in the order received. The volume and chapter number for each element are also located in the chart.

The chapters are generally organized into five sections:

1. Introduction and Standard Potentials
   - 1.1. Aqueous Solutions
   - 1.2. Nonaqueous Solvents
   - 1.3. Fused Salts
   - 1.4. Other Data
2. Voltammetric Characteristics

   Presentation of Polarographic and Other Voltammetric Results

3. Kinetic Parameters and Double-Layer Properties

Presentation of available rate constants, exchange current densities, transfer coefficients, potential of zero charge, double-layer capacities, etc.

4. Electrochemical Studies

A survey and critical review of the electrochemical reactions of the element and its compounds. Description of known mechanisms of the electrode reactions, oxidation and reduction products, current efficiencies, reaction orders, etc. Discussion of passivation phenomena, oxide films, and anodization.

5. Applied Electrochemistry

The use of electrochemistry in the isolation or purification of the element, the electrochemical production of compounds, and the use and behavior of the element and its compounds in electrochemical devices.

5.1. Electrowinning and Electrorefining
5.2. Electrodeposition, Electroplating, and Electropolishing
5.3. Electrosynthesis
5.4. Corrosion
5.5. Batteries and Cells
5.6. Other

Conventions generally follow the recommendations of IUPAC [1]. Half reactions corresponding to a standard electrode potential are uniformly written as reductions and the sign of the electrode potential follows the Gibbs-Stockholm convention [2]. Notation generally follows that in the "Manual of Symbols and Terminology for Physicochemical Quantities and Units" [3]. Modified and frequently-used symbols are given below.

1. POTENTIALS

E: Potential vs NHE, SCE, or other reference electrode. When corresponding cell is to be indicated, it may be written

$$E(Zn^{2+}/Zn) \text{ or } E(1/2)Zn^{2+} + e = (1/2)Zn$$

$E°$: standard electrode potential. It is the EMF of the cell in which the reaction is the reduction of the oxidized species by hydrogen, all reactants being in their standard state, e.g.,

$$(1/2)Zn^{2+} + (1/2)H_2 = (1/2)Zn + H^+ \qquad E°(Zn^{2+}/Zn)$$

$E°'$: formal electrode potential or conditional electrode potential corresponding to unit concentration quantities.

η: Overpotential, the deviation of the potential of an electrode from its equilibrium value. Negative values are associated with reductions and positive values are associated with oxidations.

Polarographic and Voltammetric Potentials

$E_{1/2}$: half-wave potential in dc polarography, i.e., potential at which $i = (1/2)i_d$.

$E_{1/4}$, $E_{3/4}$: 1/4 and 3/4 wave potentials in dc polarography.

E_p, $E_{p/2}$: peak and half-peak potential, e.g., in linear scan or cyclic voltammetry. Cathodic and anodic waves are indicated by E_{pc} and E_{pa}.

$E_{\tau/4}$: quarter-wave potential in chronopotentiometry.

E_Z: point of zero charge.

E_{mix}: mixed potential.

2. CURRENT

i: current, instantaneous current, total current.

i_a: anodic current.

i_c: cathodic current.

i_o: exchange current.

i_d: diffusion current.

i_l: limiting current.

i_p: peak current.

Anodic or cathodic waves are indicated by i_{pa} and i_{pc}, i_{da} and i_{dc}, etc. Average value of current is indicated by a bar over the appropriate symbol, e.g., $\bar{i}_d$: average diffusion current.

j: current density, subscripted as above to indicate j_a, anodic; j_c, cathodic; j_o, exchange, etc.

I, $\bar{I}$: diffusion current constant in dc polarography

$$I = i_d/m^{2/3}t_d^{1/6} \qquad \bar{I} = \bar{i}_d/m^{2/3}t_d^{1/6}$$

with i_d in μA, m in mg/s and t in s.

3. CONCENTRATION

C, C_O, C_R: concentration of O, R, ..., at electrode surface. Unless stated otherwise, [O], [R], ..., may be used in discussing equilibria in solution.

C°, C_R°: bulk concentrations, bulk concentration of species R.

$C_R(x, t)$: concentration of species R at distance x from electrode at time t.

4. TIME

t: time.

t_d: drop time of DME.

τ: transition time in chronopotentiometry.

5. RATE CONSTANTS

For the reaction $O + ne = R$:

$$j = i/A = nF(k_a C_R - k_c C_O)$$

n: charge number of elementary step.

k_a: formal or conditional rate constant of anodic (oxidation) reaction at potential E vs reference electrode, in cm/sec (k_a^o refers to value of k_a at E = 0 vs reference electrode).

k_c: formal or conditional rate constant of cathodic (reduction) reaction at potential E vs reference electrode, in cm/sec (k_c^o refers to value of k_c at E = 0 vs reference electrode).

The current convention corresponds to an anodic current positive.

$$k_c = k^\circ \exp\{-\alpha zF(E - E^{\circ\prime})/RT\}$$

$$k_a = k^\circ \exp\{(1 - \alpha)zF(E - E^{\circ\prime})/RT\}$$

k°: formal or conditional standard rate constant for the electrode reaction.

α: the electrochemical charge transfer coefficient for the cathodic reaction.

$1 - \alpha$: the electrochemical charge transfer coefficient for the anodic reaction.

$$j_0 = i_0/A = nFk^\circ C_O{}^{1-\alpha} C_R{}^{\alpha}$$

i_0: formal or conditional exchange current (A).

j_0: formal or conditional exchange current density (A/cm^2).

If the rate constants are determined by extrapolation to zero ionic strength so that they are written in terms of activities (a_O, a_R) in place of concentrations (C_O, C_R), the symbols are primed: k_c', k_a', $k^{\circ\prime}$, j_0', i_0'.

For rate constants corrected for double layer effects, the subscript t is used; k_t°, j_t° where $k_t^\circ = k^\circ \exp - \{(\alpha n - z)F\phi_2/RT\}$.

6. OTHER

A = electrode area (cm^2).

F = Faraday constant (C/mol).

u_i: mobility of species i (cm^2/V/sec).

z_i: charge number of species i, positive for cations and negative for anions.

t_i: transport number of species i.

D_O, D_R: diffusion coefficient of species O, R.

δ: thickness of diffusion layer (cm).

m_O, m_R: mass transport constant of species O, R (cm/sec)

$$m_R = D_R/\delta_R$$

Q = quantity of electricity, charge.

C = capacitance (μF/cm^2).

C_{dl} = double-layer capacitance.

m = rate of mercury flow in DME (mg/sec).

v = potential scan rate (V/sec).

ω = angular frequency of rotation (sec^{-1}).

Abbreviations for words and journals follow the American Chemical Society Handbook for Authors and the Chemical Abstracts Service Guide for Abbreviating Periodical Titles. Some special abbreviations used in the tables in this compilation are given below.

1. TECHNIQUES

E swp: potential sweep.
i swp: current sweep.
CV: cyclic voltammetry
E stp: potentiostatic or potential step.
V stp: voltage step.
i stp: chronopotentiometry, galvanostatic, or current step.
ac pol: ac polarography.
ac harm: ac harmonic method.
farad. imp: faradaic impedance (ac bridge) method.
chramp: chronoamperometry.
chrcoul: chronocoulometry.
coul: coulostatic method.
farad. rect: faradaic rectification method.
dc pol: dc polarography.

2. ELECTRODES

rot: rotating.
hng: hanging.
vib: vibrating.
stat: stationary.
stir: stirred.
bub: bubbling.
soln: solution
drp: drop.
dsk: disk.
wr: wire.
pl: pool.
pwd: powder.
pst: paste.
fl: foil.
rd: rod.
DME: dropping mercury electrode.
SCE: saturated calomel electrode.
NHE: normal hydrogen electrode.

Formula of element or compound is used to complete description of electrode, e.g., rot Pt wr (rotating platinum wire electrode), C pst - stir soln (carbon paste electrode in stirred solution), etc.

3. DESCRIPTIONS OF REACTIONS AND WAVES

rev: reversible.
irr: irreversible.

sl: slightly.

q: quasi.

w: well-defined.

i: ill-defined.

fw: fairly well-defined.

fi: fairly ill-defined.

do: drawn out.

mb: merges with background.

The general philosophy of this compilation was to let each author choose the extent and scope of his chapter, subject only to limitations of format and notation, since it would clearly be impossible to prescribe too closely the type of treatment appropriate for the different elements. Supplementary volumes and, eventually, new editions are planned which will contain new results and corrections. The authors and the editor would welcome comments, suggestions, and corrections for inclusion in the supplements. The editor is indebted to the members of the advisory board, his colleagues, and his students for suggestions and assistance in preparing this volume. Special thanks are due to Mrs. Gaynel Klingemann for her secretarial and administrative efforts.

REFERENCES

1. J. Electroanal. Chem., 7, 417 (1964).
2. A. J. deBethune, J. Electrochem. Soc., 102, 288C (1955); T. S. Licht and A. J. deBethune, J. Chem. Educ., 34, 433 (1957).
3. Pure Appl. Chem., 21, 3 (1970).

Allen J. Bard
Austin, Texas

Chapter VI-1

INDIUM

V. V. LOSEV and A. I. MOLODOV

Karpov Institute of Physical Chemistry
Moscow, USSR

1. STANDARD AND FORMAL POTENTIALS

1.1. AQUEOUS SOLUTIONS*

The standard potential of the couple In/In^{3+} was determined by measuring the emf of cells without transference in chloride [1, 4, 6], sulfate [2-4], and sulfate-perchlorate [7]

*See also Section 1.4.

TABLE 1.1.1. Standard and Formal Potentials in Aqueous Solutions

Half-reaction	Standard or formal potential (V)	Conditions	Refs.
$In^{3+} + 3e = In(Pt)$	-0.336		[1]
	-0.340		[2]
	-0.3358[a]		[3]
	-0.3382[b]		[4]
	-0.343	I = 3 M, pH ≥ 2	[5]
$In^{3+} + 3e = In(Hg)$	-0.334	20°C	[6]
	-0.337		[7]
$In^{2+} + 2e = In$	-1.66 ± 0.06[c]		[8]
$In^{+} + e = In$	-0.126	20°C	[9]

[a] dE°/dT = 0.965 mV/°C (18 to 60°C).

[b] dE°/dT = +40 μV/°C (0 to 60°C).

[c] Versus Cl_2/Cl^- electrode; calculated from emf of the cell $In|InCl_2|Cl_2(g)$.

solutions (Table 1.1.1). The exact determination of E° is complicated by

$$2In + In^{3+} = 3In^{+} \qquad (1.1.1)$$

for the formation of noticeable quantities of unstable monovalent ions whose concentration affects the equilibrium potential of indium [10], while they themselves are easily oxidized by hydrogen ions [3, 5, 9, 11]. Consequently, in acid solutions the steady-state concentration of In^+ and the potential of indium are lower than the equilibrium values [10]. At the same time, measurements of E° in weakly-acidic solutions could be distorted by the hydrolysis of indium ions [1, 5, 5a]. Complications arise in sulfate solutions because of the incomplete dissociation of indium sulfate [12], while in comparatively concentrated chloride solutions they are caused by the ability of In^{3+} to form complexes with Cl^- ions [13, 13a]. The use of cells without liquid junctions [2, 7, 12] can be complicated due to the oxidation of In^+ on the reference electrode if its potential is more positive than the equilibrium potential of the couple In^+/In^{3+}. Reliable results, within the technique of the experiment, are obtained in Ref. 4 which is based on the measurement of the emf of the cell

In(Pt) | $InCl_3$(m), AgCl | Ag in solutions with a sufficiently low concentration of chloride ions (2×10^{-3} to 2×10^{-2} mole/kg). Unfortunately, the authors do not give the pH of the solution so it is not clear how the hydrolysis of indium ions affects the accuracy of the determination of E°. (See Ref. 14 for the influence of the pH on E°.)

The values of the E° of the couples In/In^+, In^+/In^{2+}, and In^{2+}/In^{3+}, calculated on the basis of the titrametric determination of the concentrations of low-valent indium ions in the system In/In^{3+} + 0.1 M $HClO_4$ which are equal to -0.14, -0.40, and -0.49 V, respectively [15], should be distorted since in 0.1 M $HClO_4$ the system In/In^{3+} is not in equilibrium because of the oxidation of In^+ by H^+ ions [10]. In addition, the results of titration could be affected by the existence of minute particles of metallic indium in the solution. These shortcomings also raise doubts about the conclusion of the authors [15] on the existence in the system of a noticeable concentration of In^{2+} ions. Other criticisms of the results of Ref. 15 are also given in Ref. 5. The most reliable data on E° for the couple In/In^+ were obtained by Visco [9] (Table 1.1.1), in which the reaction of the oxidation of In^+ by H^+ ions is taken into account quantitatively. The concentration of In^+ was determined polarographically. It was assumed that $D_{In^+} = D_{Tl^+}$, while the average ionic activity coefficient of $InClO_4$ equals 0.65. The equilibrium constant of Reaction (1.1.1) was calculated as $K = 1.3 \times 10^{-11}$ mole^2/l^2 [9].

The results of measurements of E° of the couple In^+/In^{3+} by means of platinum electrodes [3] and platinum electrodes with the mercury coating Hg(Pt) [5] are questionable. As shown in Ref. 16, in view of the heterogeneous Reaction (1.1.1) on the surface of mercury, which leads to the formation of indium amalgam, the potential of the Hg(Pt) electrode actually relates to the couple $In(Hg)/In^+$. In the case of the Pt electrode, Reaction (1.1.1) apparently also takes place on its surface [16], but in acidic solutions, because of the low hydrogen overvoltage on Pt, the indium formed by Reaction (1.1.1) would be dissolved at the expense of the coupled reaction of H^+ reduction, and the electrode should acquire a mixed potential [16a].

1.2. NONAQUEOUS SOLVENTS

There are indications that in anhydrous acetic acid the potential of the couple In/In^+ does not follow the Nernst equation [17].

1.3. FUSED SALTS

See Table 1.3.1.

TABLE 1.3.1. Emf (V) of the Cell In|$InCl_n$|Cl_2(g) Calculated from Thermodynamic Data for Fused Chlorides[a] [8]

°C	100	200	300	350	400	450	500	550	600	609
n = 1 (In/In^+)	1.654	1.581	1.520	1.492	1.465	1.439	1.414	1.389	1.364	1.360 v = 609°C
n = 2 (In/In^{2+})	1.601	1.528	1.464	1.435	1.407	1.380	1.361 v = 485°C			
n = 3 (In/In^{3+})	1.588	1.519	1.451	1.417	1.384	1.352	1.321 s = 498°C			

[a] v = vaporization; s = sublimation.

1.4. EQUILIBRIUM IN INDIUM-WATER SYSTEM (POURBAIX DIAGRAMS)

The equilibrium E_0 vs pH diagram for the indium-water system (Fig. 1.4.1) is presented for 25°C for solutions in the absence of substances with which indium can form stable complexes or insoluble salts on the basis of the following equations (the numbers of the equations correspond to the numbers of the curves in Fig. 1.4.1) [18]:

1. $In^{3+} + H_2O = InOH^{2+} + H^+$ $\qquad \log\frac{[InOH^{2+}]}{[In^{3+}]} = -3.88 + pH$

2. $InOH^{2+} + H_2O = InO_2^- + 3H^+$ $\qquad \log\frac{[InO_2^-]}{[InOH^{2+}]} = -20.38 + 3\ pH$

3. $In^{2+} + e = In^+$ $\qquad E = -0.400 + 0.0591 \log\frac{[In^{2+}]}{[In^+]}$

4. $In^{3+} + 2e = In^+$ $\qquad E = -0.443 + 0.0295 \log\frac{[In^{3+}]}{[In^+]}$

5. $InOH^{2+} + H^+ + 2e = In^+ + H_2O$ $\qquad E = -0.330 - 0.0295\ pH + 0.0295 \log\frac{[InOH^{2+}]}{[In^+]}$

6. $InO_2^- + 4H^+ + 2e = In^+ + 2H_2O$ $\qquad E = 0.262 - 0.1182\ pH + 0.0295 \log\frac{[InO_2^-]}{[In^+]}$

7. $In^{3+} + e = In^{2+}$ $\qquad E = -0.489 + 0.0591 \log\frac{[In^{3+}]}{[In^{2+}]}$

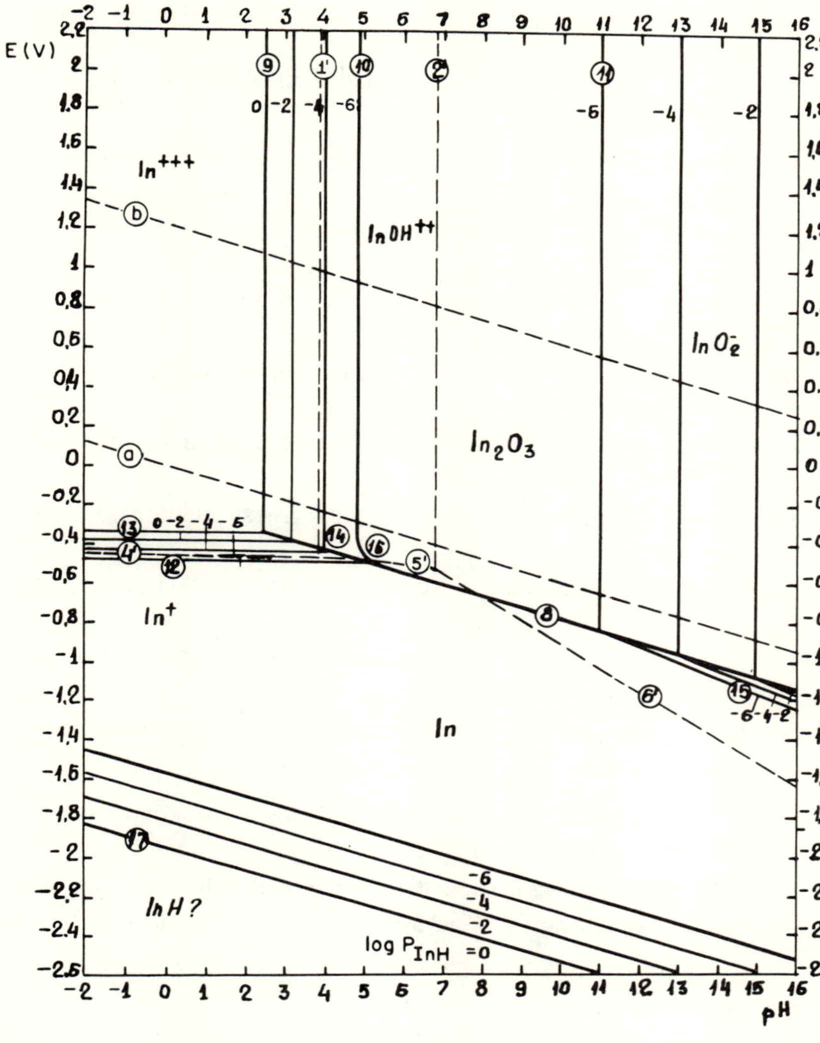

FIG. 1.4.1. Potential vs pH equilibrium diagram for the system indium-water at 25°C (reprinted from Ref. 18).

8. $In_2O_3 + 6H^+ + 6e = 2In + 3H_2O$ 8a. $E = -0.190 - 0.0591\ pH$

8b. $E = -0.172 - 0.0591\ pH$

9. $2In^{3+} + 3H_2O = In_2O_3 + 6H^+$ 9a. $\log[In^{3+}] = 7.73 - 3\ pH$

9b. $\log[In^{3+}] = 8.65 - 3\ pH$

10. $2InOH^{2+} + H_2O = In_2O_3 + 4H^+$	10a. $\log[InOH^{2+}] = 3.85 - 2\,pH$
	10b. $\log[InOH^{2+}] = 4.77 - 2\,pH$
11. $In_2O_3 + H_2O = 2InO_2^- + 2H^+$	11a. $\log[InO_2^-] = -17.05 + pH$
	11b. $\log[InO_2^-] = -16.13 + pH$
12. $In^+ + e = In$	$E = -0.139 + 0.0591 \log[In^+]$
13. $In^{3+} + 3e = In$	$E = -0.342 + 0.0197 \log[In^{3+}]$
14. $InOH^{2+} + H^+ + 3e = In + H_2O$	$E = -0.266 - 0.0197\,pH + 0.0197 \log[InOH^{2+}]$
15. $InO_2^- + 4H^+ + 3e = In + 2H_2O$	$E = 0.146 - 0.0788\,pH + 0.0197 \log[InO_2^-]$
16. $In_2O_3 + 6H^+ + 4e = 2In^+ + 3H_2O$	16a. $E = -0.216 - 0.0886\,pH - 0.295 \log[In^+]$
	16b. $E = -0.189 - 0.0886\,pH - 0.0295 \log[In^+]$
17. $In + H^+ + e = InH$	$E = -1.951 - 0.0591\,pH - 0.0591 \log p_{InH}$

where equations "a" correspond to the anhydrous oxide In_2O_3 and equations "b" to the trihydrated In_2O_3 [or to the hydroxide $In(OH)_3$] .

From Eqs. (1)-(7), assuming the activity of the species to be equal to 1, we obtain the limits of the domains of simple hydrated and hydrolyzed indium ions (dashed line in Fig. 1.4.1):

1'. $In^{3+}/InOH^{2+}$	$pH = 3.88$
2'. $InOH^{2+}/InO_2^-$	$pH = 6.79$
3'. In^+/In^{2+}	$E° = -0.400$
4'. In^+/In^{3+}	$E° = -0.443$
5'. $In^+/InOH^{2+}$	$E° = -0.330 - 0.0295\,pH$
6'. In^+/InO_2^-	$E° = 0.262 - 0.1182\,pH$
7'. In^{2+}/In^{3+}	$E° = -0.489$

The Pourbaix diagram of the active-passive state of indium (Fig. 1.4.2) is plotted in a similar manner.

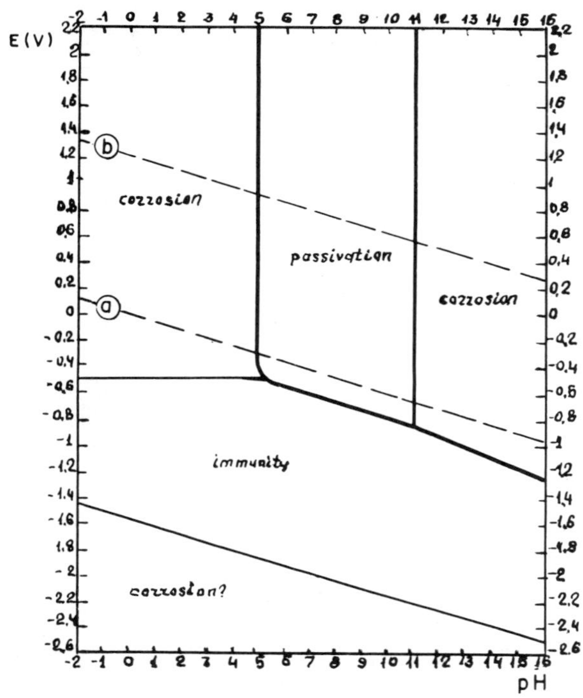

FIG. 1.4.2. Theoretical conditions of corrosion, immunity, and passivation of indium at 25°C (reprinted from Ref. 18).

In estimating the accuracy of the state of the system by the Pourbaix diagrams, it is essential to bear in mind that the values of E° used to plot them were for simple hydrated indium ions from Ref. 15 (see Section 1.1 for critical remarks on this work).

2. VOLTAMMETRIC CHARACTERISTICS

2.1. POLAROGRAPHIC CHARACTERISTICS

See Tables 2.1.1 and 2.1.2.

In acidic noncomplexing solutions, In(III) is reduced irreversibly on the dropping mercury electrode [31, 55]. Thus, in 0.1 M $HClO_4$, the irreversible wave of the indium discharge appears only at E = -1.0 V (SCE). With an increase of the pH this wave shifts to positive potentials, remaining irreversible and only at pH = 2 in the region of potentials near E = -0.55 V does a well-defined limiting current appear on it. It is of a kinetic nature and

TABLE 2.1.1. Polarographic Characteristics

Substance	Product	Conditions	$E_{1/2}$ (V vs SCE)	I	$E_{3/4}-E_{1/2}$ (mV)	Remarks	Refs.
In(III)	0	1 $\underline{M}$ $NaClO_4$ + $HClO_4$ (pH 3), 22°C	-0.52		rev		[19]
In(Hg)	In(III)	0.1 $\underline{M}$ $HClO_4$	(-0.488)			$E_{p/2}$ for E swp	[20]
In(III)	0	5 X 10^{-3} $\underline{M}$ H_2SO_4	-0.985		irr	Second wave	[21]
In(III)	0	0.124 $\underline{M}$ H_2SO_4	-1.069		irr	Second wave	[21]
In(III)	0	1 $\underline{M}$ KNO_3 + HNO_3 (pH 1.0)	-0.485		rev	Kinetic wave	[22]
In(III)	0	1 $\underline{M}$ KNO_3 + HNO_3 (pH 3.6)	-0.512		rev		[22]
In(Hg)	In(III)	2 $\underline{M}$ KNO_3	(-0.47)				[23]
In(III)	0	1 $\underline{M}$ NH_4NO_3	-0.56		irr		[23]
In(III)	0	2 $\underline{M}$ $Ca(NO_3)_2$	-0.480		irr		[24]
In(III)	0	2 $\underline{M}$ $Ba(NO_3)_2$	-0.505		irr		[24]
In(III)	0	0.5 $\underline{M}$ $Al(NO_3)_3$	-0.50		irr		[24]
In(III)	0	0.86 $\underline{M}$ H_3PO_4	-1.10		irr		[25]
In(III)	0	0.1 $\underline{N}$ HCl	-0.500 (NCE)				[26]
In(Hg)	In(III)	1 $\underline{N}$ HCl	(-0.55)			$E_{p/2}$ for E swp	[27]

In(III)	0	2 M HCl + 2 M KBr	-0.43			[28]
In(III)	0	1 M LiCl, 30°C	-0.583w	q-rev	10^{-3}% Triton X100	[29]
In(III)	0	1 N KCl	-0.636 (NCE)	rev		[30]
In(III)	0	1 N KCl	-0.597	rev		[31]
In(III)	0	1 M KBr	-0.58			[32]
In(Hg)	In(III)	1 M HBr	(-0.60)		$E_{p/2}$ for E swp	[27]
In(III)	0	1 M NaI	-0.54		$E_{1/2}$ is practically constant in the range 1 to 5 M NaI	[33]
In(Hg)	In(III)	0.5 M KI	(-0.54)			[34]
In(III)	0	2 M NaSCN	-0.6309	q-rev	0.003% gelatin	[35]
In(III)	0	0.1 M NH_4SCN + 0.1 M $HClO_4$ + $NaClO_4$ (I = 2 M), 30°C	-0.5176			[36]
In(III)	0	1.0 M NH_4SCN + 0.1 M $HClO_4$ + $NaClO_4$ (I = 2 M), 30°C	-0.5885		0.003% gelatin	[36]
In(III)	0	10 N NaOH	-1.38	irr		[37]
In(Hg)	In(III)	1 N NaOH	(-1.10)		$E_{p/2}$ for E swp	[27]
In(Hg)	In(III)	NH_4OH + NH_4Cl (pH 9.5)	(-0.76 to -0.77)		$E_{p/2}$ for E swp	[27]

(continued)

TABLE 2.1.1 (continued)

Substance	Product	Conditions	$E_{1/2}$ (V vs SCE)	I	$E_{3/4} - E_{1/2}$ (mV)	Remarks	Refs.
In(III)	0	1 N NaOH + satd $Na_4P_2O_7$	-1.13		-57	0.002% Triton X100	[37]
In(III)	0	2 M $N_2H_4 \cdot HCl$	-0.69 (NCE)				[38]
In(III)	0	Satd $N_2H_4 \cdot 2HCl$	-0.648		-25		[37]
In(III)	0	1 M N_2H_4 + 1 M NaOH	-1.15		-31		[37]
In(III)	0	0.1 M $C_2H_4(NH_2)_2$(ethylenediamine)	-1.07				[39]
In(III)	0	0.10 M $C_2H_4(NH_2)_2$ + 0.05 M $KClO_4$ + 0.05 M KOH	-1.097			Wave height increases with ethylenediamine concn from 0.01 to 0.25 M, then remains constant	[40]
In(III)	0	0.50 M $C_2H_4(NH_2)_2$ + 0.05 M $KClO_4$ + 0.05 M KOH	-1.107				[40]
In(III)	0	0.50 M $C_2H_4(NH_2)_2$ + 0.5 M K_3PO_4	-1.18				[40]
In(III)	0	0.50 M $C_2H_4(NH_2)_2$ + 0.5 M KCN + 0.05 M KOH	-1.33				[40]
In(III)	0	2 M HCOOH (pH 2.10)	-0.567				[41]
In(III)	0	2 M CH_3COOH (pH 2.23)	-0.578				[41]
In(Hg)	In(III)	0.5 M CH_3COOH	(-0.33)				[34]

In(III)	0	$CH_3COOH + CH_3COONa$ (pH 4.64) $^aCH_3CO_2^-$ ~ I^a	-0.690				[42]
In(III)	0	1 M CH_3COONH_4	-0.70 (NCE)				[43]
In(Hg)	In(III)	1 M CH_3COONH_4	(-0.64)w				[34]
In(III)	0	0.1 M $(COOH)_2$ + NaOH (pH 4.6)	-0.76	4.2	25	0.040% decylamine	[44]
In(III)	0	0.05 M $HOOC-(CH_2)_2-COOH$ (pH 3.30)	-0.581		rev		[45]
In(III)	0	0.1 M $HOOC-CHOH-CH_2-COOH$ (pH 2.70)	-0.585				[45]
In(III)	0	10% $HOOC-(CHOH)_2-COOH$	-0.63 (NCE)				[46]
In(III)	0	0.25 M $NaK(C_4H_4O_6)$	-0.68 (NCE)				[43]
In(III)	0	0.1 M $HOOC-(CHOH)_3-COOH$ (pH 1.62)	-0.527w		rev		[20]
In(III)	0	0.1 M $HOOC-(CHOH)_3-COOH$ + 1 M NaOH	-1.47		irr	Wave height decreases and $E_{1/2}$ shifts to more negative E with increasing pH	[20]
In(III)	0	0.012 M D-Na-gluconate + 0.5 M NaCl (pH 12.7)	-1.220			Kinetic wave	[47]
			-1.572			Second wave	
In(III)	0	0.2 M H_3 Cit (pH 1.09)	-0.545		irr		[48]
		0.2 M H_3 Cit (pH 3.13)	-0.637		irr		[48]
In(Hg)	In(III)	0.5 M EDTA	(-0.52)				[34]

(continued)

TABLE 2.1.1 (continued)

Substance	Product	Conditions	$E_{1/2}$ (V vs SCE)	I	$E_{3/4} - E_{1/2}$ (mV)	Remarks	Refs.
In(III)	0	0.3 M Na-sulfosalicylate + $NaClO_4$ (or KNO_3) (I = 1 M), 30°C	-0.574			0.005% Triton X100	[49]
In(III)	0	0.30 M Na-levulinate + KNO_3 (I = 1 M), 30°C	-0.6050		q-rev		[50]
In(III)	0	0.26 M Na-pyruvate + KNO_3 (I = 1 M), 30°C	-0.6250		q-rev		[50]
In(III)	0	0.50 M Na-butyrate + KNO_3 (I = 1 M), 30°C	-0.6500		q-rev		[50]
In(III)	0	0.28 M Na-glycolate + KNO_3 (I = 1 M), 30°C	-0.6462		rev		[50]
Complex of In(III) with Na-benzoyl-N-phenylhydroxylamine	0	0.1 M LiCl	-0.99			Solvent: 15 ml methanol + 10 ml isoamyl alcohol	[51]
Complex of In(III) with Na-benzoyl-N-phenylhydroxylamine	0	0.1 M LiCl	-0.73			Solvent: 15 ml methanol + 10 ml isoamyl alcohol	[51]

[a] a = activity.

TABLE 2.1.2. Polarographic Characteristics (Fused Salts)

Substance	Product	Conditions	$E_{1/2}$ (V vs SCE)	I	$E_{3/4} - E_{1/2}$ (mV)	Remarks	Refs.
In(III)	0	20 mole % $LiCH_3CO_2$ + 30 mole % $NaCH_3CO_2$ + 50 mole % KCH_3CO_2, 202°C	+0.395	9.88		$E_{1/2}$ vs $E_{1/2}$ of Zn^{2+}-reduction in the same medium	[52]
In(Hg)	In(III)	20 mole % $LiCH_3CO_2$ + 30 mole % $NaCH_3CO_2$ + 50 mole % KCH_3CO_2, 202°C	(+0.401)	9.88			[52]
In(III)	0	$NH_4SO_3NH_2$(sulfamate), 162°C	-0.190	1.22	rev	$E_{1/2}$ vs $E_{1/2}$ of Pb^{2+}-reduction in the same medium	[53]
In(III)	0	$Ca(NO_3)_2 \cdot 4H_2O$ (aqueous melt), 50°C	-0.360 Ag/AgCl (satd)		rev		[54]
In(III)	0	$LiNO_3 \cdot 3H_2O$ (aqueous melt), 47°C	-0.710 (Hg/Hg_2Cl_2) (satd)				[54]
In(III)	0	$CaCl_2 \cdot 6H_2O$ (aqueous melt), 35°C	-0.375 (Ag/AgCl) (satd)		rev		[54]

increases as the pH increases. Finally, at pH 3 a reversible wave appears on the polarogram ($E_{1/2}$ = -0.52 V) and there is a diffusion limiting current [19, 22, 23, 42, 56]. The appearance of a reversible polarographic wave at more positive potentials is also observed when polarizable anions (halide, thiocyanate, acetate, etc.) are added, as well as certain organic compounds [19, 23, 42, 55-60]. Under these conditions the polarographic behavior of indium is greatly affected by adsorption and complex formation. In the presence of Cl^-, Br^-, I^-, and SCN^- ions and certain organic compounds, a decrease of the current is observed on the polarogram of indium reduction and a minimum appears [42, 60-65]; on ac polarograms a negative faradaic admittance appears in the same region of potentials [62, 63, 66, 67]. There is no single point of view on the nature of this effect [19, 23, 42, 59-63, 67-70]. In the presence of the above-mentioned additives a catalytic current appears on the polarogram at potentials of about -0.50 V, which corresponds to the reduction of In(III). It can be used to determine the quantity of the additives with great sensitivity (for instance, I^- up to 10^{-5} M) [19, 71].

Polarographic methods for the quantitative determination of indium are reviewed in the monograph of Busev [13]. Variations of the polarographic method are used for the quantitative determination of indium, specifically in the presence of other elements: ac polarography [72-73], square-wave polarography [74], differential pulse polarography [75], and oscillographic amalgam extraction-stripping polarography [76].

3. KINETIC PARAMETERS AND DOUBLE-LAYER PROPERTIES

3.1. KINETIC PARAMETERS

When E shifts by 40 to 50 mV from E_r, or from the potential of indium discharge on mercury, the cathodic process is distorted by the preceding chemical reactions (see Section 4). Therefore Tables 3.1.1 and 3.1.2 include the kinetic parameters obtained from measurements made at potentials more positive than and close to the equilibrium potential or $E_{1/2}$.

The actual electrochemical reactions of the ionization of indium and the discharge of trivalent indium ions include rapid and slow charge transfer steps. Consequently, the values k_s and α, determined by steady-state and relaxation methods, can correspond to different steps, and this should be borne in mind when analyzing the data of Tables 3.1.1 and 3.1.2 (cf. data for steady-state and impedance measurements in Table 3.1.2). It should also be noted that unlike other methods, dc polarography often gives $\alpha > 0.5$ (see Refs. 29, 35, and 86 in Table 3.1.1). The reason for this deviation is not clear.

TABLE 3.1.1. Kinetic Parameters (Amalgam and Mercury Electrodes)

Reaction	Conditions	Electrode	Technique	Rate constant k_s (cm/sec)	α	Remarks	Refs.
In^{3+} + 3e + Hg = In(Hg)	1×10^{-3} $\underline{F}$ In^{3+} + 0.86 $\underline{F}$ H_3PO_4	DME	oscillogr pol	6.6×10^{-11} [a]	0.60[b]		[25]
	1×10^{-3} $\underline{F}$ In^{3+} + 0.48 $\underline{F}$ $HClO_4$			4.4×10^{-9} [a]	0.66[b]		
In^{3+} + 3e + Hg = In(Hg)	1×10^{-3} $\underline{F}$ In^{3+} + 0.5 $\underline{F}$ ClO_4^-	DME	dc pol	2.2×10^{-12} [a]	0.66[b]	293.2 ± 0.1K	[77]
				4.7×10^{-12} [a]	0.66[b]	299.7 ± 0.1K	
				7.3×10^{-12} [a]	0.65[b]	303.2 ± 0.1K	
In^{3+} + 3e = In(Hg)	0.1 $\underline{M}$ In(Hg); 1×10^{-2} $\underline{M}$ $In(ClO_4)_3$ + ~1×10^{-2} $\underline{M}$ $HClO_4$ + X$\underline{M}$ $NaClO_4$	In(Hg)-stir Pl	Radiotracer			20°C $\Delta H_{(20-58°C)}$ = 47 kJ/mole	[78]
	X = 1.0			$j_0 = 6.9 \times 10^{-4}$			
	X = 3.0			$j_0 = 1.52 \times 10^{-4}$			
	X = 6.1			$j_0 = 2.5 \times 10^{-4}$			
In^{3+} + 3e = In(Hg)	0.1 $\underline{M}$ In(Hg); 1×10^{-2} $\underline{M}$ $In(ClO_4)_3$ + (3 - X) $\underline{M}$ $NaClO_4$ + X$\underline{M}$ $HClO_4$	In(Hg)-stir Pl	Radiotracer and current-potential curves			20°C	[79]
	X = 7.7×10^{-3}			$j_0 = 3.5 \times 10^{-4}$	2.2[c]		
	X = 1.6×10^{-2}			$j_0 = 2.2 \times 10^{-4}$	2.2[c]		
	X = 5.0×10^{-2}			$j_0 = 5.2 \times 10^{-5}$	2.2[c]		

(continued)

TABLE 3.1.1 (continued)

Reaction	Conditions	Electrode	Technique	Rate constant k_s (cm/sec)	α	Remarks	Refs.
$In^{3+} + 3e = In(Hg)$	0.97 M In(Hg); 5.3×10^{-3} M $In(ClO_4)_3$ + 1 M $HClO_4$	In(Hg)-stir Pl	Current-potential curves	$j_0 = 4 \times 10^{-6}$		20°C	[80]
$In^{3+} + 3e = In(Hg)$	0.3 wt% In(Hg); 0.116 N $In_2(SO_4)_3$ + 0.5 N K_2SO_4; pH 2.5	In(Hg) drp	i stp		2.8^d 0.87^e	25 ± 1°C	[81]
$In^{3+} + 3e = In(Hg)$	0.1 M In(Hg); 1×10^{-2} M $In(ClO_4)_3 + HClO_4 + NaClO_4$ (I = 3 M) + X	In(Hg)-stir Pl	Radiotracer and current-potential curves			20°C	[82]
	X = 0.1 M Na_2SO_4, ~10^{-2} M $HClO_4$			$j_0 = 3.0 \times 10^{-4}$	2.2^c		
	X = 0.6 M Na_2SO_4, ~10^{-2} M $HClO_4$			$j_0 = 1.7 \times 10^{-4}$	2.2^c		
	X = 2×10^{-2} M NaF, pH 1.3			$j_0 = 9.1 \times 10^{-5}$	2.3 ± 0.4^c		
	X = 0.1 M NaF, pH 1.3			$j_0 = 1.6 \times 10^{-5}$	2.3 ± 0.4^c		
$In^{3+} + 3e + Hg = In(Hg)$	6 M HCl	Hg drp	farad. rect	0.19-0.21			[83]
$In^{3+} + 3e = In(Hg)$	0.1 M In(Hg); 1×10^{-2} M $In(ClO_4)_3$ + ~10^{-2} M $HClO_4$ + $NaClO_4$ (I = 3 M) + X	In(Hg)-stir Pl	Radiotracer and current-potential curves				[82]

	X = 0.1 M NaCl			$j_0 = 1.7 \times 10^{-4}$	2.3[c]		
	X = 0.5 M NaCl			$j_0 = 1.7 \times 10^{-4}$	2.3[c]		
In^{3+} + 3e + Hg = In(Hg)	1.5×10^{-3} M $In(ClO_4)_3$ + 3 M $NaClO_4$ + $HClO_4$ + X	DME	dc pol			Room temp	[56]
	X = 92.9 mM NaCl, pH 0.0			1.48×10^{-3} [f]	0.41		
	X = 0.6 mM NaCl, pH 1.4			6.4×10^{-4} [f]	0.41		
	X = 0.6 mM NaCl, pH 2.5			1.7×10^{-3} [f]	0.36		
In^{3+} + 3e + Hg = In(Hg)	0.43 mM In^{3+} + 1 M KNO_3 + 0.1 M HNO_3 + 12 mM KCl, pH 3	DME	farad. imp	$(3.1 \pm 0.4) \times 10^{-3}$ [g]	0.90 ± 0.05[h]		[62]
	0.4 mM In^{3+} + 0.1 N HNO_3 + 1 M KCl, pH 3			0.06 ± 0.02	0.9 ± 0.1[h]		
In^{3+} + 3e + Hg = In(Hg)	3×10^{-4} M $In(NO_3)_3$ + 1 M NaCl, pH 2 to 3	DME	farad. imp	3.4×10^{-2}			[84]
In^{3+} + 3e = In(Hg)	0.3 wt% In(Hg); 0.116 N $In_2(SO_4)_3$ + 0.5 N K_2SO_4 + 5×10^{-2} N KCl, pH 2.5	In(Hg) drp	i stp		2.81[e]		[85]
	0.3 wt% In(Hg); 0.116 N $InCl_3$ + 0.5 N KCl, pH 2.5				0.69[e]		
In^{3+} + 3e + Hg = In(Hg)	0.8 mM $InCl_3$ + X + Triton X100 (10^{-3}%)	DME	dc pol			$\Delta H_{(30-40°C)}$ = 7.3 kcal/mole	[86]
	X = 0.1 M $NaClO_4$			1.28×10^{-3} [i]	0.66	30°C	
	X = 0.1 M $NaClO_4$			1.42×10^{-3} [i]	0.67	40°C	
	X = 1.0 M $NaClO_4$			1.21×10^{-3} [i]	0.59	30°C	
	X = 0.1 M $NaNO_3$			1.285×10^{-3} [i]	0.57	30°C	

(continued)

TABLE 3.1.1 (continued)

Reaction	Conditions	Electrode	Technique	Rate constant k_s (cm/sec)	α	Remarks	Refs.
In^{3+} + 3e + Hg = In(Hg)	0.8 m$\underline{M}$ $InCl_3$ + X + Triton X100 (10^{-3}%)	DME	dc pol				[29]
	X = 0.1 $\underline{M}$ LiCl			3.147×10^{-4}	2.06[b]	30°C	
	X = 0.1 $\underline{M}$ LiCl			4.895×10^{-4}	1.93[b]	40°C	
	X = 1.0 $\underline{M}$ LiCl			6.65×10^{-4}	1.50[b]	30°C	
In^{3+} + 3e + Hg = In(Hg)	0.3 m$\underline{M}$ $In(NO_3)_3$ + 1 $\underline{M}$ NaN_3, pH 2 to 3	DME	farad. imp	5.3×10^{-2}		25 ± 0.5°C	[84]
	0.3 m$\underline{M}$ $In(NO_3)_3$ + 1 $\underline{M}$ NaBr, pH 2 to 3			0.65			
In^{3+} + 3e + Hg = In(Hg)	1 $\underline{M}$ KBr	Hg drp	farad. rect	1.33 to 1.56			[83]
In^{3+} + 3e = In(Hg)	0.1 $\underline{M}$ In(Hg); 1×10^{-2} $\underline{M}$ $In(ClO_4)_3$ + ~10^{-2} $\underline{M}$ $HClO_4$ + $NaClO_4$ (I = 3 $\underline{M}$) + X	In(Hg)-stir Pl	Radiotracer and current-potential curves				[82]
	X = 0.1 $\underline{M}$ NaBr			$j_0 = 1.1 \times 10^{-3}$	2.3[c]		
	X = 0.5 $\underline{M}$ NaBr			$j_0 = 2.9 \times 10^{-3}$	2.3[c]		
	X = 1×10^{-3} $\underline{M}$ NaI			$j_0 = 2.6 \times 10^{-4}$ j	2.3[c]		
	X = 5×10^{-3} $\underline{M}$ NaI			$j_0 = 1.26 \times 10^{-3}$ j	2.3[c]		
	X = 2×10^{-2} $\underline{M}$ NaI			$j_0 = 1.9 \times 10^{-3}$	2.3[c]		
In^{3+} + 3e + Hg = In(Hg)	0.4 m$\underline{M}$ In^{3+} + 1 $\underline{M}$ KSCN, pH 3	DME	farad. imp	0.06 to 0.09			[87]

$In^{3+} + 3e + Hg = In(Hg)$	5 m<u>M</u> $In(SCN)_3$ + 1 <u>M</u> NaSCN + HSCN, pH 3.1	DME	farad. imp	~0.3[g]		[88]
$In^{3+} + 3e + Hg = In(Hg)$	1 m<u>M</u> $In(ClO_4)_3$ + 0.1 <u>M</u> $NaClO_4$ + 5 X 10^{-2} <u>M</u> $HClO_4$	DME	dc pol			[67]
	X = 1 X 10^{-4} <u>M</u> NaSCN			~5 X 10^{-5} j		
	X = 1 X 10^{-3} <u>M</u> NaSCN			~4 X 10^{-3} j		
$In^{3+} + 3e + Hg = In(Hg)$	1 m<u>M</u> In^{3+} + ClO_4^- + 2 X 10^{-2} <u>M</u> KSCN + 5 X 10^{-3}% gelatin (I = 2 <u>M</u>)	DME	dc pol	$(0.64 \pm 0.16) \times 10^{-4}$ i 4.12×10^{-12} a	1.66[b]	[35]
$In^{3+} + 3e = In(Hg)$	3 X 10^{-4} <u>M</u> In(Hg); 3 X 10^{-4} <u>M</u> In^{3+} + 11 <u>M</u> NaOH	In(Hg)-stir Pl	Radiotracer and current-potential curves	$j° = 9.2 \times 10^{-3}$ k	0.74[b] 2.19[c]	[89]

[a] k referred to E = 0 vs NHE.

[b] Value of αn.

[c] Value of $(1 - \alpha)n$.

[d] Value of $(1 - \alpha)n$ (calculated from the Tafel slope of the anodic current-potential curve at low j_a).

[e] Value of $(1 - \alpha)n$ (calculated from the Tafel slope of the anodic current-potential curve at high j_a).

[f] k referred to the reversible $E_{1/2}$.

[g] For E more anodic than $E_{1/2}$.

[h] Value of $(1 - \alpha)$.

[i] k_s referred to E° of In.

[j] Estimated from a figure.

[k] j° calculated from the equation $j_0 = j°[In]^{0.25}[In^{3+}]^{0.73}$.

TABLE 3.1.2. Kinetic Parameters (Solid Electrodes)

Reaction	Conditions	Electrode	Technique	Rate constant	α	Remarks	Refs.
$In^{3+} + 3e = In$	10^{-2} M $In(ClO_4)_3$ + 10^{-2} M $HClO_4$ + $NaClO_4$ (I = 3 M) + X	rot plate	Radio-tracer			20°C	[90]
	X = 0 M NaCl			$j_0 = 6.7 \times 10^{-5}$			
	X = 5 × 10^{-2} M NaCl			$j_0 = 2.4 \times 10^{-4}$			
	X = 0.16 M NaCl			$j_0 = 4.2 \times 10^{-4}$			
	10^{-2} M $In(ClO_4)_3$ + 3 × 10^{-3} M $HClO_4$ + $NaClO_4$ (I = 3 M)			$j° = 2.88 \times 10^{-3}$ [a] 1.85[b]		$\Delta H_{S(5-60°C)}$ = 16 kcal/mole	
$In^{3+} + 3e = In$	10^{-2} M $In(ClO_4)_3$ + 3 × 10^{-3} M $HClO_4$ + $NaClO_4$ (I = 3 M)	wr	farad. imp	$j_0 = 8.6 \times 10^{-2}$ [c]			[91]
$In^{3+} + 3e = In$	10^{-2} M $In(ClO_4)_3$ + 10^{-2} M $HClO_4$ + $NaClO_4$ (I = 2 M)	rot disk	Current-potential curves		2.0[d]		[91a]
	X M HCl + (1 - X) M NaCl X 0.01 ÷ 0.75				2.0[d]		
$In^+ + e = In$	1 M $HClO_4$	rot disk	Current-potential curves		0.5[e]		[91a]

[a] $j°$ calculated from the equation $j_0 = j°[In^{3+}]0.62$.

[b] $3\partial \log j_0/\partial \log[In^{3+}]$.

[c] j_0 of the step $In \rightleftarrows In^+ + e$.

[d] Value of $(1 - \alpha)n$ (calculated from the Tafel slope of the anodic current-potential curve).

[e] Value of $1 - \alpha$ (calculated from the Tafel slope of the anodic current-potential curve).

3.2. DOUBLE-LAYER PROPERTIES

Table 3.2.1 includes data on the p.z.c., E_z (vs NHE), obtained chiefly by methods giving the most reliable results [92]. Data on the dependence of E_z on the concentration of the indium amalgam obtained by the streaming electrode method [93, 104] agree well with the results obtained on the dropping electrode by the null solution method [102] and with the data of electrocapillary measurements [94, 105, 106]. A comparison is made in Refs. 94 and 106.

A negative surface excess of indium is typical for the indium amalgam [93, 105], and it increases monotonically with an increase in the indium content of the amalgam [94, 106].

TABLE 3.2.1. Potential of Zero Charge

Electrode	Conditions	Technique	E_z	Refs.
Amalgam	5.95 mole % In, 0.1 M $HClO_4$, 25°C	Streaming electrode	-0.310^a	[93]
Amalgam	10.00 mole % In, 0.1 M $HClO_4$, 25°C	Streaming electrode	-0.333^a	[93]
Amalgam	20.00 mole % In, 0.1 M $HClO_4$, 25°C	Streaming electrode	-0.388^a	[93]
Amalgam	39.83 mole % In, 0.1 M $HClO_4$, 25°C	Streaming electrode	-0.480^a	[93]
Amalgam	63.02 mole % In, 0.1 M $HClO_4$, 25°C	Streaming electrode	-0.558^a	[93]
Amalgam	64.6% In, 1 N Na_2SO_4, ~ 25°C	Electrocapillary max	-0.64 ± 0.01	[92], [94]
Amalgam	51% In, diluted NaF and KF, 18-20°C	Differential capacity	-0.915^b	[95]
Metal	1.0 N K_2SO_4, room temp.	Open circuit scrape method	-0.63	[96]
Metal	0.01 N Na_2SO_4, pH 3.0	Differential capacity	-0.67 ± 0.01^c	[97]
Metal	4×10^{-3} M Na_2SO_4, pH 4.01	Differential capacity	-0.95 ± 0.02^d	[97a]
Metal	3×10^{-3} N NaAF	Differential capacity	-0.93 ± 0.01^b	[98]
			(-0.65 ± 0.02)	[92]
Metal	10^{-3} N HCl, $0 \div 10^{-2}$ N KCl		-0.9 ± 0.03^d	[99]
Fused metal	Eutectic KCl-LiCl, 450°C	Electrocapillary max	-0.52^e	[100]
Fused metal	Eutectic KCl-LiCl, 700°C	farad. imp	-0.78^e	[101]

[a]Versus rev. HE in the same solution. (E of this reference electrode is -0.065 V vs NHE [102, 103]); for 20 and 40 mole % indium amalgams, dE_z/dt is -0.18 ± 0.05 and -0.15 ± 0.03 mV/°C, respectively [104].

[b]Versus NCE.

[c]Corrected for electrolyte nonsymmetry.

[d]Versus SCE; in Ref. 99, E_z determined from the intercept of $j^{0.4}$-E curves (j photocurrent) corresponding to different inert electrolyte concentrations.

[e]Versus rev. Pb = electrode in the same electrolyte.

For 5.95 atom % indium in the amalgam (0.1 N $HClO_4$) in the region of potentials from -0.8 to -1.3 V NCE, the thickness of the surface layer of the amalgam with an indium concentration equal to zero (assuming that the entire negative adsorption occurs in this layer) is equal to 7.3×10^{-8} cm [106].

The ClO_4^- ion is specifically adsorbed on the indium [98] and its amalgam [104], with the minimum specific adsorption of ClO_4^- corresponding to 5 to 10 mole % In in the amalgam. The ion F^- is not adsorbed on indium, while the adsorption of Cl^- (in 0.01 N NaCl) shifts E_z in the negative direction by 40 mV [104]. A shift of E_z in the negative direction for 50 mole % indium amalgam is also caused by an addition of thiourea (5.2×10^{-3} M) and tetramethylthiourea (1.6×10^{-3} M) to a 0.1 M solution of $HClO_4$ [107].

For melts, the value of E_z, determined in various works for 450°C [100, 108, 109] and for 750°C [101, 108, 110], varies by 0.1 V and more. The most reliable results are apparently obtained by using the dispersion analysis of electrode impedance [101]. A review of works on E_z for indium in melts is given in Ref. 111 and partially in Ref. 101; see also Refs. 5a and 112.

4. ELECTROCHEMICAL STUDIES

A distinctive feature of the electrochemical behavior of indium is the comparatively low value of the exchange current in acid solutions in the absence of complex formation [65, 79]. In these conditions, indium exists in the solution in the form of an aquo-complex $In(H_2O)_6^{3+}$, which is characterized by a very low rate of water molecule dissociation from its hydration shell [113], which is one of the reasons for the irreversibility of electrode processes with the participation of indium ions in such solutions. In view of the slowness of the direct discharge and of the formation of stable aquo-complexes of indium, the overall electrode process, as a rule, also includes chemical steps leading to the labilization of the hydration shell of the cation as a result of the formation from the simple hydrated indium particles (prevailing in the solution) of less stable and more reactive intermediate particles to be discharged at a higher rate. Therefore the indium electrode is suitable for elucidation of general regularities and of the mechanism of the influence exerted on the discharge ionization process of the metal by activating anions, pH, and the complex formation [65] to which indium is very sensitive [13, 42].

The rate of the process of discharge and of the ionization of indium on the amalgam and solid indium electrode depends considerably on the acidity of the solution. The rate of the anodic process at a constant potential and the exchange current are inversely proportional to the concentration of hydrogen ions [79, 90, 114]. The inhibiting effect of the hydrogen ions was explained quantitatively on the basis of a stepwise mechanism (see below) and the concept of the participation in the limiting step of partially hydrolyzed indium ions of intermediate valency [65, 79, 90, 115]

$InOH^+ \rightarrow InOH^{2+} + e$ (or $InOH \rightarrow InOH^{2+} + 2e$)

with a subsequent chemical step

$$InOH^{2+} + H^+ \rightleftarrows In^{3+} + H_2O$$

During the anodic polarization of indium and its amalgam it is possible to detect $InOH^{2+}$ ions in the bulk of the solution in concentrations exceeding their equilibrium concentration, which confirms the indicated mechanism [116, 117].

The process of indium ion discharge is also retarded with an increase in acidity; this is observed both on the mercury and amalgam electrodes [19, 22, 23, 42, 79, 80] and also on the solid indium cathode [114, 118-125]. Even when there is an insignificant shift in the negative direction from the equilibrium potential, the cathode process is limited by the rate of the preceding chemical reaction of the formation of partially hydrolyzed indium ions from the prevailing simple aquo-ions In^{3+}; the particles that are directly discharged are the $InOH^{2+}$ ions [42, 56, 79, 80, 122]. (It was concluded on the basis of an analysis of polarographic data for nitrate solutions that the ion $In(OH)_2{}^+$ participates directly in the electrochemical step [22].) At much more negative potentials the direct reduction of simple aquo-complexes $In(H_2O)_6{}^{3+}$ takes place in parallel with the discharge of hydrolyzed indium ions, and the rate of this process does not depend on the pH [80, 122]. For the indium amalgam the standard exchange current density of this process ($\sim 4 \times 10^{-6}$ A/cm^2) is lower by three orders of magnitude than the corresponding value for the process with the participation of hydrolyzed ions [80]. Apparently, when the hydroxyl ion is included in the inner sphere of the aquo-complex of trivalent indium, it has a labilizing effect, accelerating the steps of further dehydration and discharge of In(III) ions [19, 22, 80].

In weakly acid, neutral, and weakly alkaline solutions in the absence of complex formation, indium salts are practically insoluble and indium is readily passivated during anodic polarization [5a, 126-131]. In more concentrated alkaline solutions, indium exists in the form of hydroxy complexes $In(OH)_m{}^{(3-m)+}$ (indates). The charge-transfer step is actually preceded by the dissociation of several hydroxyl ions from the indate ions prevailing in the solution [89]. But unlike acid solutions, this chemical reaction is reversible in alkaline solutions and has practically no effect on the kinetics of the overall electrode process [128].

The kinetics of the processes of discharge and ionization of indium depends upon the nature of the anions present in the electrolyte (Tables 3.1.1 and 3.1.2). As a rule, an increase in the concentration of anions that can be easily adsorbed (e.g., on the mercury surface), are polarizable, and can form unstable complexes with indium ions, leads to an acceleration of the electrode reactions [57, 58, 65, 82, 85, 90, 91a, 132, 133]. This also happens in the presence of certain surface-active organic substances [70]. These substances, in the absence of activating anions, lead to the retarding of the processes of the discharge and ionization of indium [70]. A review of work on the effect of anions on the kinetics of indium reduction is given in Refs. 5a, 42, 63, and 65. The catalytic action of anions increases in the order $F^- < ClO_4{}^-$, $SO_4{}^{2-} < Cl^- < Br^- < I^-$, SCN^- [42, 57, 58, 63, 65, 70, 82, 85, 132, 133] and is connected not only with their effect on the double-layer structure, since it is also observed in the region of high concentrations of anions. Moreover, in the case of SCN^- the effect of an increase in the rate of indium ion discharge with the growth of anion concentration remains even after the change of the ψ_1 (or ϕ_2^-) potential as a result of anion adsorption is quantitatively taken into account [67].

Determination of the electrode reaction orders for I^-, Br^-, Cl^-, and SO_4^{2-} anions showed that they participate in the processes of discharge and ionization of indium, forming reactive complexes with it [56, 67, 82]. The compositions of the reactive complexes often differ from those of the complexes prevailing in the bulk solution, and consequently the charge-transfer reaction is accompanied by chemical reactions of the formation and dissociation of complexes [56, 65, 82]. In the general case it is possible that parallel reactions can take place with the participation of complexes of different compositions [65, 82]. In the presence of thiocyanate ions the indium complex discharging near $E_{1/2}$ is adsorbed on the mercury surface [62, 67, 87, 88, 134, 135], and the formation of the complex apparently proceeds at a much lower rate than its discharge [67]. Moreover, the thiocyanate complexes formed on the mercury surface are discharged more easily than complexes formed in the bulk of the solution; this effect might possibly depend upon which of the atoms in the SCN^- particle form a bond between the indium ion and ligand, and which form a bond between the ligand and the surface of the mercury electrode [67]. In halide solutions the discharging particles are formed as a result of chemical reactions in the solution, since halide complexes of indium are not adsorbed on the mercury surface [62, 135, 136]. The absence of adsorbed indium complexes which are directly formed in the electrode reaction makes it difficult to explain the participation of halide ions in the anodic process. The halide ions apparently form complexes with intermediate low-valent indium ions that participate directly in the limiting step of the ionization of indium [65]. The catalytic effect of the anions could possibly be confined to the labilization of a solvation shell of the indium ion, thereby facilitating its desolvation [22, 63, 65], and is due to the anions ability to act as electron bridges between the indium ions and the electrode [63, 65, 82, 132].

F^- ions are not specifically adsorbed on the mercury surface and unlike other halide ions have no accelerating effect on the electrode process of the indium amalgam [60, 82]. The decrease in the exchange current observed under these conditions is due to a decrease in the concentration of noncomplexed indium ions as a result of the bonding of the greater part of the indium ions into electrochemically less active stable fluoride indium complexes [82].

Another specific feature of indium is its ability to form, apart from stable trivalent ions, unstable monovalent ions in solution. In^+ ions are formed both when metallic indium comes into contact with the solution of a trivalent indium salt [3, 15] and during the anodic polarization of indium and its amalgam. This is evident from the results of identification of low-valent indium particles by means of various indicator electrodes [9, 116, 117, 137, 138], and specifically by the ring disk electrode [139, 140], as well as from values of anodic current efficiencies that considerably exceed 100% [3, 133, 137, 141-143]. By changing the electrolysis conditions it is possible to obtain as the final product of the anodic reaction either only In^{3+} ions or predominantly In^{+} ions [65, 143]. Increased values of anodic current efficiency were likewise observed in nonaqueous solutions [138, 144-147] and were explained by the formation of In^{+}.

The In^{+} ions are unstable and are oxidized by hydrogen ions [9, 11]; with I^{-} ions they form an insoluble deposit of InI [143, 148], while in the presence of metallic surfaces they are subject to heterogeneous disproportionation which takes place as a combination of the coupled charge transfer steps $In^{+} + e \rightarrow In$ and $In^{+} \rightarrow In^{3+} + 2e$ [16, 143].

In perchlorate and sulfate solutions the electrode process on indium is irreversible. Judging from the low values of the anodic Tafel slopes that amount to 21 to 26 mV for indium amalgam [81, 149, 150] and 26 to 31 mV for solid indium [90, 91a, 114], as well as by the results of impedance measurements [91, 91a], the process takes place as a combination of successive charge-transfer steps, with the equilibrium of the step $In \rightleftharpoons In^{+} + e$ being preserved. This last conclusion was confirmed by the results of studies of the effect of various factors on the concentration of In^{+} ions that are formed during the electrolysis. Thus, during the anodic polarization of indium the concentration of In^{+} ions at the electrode increases in quantitative agreement with the Nernst equation for this step [91, 116, 133, 137, 139, 140].

The stepwise mechanism of the electrode process on indium with the formation of intermediate low-valent particles affects the whole electrochemical behavior of indium [5a]. For one thing, bearing in mind the stepwise mechanism and the oxidation of low-valent indium particles by hydrogen ions, it is possible to explain such unusual effects observed in acid solutions as the deviation of the In-In^{3+} system from the equilibrium state (the shift of the indium potential in the negative direction with the increase of acidity and with stirring as well as the deviation from the Nernst equation) [10] and the corrosion of indium at a rate greatly exceeding that calculated by usual methods through the extrapolation of the cathodic polarization curve of oxidant reduction to the mixed potential [151].

5. APPLIED ELECTROCHEMISTRY

Electrochemical methods find wide application in the recovery and refining of indium. The method of cementation by zinc and aluminum [152-155] is used for the recovery of indium from sulfate solutions containing approximately 0.1 g/l of indium and high concentrations of zinc, copper, cadmium, aluminum, arsenic, and other elements (subproducts of lead and zinc production). The amalgam method is also used to recover indium from solutions: indium is transferred from the solution in an amalgam together with certain other elements by cementation with a zinc amalgam or by electrolysis with a mercury cathode

[156, 157]. The subsequent selective anodic dissolution of the amalgam is carried out at a current density of 0.05 to 0.1 A/cm^2 in a hydrochloric acid medium [158].

Several consecutive cementation operations are carried out to obtain high purity indium. First, impurities are removed by cementation from a sulfuric acid solution, then indium is cemented on aluminum. Chloride solutions are used for the electrolytic refining of indium at a current density of 100 to 200 A/m^2 and 40°C [153, 159, 160]. The efficiency of this refining is enhanced if impurities are simultaneously and continuously removed from the electrolyte by cementation on metallic indium [161]. Extra-pure indium can be obtained by repeated amalgam refining using electrolyzers with several cells and bipolar mercury electrodes [162-164].

An important feature of indium electrodeposits is the possibility of obtaining surface alloy layers at relatively low temperatures. A lead-indium coating on silver possesses excellent protective properties against corrosion and seizing. It is used as a bearing alloy. Indium coatings on zinc, lead, and copper are also used. The possibility of obtaining thin electrodeposits of In-Sb is of considerable interest.

Sulfate, fluoroborate, sulfamate, and alkaline cyanide baths are used to obtain indium electrodeposits [5a, 165-168]. Trilonate [169] and polyethylenepolyamine baths [170] have also been used. From tartrate, ammonium-tartrate, and pyrophosphate-tartrate baths it is possible to obtain not only indium electrodeposits, but also galvanic alloys of indium with antimony, copper, tin, zinc, and nickel [171, 172].

The main field in which indium is used is in the production of semiconductors. Indium's ability to impart hole conduction to germanium is used for p-n junctions in various semiconductor devices. Indium compounds with the elements of the V group of the periodic system — antimony, arsenic, phosphorus — having semiconductor properties have found wide application. Indium antimonide possesses photo-conductivity in the infrared region, and it is used as an infrared detector as well as a Hall-effect indicator. Indium arsenide and indium phosphide are used for high-temperature thermoelements.

In radio engineering and electronics, indium oxide is used to make contacts and transparent conducting films on dielectric materials, while indium is used as a neutron indicator in nuclear technology.

REFERENCES

1. S. Hakomori, J. Amer. Chem. Soc., 52, 2372 (1930).
2. E. M. Hattox and T. de Vries, ibid., 58, 2126 (1936).
3. W. Kangro and F. Weingärtner, Z. Elektrochem., 58, 505 (1954).
4. A. K. Covington, M. A. Hakeem, and W. F. K. Wynne-Jones, J. Chem. Soc., 1963, 4394.

5. G. Biedermann and T. Wallin, Acta Chem. Scand., 14, 594 (1960).

5a. R. Piercy and N. A. Hampson, J. Appl. Electrochem., 5, 1 (1975).

6. N. Sunden, Z. Elektrochem., 57, 100 (1953).

7. J. N. Butler, J. Phys. Chem., 68, 1828 (1964).

8. W. Hammer, M. Malmberg, and B. Rubin, J. Electrochem. Soc., 103, 8 (1956).

9. R. E. Visco, J. Phys. Chem., 69, 202 (1965).

10. V. V. Losev and A. P. Pchelnikov, Elektrokhimiya, 4, 264 (1968).

11. A. P. Pchelnikov and V. V. Losev, Zh. Fiz. Khim., 41, 971 (1967).

12. M. H. Lietzke and R. W. Stroughton, J. Amer. Chem. Soc., 78, 4520 (1956).

13. A. I. Busev, Analiticheskaya Chimiya Indiya, Izd. AN SSSR, Moscow, 1958; A. I. Busev, Analytical Chemistry of Indium, Macmillan, New York, 1962.

13a. D. Ferri, Acta Chem. Scand., 26, 733, 747 (1972).

14. N. A. Hampson and R. Piercy, J. Electroanal. Chem., 51, 91 (1974).

15. L. G. Hepler, Z. Z. Hugus, and W. M. Latimer, J. Amer. Chem. Soc., 75, 5652 (1953).

16. A. I. Molodov, G. N. Markosyan, A. P. Pchelnikov, and V. V. Losev, Elektrokhimiya, 4, 1370 (1968).

16a. L. Kiss, A. Körösi, and J. Farkas, Ann. Univ. Sci. Budapest, Sect. Chim., 9, 3 (1967).

17. J.-C. Touller and B. Trémillon, Anal. Chim. Acta, 49, 115 (1970).

18. M. Pourbaix, Atlas of Electrochemical Equilibria in Aqueous Solutions, Cebelcor, Brussels, 1966, p. 436.

19. A. J. Engel, J. Lawson, and D. A. Aikens, Anal. Chem., 37, 203 (1965).

20. E. N. Vinogradova and An Tszun-zhu, Zh. Anal. Khim., 16, 555 (1961).

21. N. V. Akselrud, Ukr. Khim. Zh., 21, 457 (1955).

22. J. G. Lawson and D. A. Aikens, J. Electroanal. Chem., 15, 193 (1967).

23. E. D. Moorhead and W. M. MacNevin, Anal. Chem., 34, 269 (1962).

24. P. S. Jain and S. Lal, Monatsh. Chem., 103, 751 (1972).

25. H. Imai, J. Sci. Hiroshima Univ., A22, 291 (1958).

26. S. Takagi, J. Chem. Soc., 1928, 301.

27. M. Breant and G. C. Merlin, C.R. Acad. Sci., Paris, 254, 665 (1962).

28. M. Kvaček and P. Kühn, Chem. Listy, 55, 1296 (1961).

29. D. S. Jain and J. N. Gaur, Electrochim. Acta, 12, 413 (1967).

30. J. Tomes, Collect. Czech. Chem. Commun., 9, 12 (1937).

31. J. J. Lingane, J. Amer. Chem. Soc., 61, 2099 (1939).

32. N. V. Akselrud and V. B. Spivakovskii, Zh. Anal. Khim., 12, 78 (1957).

33. L. Treindl, Chem. Listy, 50, 534 (1956).

34. K. Zh. Sagadieva and M. T. Kozlovskii, Izv. Akad. Nauk Kaz. SSR, Ser. Khim., 1965(1), 3.

35. Y. Narusawa, J. Hashimoto, and H. Hamaguchi, Bull. Chem. Soc. Japan, 38, 234 (1965).

36. T. P. Radhakrishnan and A. K. Sundram, J. Electroanal. Chem., 5, 124 (1963).

37. J. W. Grenier and L. Meites, Anal. Chim. Acta, 14, 482 (1956).

38. S. I. Sinjakova, Zh. Anal. Khim., 1, 241 (1946).

39. J. Doležal, Chem. Listy, 49, 1237 (1955).

40. J. Doležal, ibid., 50, 54 (1956).

41. H. Nakatani, J. Sci. Hiroshima Univ., A19, 183 (1955).

42. D. Cozzi and S. Vivarelli, Z. Elektrochem., 57, 408 (1953).

43. S. I. Sinjakova, Dokl. Akad. Nauk SSSR, 29(5-6), 375 (1940).

44. E. Jacobsen and G. Tanberg, J. Electroanal, Chem., 30, 161 (1971).

44. D. Cozzi and S. Vivarelli, Z. Elektrochem., 58, 907 (1954).

46. G. Rienäcker and E. Hoschek, Z. Anorg. Chem., 268(4-6), 260 (1952).

47. M. Komatsu, Y. Masuda, and S. Misumi, J. Chem. Soc. Japan, Chem. Ind. Chem., 1973(8), 1461.

48. I. A. Sheka and G. V. Lavrova, Ukr. Khim. Zh., 29, 819 (1963).

49. J. N. Gaur, R. C. Mehrotra, M. M. Palrecha, A. Kumar, D. S. Jain, and C. M. Gupta, J. Polarogr. Sci., 14, 122 (1968).

50. O. N. Shrivastava, J. K. Gupta, and C. M. Gupta, Monatsh. Chem., 104, 57 (1973).

51. I. V. Pjatnitskii and R. P. Ruzhanskaja, Zh. Anal. Khim., 26, 1475 (1971).

52. R. Marassi, V. Bartocci, and M. Fiorani, Chim. Ind. (Milan), 52, 365 (1970).

53. V. Bartocci, R. Marassi, and F. Pucciarelli, ibid., 52, 1201 (1970).

54. D. G. Lovering, Collect. Czech. Chem. Commun., 37, 3697 (1972).

55. I. M. Kolthoff and J. Lingane, Polarography, Vol. 2, 2nd ed., Interscience, New York, 1952, pp. 519-520.

56. S. A. Levitskaya, A. I. Zebreva, L. S. Palagutina, and L. E. Vokhrysheva, Izv. Vyssh. Ucheb. Zaved., Khim. Khim. Tekhnol., 14, 1468 (1971).

57. M. Bulovova, Chem. Listy, 48, 655 (1954).

58. M. Bulovova, Collect. Czech. Chem. Commun., 19, 1123 (1954).

59. N. Tanaka and R. Tamamushi, Proceedings of the 1st Australian Congress on Electrochemistry, Pergamon, London, 1965, p. 248.

60. A. M. Bond, J. Electrochem. Soc., 119, 1503 (1972).

61. A. G. Stromberg and Kh. Z. Brainina, Zh. Fiz. Khim., 35, 2016 (1961).

62. B. Timmer, M. Sluyters-Rehbach, and J. H. Sluyters, J. Electroanal. Chem., 19, 73 (1968).

63. R. De Levie, J. Electrochem. Soc., 118, 185C (1971).

64. Ja. I. Turyan, N. K. Strizhov, and B. M. Vasilenko, Elektrokhimiya, 6, 1560 (1970).

65. V. V. Losev and A. I. Molodov, Itogi Nauki, 8, 25 (1972).

66. N. Tanaka, T. Takeuchi, and R. Tamamushi, Bull. Chem. Soc. Japan, 37, 1435 (1964).

67. L. Pospíšil and R. De Levie, J. Electroanal. Chem., 25, 245 (1970).

68. P. N. Zabotin, S. P. Buchman, and G. Z. Kiryakov, Trudy 4 Elektrokhimicheskogo Soveshchaniya (Lectures from the Fourth Electrochemical Conference), Moscow, 1959, p. 179,

69. R. Visco, J. Electroanal. Chem., 10, 82 (1965).

70. M. A. Loshkarev and A. A. Kazarov, Ukr. Khim. Zh., 33, 568 (1967); Elektrokhimiya, 3, 39 (1967); A. A. Kazarov and M. A. Loshkarev, ibid., 3, 681 (1967).

71. Ja. I. Turyan and N. K. Strizhov, Zh. Anal. Khim., 27, 1423 (1972); 28, 1615 (1973).

72. R. G. Pats and S. B. Tsfasman, Zavod. Lab., 27, 266 (1961).

73. J. Devay, T. Garai, B. Palagyi-Fenyes, L. Meszaros, and A. R. S. Sayed, Acta Chim. (Budapest), 76(1), 29 (1973).

74. O. Kammori and I. Inamoto, Japan Analyst, 16, 1324 (1967).

75. E. Temmerman and F. Verbeek, Z. Anal. Chem., 244, 25 (1969).

76. I. P. Alimarin, F. A. Abduel Razak, E. N. Vinogradova, and A. I. Kamenev, Izv. Akad. Nauk SSSR, Ser. Khim., 1969(10), 2101.

77. S. Inouye and H. Imai, Bull. Chem. Soc. Japan, 33, 149 (1960).

78. A. I. Molodov and V. V. Losev, Zh. Fiz. Khim., 38, 1481 (1964); Elektrokhimiya, 1, 53 (1965).

79. V. V. Losev and A. I. Molodov, Dokl. Akad. Nauk SSSR, 135, 1432 (1960); Electrochim. Acta, 6, 81 (1962).

80. A. I. Molodov and V. V. Losev, Elektrokhimiya, 1, 651 (1965).

81. V. Markovac and V. Lovrecek, J. Electrochem. Soc., 112, 520 (1965).

82. V. V. Losev and A. I. Molodov, Dokl. Akad. Nauk SSSR, 130, 111 (1960); 148, 1114 (1963).

83. Yu. S. Lyalikov, L. N. Vasilieva, and L. P. Chernega, Zh. Anal. Khim., 26, 278 (1971).

84. A. A. Moussa, M. M. Abou Romia, and F. El-Taib Heakal, Electrochim. Acta, 15, 1391 (1970).

85. V. Markovac and B. Lovrecek, J. Electrochem. Soc., 113, 838 (1966).

86. J. N. Gaur and D. S. Jain, Electrochim. Acta, 11, 1661 (1966).

87. B. Timmer, M. Sluyters-Rehbach, and J. H. Sluyters, J. Electroanal. Chem., 15, 343 (1967).

88. R. De Levie and A. A. Husovsky, ibid., 22, 29 (1969).

89. E. V. Leontovich, Sbornik Issledovaniya po Elektroosazhdeniyu i Rastvoreniyu Metallov, Nauka, Moscow, 1971, p. 99.

90. V. V. Losev and A. P. Pchelnikov, Elektrokhimiya, 6, 41 (1970).

91. D. P. Aleksandrova, R. A. Slivka, and D. I. Leikis, ibid., 12 (1976).

91a. R. Piercy and N. A. Hampson, J. Electroanal. Chem., 57, 297 (1974); 59, 261 (1975).

92. A. N. Frumkin, Dvoinoi Sloi i Adsorbtsiya na Tverdych Elektrodach, Materialy III Sympoziuma v Tartu, Tartu, 1972, p. 5.

93. I. N. Butler, J. Phys. Chem., 69, 3817 (1965).

94. A. Frumkin, N. Polianovskaya, I. Bagotskaya, and N. Grigoryev, J. Electroanal. Chem., 33, 319 (1971).

95. E. V. Yakovleva and N. V. Nikolaeva-Fedorovich, Elektrokhimiya, 6, 35 (1970).

96. R. Perkins and T. Andersen, Modern Aspects of Electrochemistry, Vol. 5 (J. O'M. Bockris, ed.), Plenum, New York, 1969, p. 203; R. Perkins and T. Andersen, Sovremennye Problemy Elektrokhimii, Mir, Moscow, 1971, p. 194.

97. E. D. Levin and A. A. Rotinyan, Elektrokhimiya, 7, 372 (1971).

97a. R. Piercy and N. A. Hampson, J. Electroanal. Chem., 53, 271 (1974).

98. N. B. Grigoriev, I. A. Gedvillo, and N. G. Bardina, Electrokhimiya, 8, 409 (1972).

99. Z. A. Rotenberg and Yu. V. Pleskov, ibid., 9, 1419 (1973).

100. V. A. Kuznetsov and L. S. Zagainova, Zh. Fiz. Khim., 35, 1640 (1961).

101. R. A. Alekseeva, N. G. Bukun, V. A. Kuznetsov, and E. A. Ukshe, Elektrokhimiya, 7, 1357 (1971).

102. J. N. Butler and A. C. Makrides, Trans. Faraday Soc., 60, 1664 (1964).

103. J. N. Butler, M. L. Meehan, and A. C. Mackrides, J. Electroanal. Chem., 9, 237 (1965).

104. J. N. Butler, J. Phys. Chem., 70, 2312 (1966).

105. N. S. Polyanovskaya and A. N. Frumkin, Elektrokhimiya, 1, 538 (1965).

106. N. S. Polyanovskaya, ibid., 4, 549 (1968).

107. R. Narayan and N. Hackerman, J. Electrochem. Soc., 118, 1426 (1971).

108. T. N. Prisekina and V. A. Kuznetsov, Elektrokhimiya, 2, 1307 (1966).

109. R. A. Alekseeva and V. A. Kuznetsov, ibid., 5, 1475 (1969).

110. T. N. Prisekina, V. A. Kuznetsov, and I. V. Nefedova, ibid., 3, 1385 (1967).

111. E. A. Ukshe and N. G. Bukun, Dvoinoi Sloi i Adsorbtsiya na Tverdych Elektrodach, Materialy III Symposiuma v Tartu, Tartu, 1972, p. 264.

112. A. D. Graves, J. Electroanal. Chem., 25, 349 (1970).

113. M. Eigen, Ber. Bunsenges. Phys. Chem., 67, 753 (1963); Pure Appl. Chem., 6, 97 (1963).

114. A. D. Styrkas, A. I. Ivanov, and V. V. Ostroumov, Vestn. Leningrad. Univ., Ser. Fiz. Khim., 1965(4), 106.

115. A. I. Molodov and V. V. Losev, Elektrokhimiya, 1, 1253 (1965).

116. A. I. Molodov and V. V. Losev, ibid., 4, 835 (1968).

117. A. I. Molodov, G. N. Markosyan, and V. V. Losev, ibid., 9, 1368 (1973).

118. V. M. Kochegarov, F. I. Zaburdaeva, and E. N. Zjablova, Zh. Prikl. Khim., 35, 1376 (1962).

119. G. Serravalle, Rend. Ist. Lombardo Sci., Lettere, A, 93, 653 (1959).

120. R. Piontelli, G. Poli, and G. Serravalle, ibid., 93, 301 (1959).

121. G. Serravalle and B. Mazza, Electrochim. Acta, 8, 313 (1963).

122. A. P. Pchelnikov and V. V. Losev, Elektrokhimiya, 1, 1058 (1965).

123. A. A. Voronko, R. M. Vishomirskis, and A. M. Molchadskii, Zashch. Met., 1, 703 (1965).

124. R. Piontelli and G. Serravalle, Electrochim. Metal., 1, 149 (1966).

125. A. P. Pchelnikov and V. V. Losev, Elektrokhimiya, 9, 1031 (1973).

126. T. I. Popova and N. A. Simonova, Izv. Akad. Nauk SSSR, 1963(7), 1187.

127. F. F. Faizullin and N. A. Amirhanova, Elektrokhimiya, 2, 805, 1383 (1966).

128. R. D. Armstrong, A. B. Suttie, and H. R. Thirsk, Electrochim. Acta, 13, 1 (1968).

129. T. M. Salem and A. A. Ismail, J. Chem. Soc., A, 14, 2415 (1970).

130. L. I. Filippova and Kuzmin, Izv. Vyssh. Ucheb. Zaved. Khim. Khim. Teckhol., 12, 1199 (1969).

131. R. Piercy, P. D. Stokes, and N. A. Hampson, Brit. Corr. J., 8, 275 (1973).

132. J. Heyrovsky, Discuss. Faraday Soc., 1, 212 (1947); Z. Elektrochem., 59, 802 (1955).

133. R. E. Visco, J. Electrochem. Soc., 112, 932 (1965).

134. T. Takahashi and H. Shirai, Rev. Polarogr., 11(3-4), 155 (1963).

135. G. W. O'Dom and R. W. Murray, J. Electroanal. Chem., 16, 327 (1968).

136. G. C. Barker and R. L. Faircloth, Advan. Polarogr., 1, 313 (1960).

137. A. P. Pchelnikov and V. V. Losev, Zashch. Met., 1, 482 (1965); V. V. Losev and A. P. Pchelnikov, Proceedings of the Third International Congress on Metallic Corrosion, Mir, Moscow, 1966, Vol. 1, Moscow, 1968, p. 111.

138. I. G. Erusalimchik, E. A. Efimov, T. N. Prilepina, and T. V. Gerish, Sbornik Elektrokhimicheskie Processy pri Electroosazhdenii i Anodnom Rastvorenii Metallov, Nauka, Moscow, 1969, p. 162.

139. B. Miller and R. E. Visco, J. Electrochem. Soc., 115, 251 (1968).

140. L. Kiss, J. Farkas, and A. Körösi, Magy. Kem. Foly., 75, 66 (1969); L. Kiss and J. Farkas, Ann. Univ. Sci. Budapest. Sect. Chim., 12, 123 (1971).

141. L. F. Kozin and N. N. Tananaeva, Tr. Inst. Khim. Nauk Akad. Nauk Kaz. SSR, 9, 143 (1962).

142. A. I. Molodov, G. N. Markosyan, and V. V. Losev, Elektrokhimiya, 5, 918 (1969).

143. V. V. Losev, Electrochim. Acta, 15, 1095 (1970).

144. A. W. Davidson and F. Iiric, J. Amer. Chem. Soc., 72, 1700 (1950).

145. A. D. McElroy, J. Kleinberg, and A. W. Davidson, ibid., 74, 736 (1952).

146. I. Epelboin, Z. Electrochem., 59, 689 (1955).

147. J. B. Headbridge and D. Pletcher, Inorg. Nucl. Chem. Lett., 3, 475 (1967).

148. A. P. Pchelnikov and V. V. Losev, Elektrokhimiya, 5, 284 (1969).

149. G. M. Budov and V. V. Losev, Dokl. Akad. Nauk SSSR, 129, 1321 (1959); Zh. Fiz. Khim., 37, 1023, 1230 (1963).

150. B. Lovrecek and V. Markovac, J. Electrochem. Soc., 109, 727 (1962).

151. V. V. Losev and A. P. Pchelnikov, Electrochim. Acta, 18, 589 (1973).

152. T. R. Wright, Mining Met., 26, 559 (1945).

153. T. R. Mills, B. G. Hunt, and G. H. Turner, J. Electrochem. Soc., 100, 136 (1953); T. R. Mills, R. A. King, and C. E. T. White, Rare Metals Handbook, (C. A. Hampel, ed.), Reinhold, New York, 1961, pp. 220-238.

154. B. G. Hunt, C. E. White, and R. A. King, Can. Mining Metal. Bull., 52(566), 359 (1959).

155. I. S. Levin and T. G. Azarenko, Bull. Tsvet, Metal., 19, 29 (1963).

156. A. T. Nizhnik, Dokl. Akad. Nauk SSSR, 137, 366 (1961).

157. S. P. Buchman, Tr. Inst. Kh. Nauk Akad. Nauk Kaz. SSR, 9, 135 (1962).

158. P. P. Tsyb, Tsvet. Metal., 8, 40 (1963).

159. L. Müller, Freiberger Forschungsh, B, 90, 105 (1963).

160. A. I. Zhurin and Li Han-guan, Tr. Leningr Politekh. Inst., 223, 69 (1963).

161. O. S. Druzhinina, G. P. Myakinkova, N. E. Galkina, and I. P. Shepshakova, Tr. Giredmeta, 6, 136 (1962).

162. P. P. Tsyb and V. I. Maltsev, Zh. Prikl. Khim., 35, 1565 (1962).

163. W. Morawietz, Chem.-Ing.-Tech., 36, 638 (1964).

164. L. F. Kozin, Amalgamnaya Metallurgiya, Technika, Kiev, 1970.

165. F. A. Lowenheim (ed.), Modern Electroplating, Wiley, New York, 1967.

166. Metal Finishing Guidebook, 38th ed., Metal and Plastic Publications, New York, 1970, p. 304.

167. P. M. Vyacheslavov, S. Ya. Grilihes, G. K. Burkat, and E. G. Kruglova, Galvanotehnika Blagorodnyh Metallov, Izd. Mashinostroenie, Leningrad, 1970.

168. L. I. Kadaner, ed., Elektroosashdenie Blagorodnyh i Redkih Metallov, Gosinti, Moscow, 1968.

169. M. A. Shluger and E. G. Chmyhalo, Zashch. Met., 9, 223 (1973).

170. N. R. Kokarev, A. V. Ryabchenkov, and A. V. Rykova, ibid., 4, 116 (1968).

171. I. A. Marchenko and L. S. Terehova, ibid., 2, 9 (1966).

172. O. P. Bespalko and I. D. Vdovenko, Elektroosazhdenie Metallov i Splavov iz Tartratnyh Elektrolitov, Naukova Dumka, Kiev, 1971.

Chapter VI-2

SCANDIUM, YTTRIUM, AND THE LANTHANIDES

HARVEY B. HERMAN

Department of Chemistry
University of North Carolina at Greensboro
Greensboro, North Carolina

and

J. R. RAIRDEN

General Electric Research and Development Center
General Electric Company
Schenectady, New York

1. STANDARD AND FORMAL POTENTIALS

1.1. AQUEOUS SOLUTIONS

The chemistry of Sc, Y, and the lanthanides is primarily concerned with the 3+ oxidation state. The standard potentials of the reactions $M^{3+} + 3e = M$ have been calculated by

TABLE 1.1.1. Standard Potentials in Aqueous Acid Solutions

Half-reaction	E_0 (25°C)	Conditions	Refs.
$Sc^{3+} + 3e = Sc$	-2.08	Free energy calcd	[1]
$Y^{3+} + 3e = Y$	-2.37	Free energy calcd	[1]
$La^{3+} + 3e = La$	-2.52	Free energy calcd	[1]
	-2.4	calcd	[3]
$Ce^{3+} + 3e = Ce$	-2.48	Free energy calcd	[1]
	-2.335	calcd	[4]
$Pr^{3+} + 3e = Pr$	-2.47	Free energy calcd	[1]
	-2.2	calcd	[3]
$Nd^{3+} + 3e = Nd$	-2.44	Free energy calcd	[1]
	-2.246	calcd	[4]
$Pm^{3+} + 3e = Pm$	-2.42	Free energy calcd	[1]
$Sm^{3+} + 3e = Sm$	-2.41	Free energy calcd	[1]
	-2.2	calcd	[3]
$Eu^{3+} + 3e = Eu$	-2.41	Free energy calcd	[1]
	-2.2	calcd	[3]
$Gd^{3+} + 3e = Gd$	-2.40	Free energy calcd	[1]
	-2.2	calcd	[3]
$Tb^{3+} + 3e = Tb$	-2.39	Free energy calcd	[1]
	-2.2	calcd	[3]
$Dy^{3+} + 3e = Dy$	-2.35	Free energy calcd	[1]
	-2.2	calcd	[3]
$Ho^{3+} + 3e = Ho$	-2.32	Free energy calcd	[1]
	-2.1	calcd	[3]
$Er^{3+} + 3e = Er$	-2.30	Free energy calcd	[1]
	-2.1	calcd	[3]
$Tm^{3+} + 3e = Tm$	-2.28	Free energy calcd	[1]
	-2.1	calcd	[3]
$Yb^{3+} + 3e = Yb$	-2.27	Free energy calcd	[1]
	-2.1	calcd	[3]
$Lu^{3+} + 3e = Lu$	-2.25	Free energy calcd	[1]
	-2.1	calcd	[3]

Latimer [1] from free energy of formation values and are shown in Table 1.1.1. Hampel [2] has summarized the electrode potentials vs the hydrogen electrode as determined by Yost et al. [3], and Spedding and Miller [4]. These data are also shown in Table 1.1.1.

Standard potentials for basic solutions for some of the elements have been calculated by Latimer [1] and are summarized in Table 1.1.2.

TABLE 1.1.2. Standard Potentials in Aqueous Base Solutions (Latimer [1])

Half-reaction	E_B°
$Y(OH)_3 + 3e = Y + 3OH^-$	-2.8
$La(OH)_3 + 3e = La + 3OH^-$	-2.90
$Lu(OH)_3 + 3e = Lu + 3OH^-$	-2.72

Oxidation states other than 3+ have been identified for a number of the lanthanides. Ce^{4+} has been well characterized and has been the subject of several studies of the redox reaction

$$Ce^{4+} + e = Ce^{3+}$$

The standard potential values for these studies have been summarized by Charlot [5] and are shown in Table 1.1.3 along with other values [6-9].

TABLE 1.1.3. Standard Potentials for the Reaction of $Ce^{4+} + e = Ce^{3+}$ [5]

E_0 (25°C)	Solution	Refs.
+1.70	$HClO_4$	[6]
+1.61	HNO_3	[7]
+1.60	HNO_3	[6]
+1.44	H_2SO_4	[8],[9]
+1.28	HCl	[6]

The elements Eu, Yb, and Sm are known to exist as 2+ ions. A value of $E_0 = -0.35$ V was determined in perchlorate solutions by Anderson and Macero [10] for the reaction

$$Eu^{3+} + e = Eu^{2+}$$

This value has been confirmed recently in two reports. Morss and Haug [11] suggest a recommended result of -0.35 ± 0.03 V for the standard potential of the europium(III, II) couple. Biedermann and Silber [12], using a coulometric approach, determined this value to be -0.379 ± 0.001 V in 1 M $LiClO_4$.

A value for E_0 of -0.58 was determined by Walters and Pearce [13] for the reaction

$Yb^{3+} + e = Yb^{2+}$

Latimer [1] estimated a value for $E_0 > -0.9$ V for the reaction

$Sm^{3+} + e = Sm^{2+}$

A study of the lanthanons was made by Onstott [14] using amalgam cathodes. He pointed out that solvent degradation takes place rapidly for the 3+ to 2+ reactions for Yb and Sm. Therefore, he suggested that the value of $E_0 = -1.15$ V for Yb as obtained by Laitinen [15] is more nearly correct. Further, he suggested a value of $E_0 = 1.2$ to 2.0 V for Sm. Recently a new correlation technique has been described [16, 17] which will allow estimation of the standard potentials [E°M(IV-III) and E°M(III-II)] of the lanthanide series using results from atomic spectroscopy.

1.2. NONAQUEOUS SOLVENTS

The nonaqueous results are reported in Section 2.1 on polarography.

1.3. FUSED SALTS

Johnson and MacKenzie [18] recently studied the electrode potentials of Sm, Eu, and Yb in a eutectic melt of LiCl-KCl. They found that these elements displace the alkali metals from their molten chlorides at 450°C. The redox potentials measured versus a Pt(III)-Pt(0) reference electrode are

$Sm^{3+} + e = Sm^{2+}$; -1.73 V

$Eu^{3+} + e = Eu^{2+}$; -0.55 V

$Yb^{3+} + e = Yb^{2+}$; -1.37 V

A decomposition potential of 3.05 V was found for $LaCl_3$ at 900°C by Kuroda [19]. A value of 2.95 V was found for the decomposition of $CeCl_3$ at 700°C by Kojima and Kikuchi [20]. The decomposition of $NdCl_3$ occurs at 1.74 V at 700°C according to Neumann and Richter [21].

A comprehensive summary of theoretical decomposition potentials has been prepared by Delimarskii and Markov [22] as shown in Table 1.3.1. These were based on adaptations of the data of Hamer et al. [23]. These calculations are based on values of heats of formation and heat capacities.

On the basis of the decomposition potentials of pure fused salts, it is possible to calculate the values of single electrode potentials of metals relative to the electrode potential of sodium, arbitrarily set at zero. These data are tabulated in Table 1.3.2.

TABLE 1.3.1. Calculated Decomposition Potentials (V) of Fused Salt Chlorides [22]

Cation	Temperature (°C)		
	700	800	1000
La^{3+}			2.876
Ce^{3+}			2.821
Pr^{3+}		2.911	2.795
Nd^{3+}		2.856	2.736
Sm^{3+}		2.861	2.763
Eu^{3+}		2.828	2.815
Gd^{3+}		2.807	2.709
Tb^{3+}	2.858	2.754	2.657
Dy^{3+}		2.690	2.599
Y^{3+}		2.643	2.548
Ho^{3+}		2.610	2.511
Er^{3+}		2.589	2.488
Tm^{3+}		2.553	2.447
Yb^{3+}			2.434
Lu^{3+}			2.356

TABLE 1.3.2. Calculated Electrode Potentials in Fused Chlorides $E_{Na} = 0$ [22]

Electrode	Temperature (°C)		
	700	800	1000
La/La^{3+}			0.143
Ce/Ce^{3+}			0.198
Pr/Pr^{3+}		0.329	0.224
Nd/Nd^{3+}		0.384	0.283
Sm/Sm^{3+}		0.389	0.256
Eu/Eu^{3+}		0.412	0.204
Gd/Gd^{3+}		0.433	0.310
Tb/Tb^{3+}	0.474		0.462
Dy/Dy^{3+}		0.550	0.420
Y/Y^{3+}			0.471
Ho/Ho^{3+}		0.630	0.508
Er/Er^{3+}		0.651	0.631
Tm/Tm^{3+}		0.687	0.572
Yb/Yb^{3+}			0.585
Lu/Lu^{3+}			0.763

1.4. EQUILIBRIUM DATA

On the basis of species considered likely to be present in the metal-aqueous solution system, potential-pH diagrams have been constructed by Van Muylder for Sc and Y, and by de Zoubov and Van Muylder for the lanthanides [24]. Equilibrium formulas are given for the calculations for each element. The plots shown in Ref. 24 are valid only in the absence of substances with which soluble complexes or insoluble salts can form.

These elements all have a very large negative free energy for the oxidation reaction and are very unstable in the presence of water.

2. POLAROGRAPHIC CHARACTERISTICS

The polarography of Eu^{3+}/Eu^{2+} has been studied by several authors. This system is of particular interest since it exhibits a relatively stable-state oxidation state of 2+ in aqueous solutions. Vlcek [25] was the first to clearly point out the irreversibility of the reduction of Eu^{3+} in many supporting electrolytes. Several of his conclusions, however, have since been explained by other theories (see Section 3.2 on kinetic parameters and double-layer properties).

Vlcek measured the dependence of the cathodic and anodic waves of Eu on drop time, on gelatin concentration, and upon temperature. He concluded that each process occurred with a different mechanism. He observed two incisions using the oscillopolarographic technique and concluded that two forms of Eu^{2+} were formed. The above correlations and other electronic structural arguments led him to postulate two parallel mechanisms for the reduction process. He concluded that there was only one energetically favored oxidation path. Eu^{3+} can orient itself with either the 5d orbitals of the metal ion in contact with the electrode surface or the ligand orbitals. In the former case, reversible electron transfer is followed by an irreversible electronic rearrangement requiring a high activation energy. The latter case involves a preceding electronic rearrangement with a subsequent reversible electron transfer. The overall process is moderately irreversible even though the electron transfer step is probably facilitated by the ligand bridge. The oxidation of Eu^{3+} occurs principally through the ligand bridge mechanism. Here the mechanistic steps occur in reverse order and the high activation energy for the 5d orbital rearrangement makes this process prohibitive. He also postulated that the rearrangement step was not necessary for EDTA complexes, thereby explaining their observed reversibility.

Kinard and Philp [26] have studied the Eu^{3+}/Eu^{2+} reaction by conventional and Kalousek polarography. The polarograms are shown in Fig. 2.1. The waves fall in two groups. The waves with more positive half-wave potentials include only anions that form relatively weak outersphere complexes with Eu^{3+}. Sulfate and possibly acetate form stronger innersphere complexes and appear in the other group. The shapes and positions of these curves and

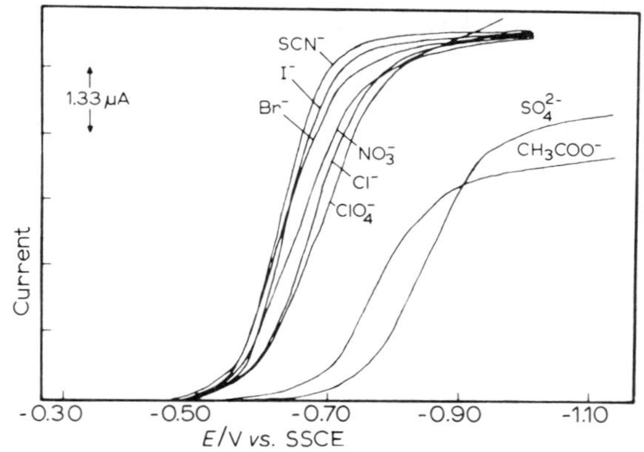

FIG. 2.1. Polarograms of 1.75×10^{-3} M Eu^{3+} in 1 M concentration of various supporting solutions [26]. Reproduced from the Journal of Electroanalytical Chemistry by permission of Elsevier Scientific Publishing Company.

others in this article were explained earlier by Gierst and Cornelissen [27] (see Section 3.2). The diffusion coefficient calculated from the average limiting current of the group is 8.6×10^{-6} cm^2/sec. It was observed that the diffusion currents for SO_4^{-2} and OAc^- are considerably smaller. This study also demonstrated the utility of Kalousek polarography for the study of unstable and irreversible systems.

The polarography of ethylenediamine tetraacetate complexes of europium was investigated by Onstott [28]. He found that in the presence of a large excess of anions of ethylenediamine tetraacetic acid (H_4Y), at 25°C, Eu^{3+} is reversibly reduced to Eu^{2+} at the DME in the pH range of 6 to 13.6. Complex ions identified from half-wave potential and diffusion current data are: pH 6 to 8 $[EuHY]^{-1}$; pH 11 to 13.6, $[EuY]^{-2}$; pH 6 to 8, $[Eu(HY)_2]^{-3}$; pH 11 to 12, $[EuY_2]^{-5}$; and pH 13 to 13.6, $[EuY(OH)]^{-2}$.

The polarography of the other rare earths has exhibited more controversy. Europium(III), as described, has a clear-cut one-electron reduction to europium(II). It is generally accepted that in water and some other solvents, ytterbium and possibly samarium also have a stable two-oxidation state. Conflicting results have been reported on the reduction of the stable two state and three-electron reductions of the other rare earths. The next few paragraphs will attempt to bring the subject up to date from the previous reviews of the subject.

An early study of several rare-earth elements was made by Noddack and Bruckl [29] in the absence of supporting electrolyte. They concluded that the reduction took place in two steps, first to the 2+ state and then to the metal. Kolthoff and Lingane [30] have stated, based on the work of Glocker for praseodymium, gadolinium, neodymium, and samarium, among others, that the first wave was probably due to the discharge of hydrogen ion and that the second wave corresponded to the ions being reduced directly to the metal. The polarographic data for the second waves obtained by Noddack and Brukl in 0.01 M solutions of the rare earth sulfates are included in Table 2.1.

TABLE 2.1. Polarographic Characteristics of Aqueous Solutions[a]

Substance	Supporting electrolyte	$E_{1/2}$ (V vs SCE)	Remarks	Refs.
Sc^{3+}	0.35 M LiCl	-1.74	pH 5.7	[37]
	0.1 M LiCl	-1.80		[30]
Y^{3+}	0.35 M LiCl	-1.92	pH 4.1	[37]
La^{3+}	0.10 M LiCl	-1.82	pH 5.7	[41]
	0.10 M LiCl	-1.91	pH 2.8	[114]
Ce^{3+}	0.01 M sulfate soln	-1.97		[29]
Pr^{3+}	0.10 M LiCl	-1.85	pH 5.9, 6.8 mM	[30]
Nd^{3+}	0.35 M LiCl	-1.80	pH 5.7	[37]
	0.10 M LiCl	-1.76	pH 5.7	[41]
Sm^{3+}	0.35 M LiCl	-1.78/-2.01	pH 5.7	[37]
	0.10 M Me_4NI	-1.80/-1.96		[30]
Eu^{3+}	0.35 M LiCl	-0.67/2.00	pH 5.7	[37]
	0.05 M Me_4NI	-0.67/-1.96	pH 6	[41]
Gd^{3+}	0.35 M LiCl	-1.79	pH 5.7	[37]
	0.10 M LiCl	-1.77		[30]
Tb^{3+}	0.10 M LiCl	-1.85		[115]
Dy^{3+}	0.35 M LiCl	-1.78	pH 5.7	[37]
	0.10 M LiCl	-1.80	pH 3.0	[111]
Ho^{3+}	0.10 M sulfate soln	-1.89		[29]
Er^{3+}	0.35 M Me_4NI	-1.77	pH 5.7	[37]
	0.10 M LiCl	-1.69	pH 5.7	[41]
Tm^{3+}	0.01 M sulfate soln	-1.85		[29]
Yb^{3+}	0.35 M LiCl	-1.46/-1.95	pH 5.7	[37]
	0.05 M Me_4NI	-1.43/-1.93	pH 6	[41]
Lu^{3+}	0.01 M sulfate soln	-1.82		[29]

[a]The products of the electrochemical reaction in all cases, except for the first step in Eu^{3+}, Yb^{3+}, and Sm^{3+}, are most probably H_2 and a hydroxide of some indeterminate type. Other waves are sometimes observed before and after the ones listed due to H^+ reduction and further hydroxide formation, respectively. The first wave for Eu^{3+}, Yb^{3+}, and Sm^{3+} is due to the one electron reduction to the II state with possible catalytic regeneration reactions with the solvent.

Doubts began to be raised about the simple reduction of the lanthanides directly to the metal. Laitinen and Mohr [31] were quoted as saying the reduction of the lanthanides was accompanied by hydrogen gas and hydrous oxide formation. Misumi and Ide [32] reported similar phenomena for yttrium. A neutron activation study [33] appeared that showed holmium was not incorporated into the mercury drop during polarography. In one study on the lanthanides in the presence of lithium, Treindl [34] reported hydrogen gas formation and hydride products. In a subsequent study [35] on scandium, the same author reported observations of hydrous oxide formation. Large and Timnick [36] reported some interesting observations on the polarography of dysprosium. They observed a two-step reduction for dysprosium in 0.1 M lithium chloride (discounting an accompanying wave due to hydrogen ion reduction). The first wave was proportional to concentration. Variation of the solution pH affected the $E_{1/2}$ of the first wave. The first wave was shifted and the second wave disappeared in D_2O solutions. Polarograms of the precipitated hydroxide gave only one wave which coincided with the second wave in dysprosium(III) solutions. They concluded on the basis of these observations that the first wave is due to the reduction of the hydrated water in the coordination sphere of dysprosium(III) and that the second wave was due to the reduction of the hydrous oxide produced during the first process.

Sevryukov and Hassan [37], in an article on the polarographic properties of the lanthanides, reviewed the recent literature. Their own results confirmed earlier studies which observed hydrogen gas formation and two reduction waves (again not counting the wave due to hydrogen ion reduction). They also observed a "grey" deposit form during the reduction and that the second wave appeared to have a higher rate of reduction. They concluded that both waves were due to the reduction of coordinated water. The second wave, however, was catalyzed by intermediate hydride formation. The authors felt that these hydrides would be stable at the applied potentials used and noted the earlier work by J. Heyrovsky [38], who concluded $LiAlH_4$ was formed in the reduction of Al^{+3} in the presence of Li^+ salts. However, Sevryukov and Hassan did not refer to later work by M. Heyrovsky [39], who concluded that lithium aluminate was the substance formed.

Sevryukov and Hassan also speculated on the reduction mechanism of the three rare earths which are known to have a stable 2+ state. Europium, ytterbium, and samarium are probably reduced in a one-electron process. A subsequent chemical reoxidation with the solvent predominates in the case of samarium. The one-electron reduction of samarium(III) occurs in the same potential region as the other rare earths undergoing three-electron reductions.

In a more recent article by Leveque and Rosset [40], work by J. Grosbois [41] is discussed which indicates a conclusion which differs slightly from either of the above. The two waves for erbium are interpreted as being due to the reduction of coordinated water and the soluble hydroxide formed in the first step. In this view, the hydroxide precipitate would be formed as the product of the electrode process during the second step. This conclusion was based on the critical evaluation of other work.

On the basis of the results quoted above, and others too numerous to mention, it would appear that the mechanism of the reduction of the lanthanons in water is still not absolutely clear. Perhaps work done in nonaqueous solvents can shed more light on the question.

Yttrium, lanthanum, and the rare-earth metal ions were studied by Kolthoff and Coetzee [42] in acetonitrile and water. They state that europium, ytterbium, and samarium in both solvents are reduced first to the 2+ state and then to the metal. Their results, obtained with perchlorates in 0.1 N tetraethylammonium perchlorate as supporting electrolyte, are summarized in Table 2.2. The relatively positive potential for the production of the 2+ state in acetonitrile is explained by changes in solvation as compared with water. The 3+ state is highly solvated in water and would be expected to be more strongly affected than the 2+ state by the low solvating power of acetonitrile. Europium and samarium also gave a second wave at more negative potentials, corresponding to the production of the metal amalgam. Other ions of the rare earths were observed to give a single irreversible wave at -1.5 V, the height of which corresponds to a three-electron reduction to the amalgam.

In this article the authors also report erratic behavior at some potentials which may be due to precipitation and hydrous oxide film formation. Large-scale electrolysis at those potentials confirmed this observation. Further confirmation is by a report by Laitinen and Mohr [31], mentioned earlier.

Cokal and Wise [43] reported on the polarography of neodymium, samarium, ytterbium, and europium in acetonitrile. They concluded that it was possible to analyze mixtures of these ions in this solvent. They noted a precipitation of ytterbium, but only on standing. Apparently this was not an electrochemically induced precipitation.

A study on the polarography of several rare earths in benzonitrile has also been reported. Headridge and Pletcher [44] predicted greater stability of the 2+ state of europium, ytterbium, and samarium because of the low solvating power of benzonitrile. However, they observed a smaller difference in the half-wave potentials between the first and second waves in this solvent than in acetonitrile. They proposed possible ion-pairing between perchlorate ion and the rare-earth ions or residual water complexes to explain the effect. Both possibilities would be more important in benzonitrile.

Cohen, Iwamoto, and Kleinberg [45] studied the reduction of praseodymium in ethanol. They observed irreversible reductions with several steps. The step ratios varied with changes in the concentration of the supporting electrolyte. They observed spectral changes which paralleled the polarographic results. The observations were attributed to formation of complexes, similar to the effects noted for nickel and indium. The effect of the supporting electrolyte then must be carefully considered before any general conclusions are drawn about the mechanism of the electrode process.

A study in anhydrous ethylenediamine has been reported. Hall and Flanigan [46] studied the ions lanthanum, praseodymium, and ytterbium. As expected, lanthanum and praseodymium gave one wave and ytterbium was reduced in a two-step process. They were one of the only studies which attempted to analyze for the reaction products after controlled

TABLE 2.2. Polarographic Characteristics of Nonaqueous Solvents

	$E_{1/2}$ (V vs aq SCE)					
Substance	Acetonitrile[a] [51]	Acetone [51]	DMF [40]	Ethylenediamine[b] [46]	DMA [54]	DMSO [116]
La^{3+}	-1.5		-2.0	-2.46	-2.11	-2.26
Pr^{3+}	-1.50	-1.47	-2.09	-2.47	-2.06	-2.20
Nd^{3+}	-1.45	-1.49	-2.09		-2.07	-2.20
Sm^{3+}	-1.03/-1.62	-1.13/-1.70	-1.86/-2.04		-1.71/-2.08	2.02/-2.12
Eu^{3+}	+0.15/-1.65	-0.03/-1.79	-0.67/-2.07	-2.1/-2.47	-0.60/-2.10	-0.81/-2.12
Gd^{3+}	-1.5	-1.49	-2.16		-2.18	-2.16
Tb^{3+}		-1.38	-2.18		-2.20	-2.19
Dy^{3+}		-1.44	-2.18		-2.24	-2.08
Ho^{3+}		-1.44	-2.20		-2.25	-2.09
Er^{3+}		-1.43	-2.22		-2.27	-2.09
Tm^{3+}	-1.50	-1.44	-2.21			
Yb^{3+}	-0.58/-1.68	-0.71/-1.71	-1.40/-2.14	-2.31/-2.65	-1.27/-2.83	-1.48/-2.21

[a]The $E_{1/2}$ values for acetonitrile are averaged from several sources.

[b]The values for ethylenediamine were referenced to a zinc reference in the original article and are recalculated here.

potential analysis. A positive qualitative chemical test was reported for each of the metals indicating amalgam formation. The polarographic reduction of europium(III) acetylacetonate complexes has been studied in DMF [47] by Misumi et al. A positive shift in the half-wave potential on addition of lithium perchlorate to the supporting electrolyte tetraethylammonium perchlorate was observed. This was attributed to coordinative relaxation of the divalent reduction product. Thus

$$Eu(acac)_3 + e^- \rightleftarrows Eu(acac)_3^-$$

$$Eu(acac)_3^- + 3Li^+ \rightleftarrows Eu^{2+} + 3Li(acac)$$

The relaxation of three ligands is inferred by the slope of a plot of $\Delta E_{1/2}$ vs $\Delta \log [Li^{2+}]$ which has been shown by Murray and Hiller [48] to be equal to 0.059q/n. In this case q, the number of ligands, was equal to three. These results contrast with the results in the same study of q equal to one for $Fe(acac)_3$ reductions and q equal to two for $Fe(acac)_3$ reductions in acetonitrile by Murray and Hiller. The later authors concluded that the reaction

$$Li^+ + Li(acac) = Li_2(acac)^+$$

was important in acetonitrile when the ratio $[LiClO_4]/[Fe(acac)_3] \gg 1$. Misumi et al. did not consider this reaction for $Fe(acac)_3$ and $Eu(acac)_3$ reductions in DMF.

The polarographic reduction of europium(III) was carried out in various nonaqueous solvents by Gutmann and Peychal-Heiling [49]. Electrochemical parameters are tabulated. A correlation was noted between the half-wave potential and the donor number of the solvent used. The donor number is defined by Gutmann and Wychera [50] as a measure of the basicity or donor ability of the solvent. Its numerical value corresponds to the negative enthalpy of the reaction between the solvent base and antimony pentachloride.

Several other studies in nonaqueous solvent have also been reported. A partial list is included in the bibliography [51-60]. A summary of the polarographic behavior of cerium, ytterbium, and the rare earths in aqueous and nonaqueous solutions is shown in Tables 2.1 and 2.2.

A recent article described a carefully thought-out series of experiments on the lanthanides in N, N-dimethylformamide. Leveque and Rosset [40] initially gave a critical review of the literature in their article. Then, experiments are described which clearly differentiate between the behavior of the ceric group and yttric group rare earth ions in DMF. The Ilkovic equation was followed by many combinations of rare-earth ions. However, mixtures of a ceric earth (lathanum, cerium, neodymium, praseodymium) and yttric earths (gadolinium through lutetium) did not show strict proportionality of current with concentration. Mixtures of Cd^{2+} and IO_3^- show very similar behavior. In this case a precipitate of $Cd(OH)_2$ is formed due to the production of OH^- in the reduction of IO_3^- in unbuffered media.

Mixtures of erbium (an yitric earth) and calcium (reduced at more negative potentials) also show interesting behavior. The Ca^{2+} wave is diminished in the presence of erbium. $Ca(OH)_2$ is known to be insoluble in DMF and apparently precipitates because of the production of OH^- ion.

These observations and others like the effect of water, pulse polarographic results, and anodic stripping (they could not make yttric earth amalgams) led them to postulate the following mechanisms for the different categories of rare earth ions:

$$Eu^{3+} + e = Eu^{2+}$$ 1) Ions that form a stable 2+ state

$$Eu^{2+} + 2e = Eu(Hg)$$

$$La^{3+} + 3e = La(Hg)$$ 2) Ceric earth ions

$$Er(DMF\text{-}H_2O)_n^{3+} + pe = Er(OH)_3 + (n - p)(DMF\text{-}H_2O) + (p - 3)OH^- + pDMF + \frac{p}{2}H_2$$ 3) Yttric earth ions

Looking back at the recent literature lends support to the conclusions of Leveque and Rosset. Similar effects undoubtedly occur in the reductions of the lanthanons in water solutions. For example, the yttric rare earths are not extracted into mercury by sodium amalgams in pH 4 acetate solutions. Barrett, Sweasey, and Topp [61] explain this by suggesting that hydrogen ion is reduced from the coordinate sphere of the more easily hydrolyzed rare earths, i.e., those with larger atomic numbers (the yttric earths). In solutions of effective complexing ligands, where a tendency to hydrolyze would be less, Shvedov [62] has shown most of the rare earths can be extracted by electrolysis.

A good bit is now known about the polarography of the rare earths. Much of it was revealed piecemeal after some false starts. The complexing effect of supporting electrolyte and inherent differences among the rare earths tend to cloud the issue. Undoubtedly more work will be forthcoming. It will be interesting to see if newer studies can provide more evidence for hydride formation under usual polarographic conditions.

3. KINETIC PARAMETERS AND DOUBLE-LAYER PROPERTIES

3.1. CERIUM

The reaction $Ce^{4+} \rightleftarrows Ce^{3+}$ has been studied by several investigators. Petrocelli and Paolucci [63] examined the reaction experimentally and presented data which indicate that the overvoltage measured can include an activation overvoltage in addition to that due to concentration polarization. This study was done using both dilute solutions [0.01 F in $H_4Ce(SO_4)_4$, 0.005 F in $Ce_2(SO_4)_3$, and 0.5 to 5.0 N H_2SO_4] and concentrated solutions (0.1 to 0.5 F in cerium salts and 5.0 N H_2SO_4). For platinized platinum electrodes, an exchange current density (i_0) of ~ 18 mA/cm^2 and a transfer coefficient (α) of 0.5 were found. On bright platinum the values were $i_0 \cong 0.2$ to 1.5 mA/cm^2 and $\alpha = 0.25$ to 0.50, with the majority of the values near 0.50. It was concluded that in dilute solutions of cerium sulfate with bright platinum, and in concentrated solutions with platinized platinum, the reaction is predominantly under diffusion control. However, in concentrated solutions with bright platinum the reaction is predominantly under activation control.

The cerous-ceric redox system was studied by Greef and Anlich [64] to determine the kinetics of the reaction at a platinum electrode. The rotating-disk electrode technique was used with an electrode area of 0.138 cm^2. The reference was Pd/H_2 and the potentials were referred to the hydrogen electrode. Cerium sulfate solutions in perchloric acid, sulfuric acid, and nitric acid were used. The Ce^{3+} and Ce^{4+} concentrations were in the range of 5×10^{-4} to 10^{-1} M.

Current voltage curves were obtained with constant and equal potential scans in the anodic and cathodic directions (triangular sweep method). The reduction curves for the various electrolytes were found to be significantly different. This was attributed to the tendency for perchloric and nitric acids to be absorbed on the platinum surface and the complex nature of the cerium species. They point out that retardation of an oxidizing species by adsorption of the reduced species on platinum may be a rather general phenomenon.

In a study of "oxygen" uptake on Au electrodes, Bonewitz and Schmid [65] obtained kinetic parameters for the Ce^{3+}/Ce^{4+} reaction in 1 N $HClO_4$ and 2 N H_2SO_4 for oxygen coverages of $\theta = 1.5$ to 2.0 and $\theta = 0.7$ to 2.0, respectively. An all-Pyrex cell was used with a saturated calomel reference electrode. The test electrode was a gold disk of 0.071 cm^2. Solutions were made with either the cerous or ceric concentration constant at 0.01 N, varying the other between 0.01 and 0.0008 N. The cerium redox couple shows different apparent standard potentials in solutions containing various electrolytes. Thus, in 2 N sulfuric acid a formal potential of 1.44 V is found. On the other hand, in 1 N perchloric acid the formal potential is about 1.70 V. This higher value is due to the fact that cerium ion complexes with the perchlorate ion do not form. This difference in the formal potentials between a sulfuric acid and a perchloric acid solution makes reaction kinetic studies possible in various regions of oxygen coverage.

The polarization resistance, R_p, was obtained as the slope of the potential-current curves at zero overvoltage. It is thought to be composed of a charge transfer resistance, R_t, and a diffusion resistance, R_d, according to

$$R_p = R_t + R_d$$

$$= \frac{RT}{nF}\left(\frac{1}{i_0{}^a}\right) + \frac{RT}{nF}\sum^{j}\left|\frac{\nu_j}{i_{d,j}}\right|$$

where ν_j and $i_{d,j}$ are the stoichiometric factor and diffusion limiting current of species j. From this apparent exchange current, $i_0{}^a$, was found as a function of oxygen coverage of the electrode by appropriate selection of the initial anodic potential of the electrode.

The anodic charge transfer coefficient, α, was obtained by using the expressions

$$(\partial \log i_0{}^a/\partial E_e)_{c_r} = \alpha F/2.3RT$$

and

$$(\partial \log i_0{}^a/\partial E_e)_{c_0} = -(1 - \alpha)F/2.3RT$$

where c_0 and c_r are the concentrations of the oxidized and the reduced species, respectively. Thus α is obtained from the slope of plots of $\log(i_0{}^a)$ vs E_e, keeping either c_r or c_0 and the coverage with adsorbed oxygen constant. One such plot is shown in Fig. 3.1.1 for $\theta = 1.3$. A charge transfer coefficient, $\alpha = 0.65$, was determined. The α value can be compared with the charge transfer coefficients for the Ce^{4+}/Ce^{3+} reaction that have been reported by others. Galus and Adams [66] measured 0.79 on platinum and 0.72 on carbon paste. A value of 0.75 on platinum was found by Vetter [67]. The similarity of α values indicates

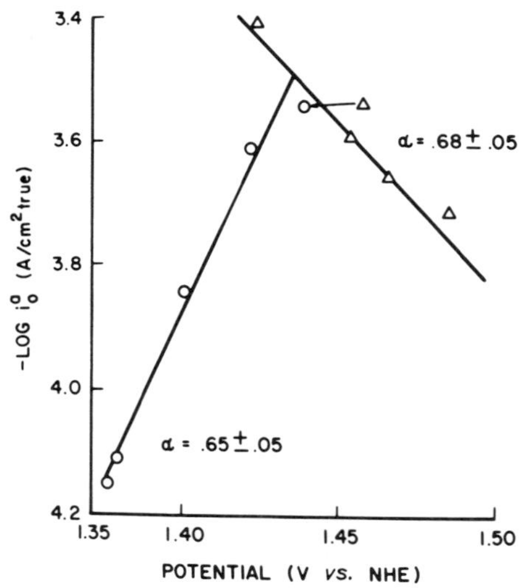

FIG. 3.1.1. Plot of log $i_0{}^a$ vs equilibrium potential, Au in 2 N H_2SO_4 with Ce(III)/Ce(IV) sulfate, oxygen coverage θ = 1.3 [65]. Reproduced from the Journal of the Electrochemical Society by permission of The Electrochemical Society.

to the authors that the mechanisms of the cerium reaction are apparently independent of the electrode material. The apparent rate constant was calculated to be $k_0{}^a = \sim 3 \times 10^{-4}$ cm/sec from the data for $i_0{}^a$ and α. Extrapolation of the data to $\theta = 0$ gives

$$k_0{}^a = \sim 12 \times 10^{-4} \text{ cm/sec}$$

It was noted that this $k_0{}^a$ value is very close to the number reported for other electrodes. This was interpreted as further proof that the activation energy (and the mechanism) for this reaction must be independent of the electrode material.

The authors list the possible effects of adsorbed oxygen on the electrochemical process. Since they found α independent of oxygen coverage, all effects except one could be eliminated; adsorbed oxygen can cause a potential drop across the diffuse double-layer change. They estimate this change to be small, about 8 mV, indicating that the adsorbed oxygen is a good electronic conductor and simply serves as an extension of the electrode surface.

The kinetics of the electroreduction of Ce^{3+} and Ca^{3+} have been studied by Sancho [68]. The irreversible reaction $Ce^{3+} + 3e = Ce$ is described by the standard equation

$$k = k_0 \exp[-\alpha nF/RT)(E - E_0)]$$

where k is the velocity constant for each value of overvoltage $E - E_0$, k_0 is k at the equilibrium potential, and α is the transfer coefficient.

Values for k_0 of 1.06×10^{-4} to 1.08×10^{-6} cm/sec were found in various media (aqueous and aqueous-alcohol, with and without gelatin). Values for the solution concentrations were between 1 and 3 mM.

3.2. EUROPIUM

Delahay [69] has discussed the Eu^{3+}/Eu^{2+} redox couple. He has pointed out that this system is important because the standard potential in perchloric acid is close to the point of zero charge on Hg (E_Z = -0.15 V vs NHE) and double-layer effects can be examined on both sides of E_Z.

Gierst [70] and Gierst and Cornelissen [27] found that the shape of the polarographic wave for Eu^{3+} and Eu^{2+} was profoundly influenced by the potential at the outer Helmholtz plane (ϕ_2) of the double layer. The step-by-step reasoning in the later paper is cogently outlined in their discussion.

1. The Eu^{3+}/Eu^{2+} system is apparently not complicated by a preceding chemical reaction as the chronopotentiometric transition times followed the Sand equation [71].
2. It was then possible to use Koutecky's polarographic method [72] to evaluate the apparent standard heterogeneous rate constant ($k_{sh}{}^{a}$).
3. The behavior of $k_{sh}{}^{a}$ follows the relationship first proposed by Frumkin [73] for an irreversible electron transfer process.
4. Complexation of Eu^{3+} and Eu^{2+} can be ruled out because the calculated values of the standard potentials were independent of solution composition (however, see below).
5. The minimum in the anodic wave can be fully explained by potential-dependent specific adsorption of anions.
6. The considerable effect of the tetraethylammonium cation can be explained partly by specific adsorption but also partly by surface blockage effects.

The rates of electron exchange reactions at an Hg electrode were examined by Randles and Somerton [74]. They calculated rate constants for the Eu^{2+}/Eu^{3+} reaction in inorganic solutions as summarized in Table 3.2.1. The rate constant k is a measure of the magnitude of the exchange reaction at the electrode surface at the equilibrium potential.

Timmer et al. [75] have studied the potential dependence of the faradaic impedance for the dc current irreversible electrode reaction Eu^{3+}/Eu^{2+} in 1 M $NaClO_4$. Solutions of 20 mM Eu^{3+} were prepared by dissolving Eu_2O_3 in a small excess of $HClO_4$ and adding $NaClO_4$. The experiments were performed at 25°C using a dropping mercury electrode. The ac polarograms were obtained by the complex plane method by which the real and imaginary components of cell impedance are measured as a function of dc potential.

TABLE 3.2.1. Rate Constants for Eu^{2+}/Eu^{3+} Reaction [74]

Base solution	E, potential against null point of Hg (V)	k (20°C) (cm/sec)
1 M KCl	-0.38	2.1×10^{-4}
1 M KI	-0.38	1.6×10^{-3}
1 M KCNS	-0.38	8.0×10^{-3}

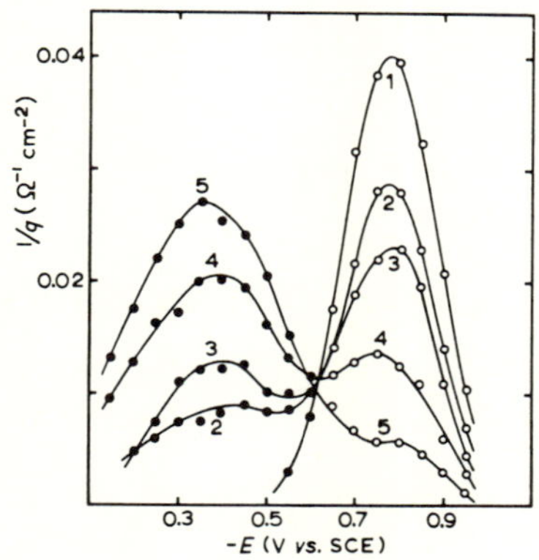

FIG. 3.2.1. The real component of the faradaic admittance (I/q) as a function of potential at 480 Hz (DME). (●) potentials anodic to E_0; (O) cathodic potentials concentrations: (1) 20 mM Eu^{3+}; (2) 14 mM Eu + 6 mM Eu^{2+}; (3) 11.7 mM Eu^{3+} + 8.3 mM Eu^{2+}; (4) 6.5 mM Eu^{3+} + 13.5 mM Eu^{2+}; (5) 2.6 mM Eu^{3+} + 17.4 mM Eu^{2+} [75]. Reproduced from the Journal of Electroanalytical Chemistry by permission of Elsevier Scientific Publishing Company.

From dc polarograms for different ratios of Eu^{3+}/Eu^{2+}, they found an average value of 0.40 for the transfer coefficient (α) and of 2.5×10^{-4} cm/sec for the rate constant (k_{sh}). The impedance measurements were used to calculate the quantity $1/q = Y_F$ (real component of the faradaic admittance) as a function of the electrode potential q where

$$q = \theta + \sigma\omega^{-1/2} + \frac{\sigma^2\omega^{-1}}{\theta + \sigma\omega^{-1/2}}$$

where θ is the activation polarization resistance, σ is the Warburg coefficient, and ω is the frequency. These plots are shown in Fig. 3.2.1 for different concentrations of Eu^{3+} and Eu^{2+}.

It was concluded from these studies that the experimental results for the faradaic impedance of the Eu^{3+}/Eu^{2+} system are in good accordance with the theory reported in Ref. 76.

The electrode kinetics and double-layer structure of the Eu^{2+}/Eu^{3+} electrode reaction at the dropping mercury electrode were studied by de Kreuk et al. [77]. This study was made using solutions of 10 m$\underline{M}$ Eu^{3+} in 1 $\underline{M}$ KCl and 1 $\underline{M}$ KI because the couple is electroactive in a potential region where chloride and iodide are strongly adsorbed.

From the limiting currents of dc polarograms, the following diffusion coefficients were calculated:

$$\left.\begin{aligned} D_{Eu^{3+}} &= 8.5 \times 10^{-6}\ cm^2/sec \\ D_{Eu^{2+}} &= 11.7 \times 10^{-6}\ cm^2/sec \end{aligned}\right\}\ 1\ \underline{M}\ KCl$$

$$\left.\begin{aligned} D_{Eu^{3+}} &= 6.3 \times 10^{-6}\ cm^2/sec \\ D_{Eu^{2+}} &= 8.3 \times 10^{-6}\ cm^2/sec \end{aligned}\right\}\ 1\ \underline{M}\ KI$$

The standard potential in 1 $\underline{M}$ KCl and 1 $\underline{M}$ KI as determined from equilibrium potentials at different Eu^{3+}/Eu^{2+} ratios is

$$E_0 = -625\ mV\ (vs\ SCE)$$

The apparent standard rate constant (k_{sh}^a) was strongly potential dependent, as expected. Several different calculation procedures gave good internal agreement. The true formal rate constant (k_{sh}^f), however, was largely potential independent. The electroactive species, Eu^{+3} and/or Eu^{+2}, affected the value for ϕ_2, and it was necessary to apply a modified version of the Frumkin correction. The potential independence of k_{sh}^f is similar to the results reported by Gierst and Cornelissen [27]. The transfer coefficients found were

0.59 in 1 $\underline{M}$ KCl

0.52 in 1 $\underline{M}$ KI

The mean values of k_{sh}^f were

1.4×10^{-5} cm/sec for 1 $\underline{M}$ KCl

0.6×10^{-5} cm/sec for 1 $\underline{M}$ KI

These values were recalculated in Ref. 74 to correct for the effect of complex formation:

2.1×10^{-5} cm/sec in 1 $\underline{M}$ KCl

0.75×10^{-5} cm/sec in 1 $\underline{M}$ KI

These authors concluded that, since k_{sh}^f was not particularly potential dependent for KI and KCl solutions and furthermore gave values consistent within an order of magnitude, that specific adsorption had no influence on the charge transfer rate. This conclusion is, of course, directly contradictory to the results of Gierst above. In addition, since there was a small potential dependence at negative potentials, de Kreuk et al. [77] also concluded that the electron transfer took place farther away from the electrode than the outer Helmholtz plane. The reactants are assumed to remain hydrated before or even during the reaction. The authors suggest that this hypothesis might be generalized to other solution-soluble electroactive pairs.

In a later study, de Kreuk et al. [78] investigated the europium reaction in KSCN. The stability of the europium thiocyanate complex is such that at the outer Helmholtz plane both the oxidized and reduced components are mainly present as uncomplexed, i.e., hydrated, ions. This makes it reasonable to assume that the reaction takes place via the hydrated species and that thus it is useful to compare the kinetic parameters with those in 1 M KCl and 1 M KI.

The solutions used were 2, 5, and 10 mM Eu^{3+} in 1 M KSCN at pH 2. Dc polarograms and impedance components at the DME were measured at 25°C.

The diffusion coefficients determined were

$$D_{Eu^{3+}} = 5.6 \times 10^{-6}\ cm^2/sec$$

$$D_{Eu^{2+}} = 7.3 \times 10^{-6}\ cm^2/sec$$

The standard potential is -640 mV vs SCE.

The ϕ_2 potentials at several Eu^{3+} concentrations were determined without consideration of the effects of complex formation as shown in Fig. 3.2.2. Complex formation of the europium ions with thiocyanate will dminish the amount of highly charged particles and thus diminish the change in ϕ_2 potential. The ϕ_2 potentials, as recalculated for complex formation for 2, 5, and 10 mM Eu^{3+} with dc polarization, are shown in Fig. 3.2.3.

It was determined that the "true" transfer coefficient $\alpha = 0.61$. The true rate constant was potential independent with a mean value of $k_{sh}^f = 3.2 \times 10^{-4}$ cm/sec.

The conclusions were much the same as before for KI and KCl solutions. Specific adsorption of the anion SCN^- had no influence on the charge transfer rate. However, several hypotheses were advanced to explain the order of magnitude difference in the true rate constant in KSCN compared with KI and KCl. The electron transfer could take place inside the outer Helmholtz plane. Using this model, the distance of the outer Helmholtz plane from the electrode surface is not constant and increases in the order $I^- < Cl^- < SCN^-$. On this assumption, a reasonable distance was calculated. Alternatively, electron transfer could take place through the complexed species rather than through the hydrated ion. The authors conclude by admitting that the problem is not yet settled but lean toward the former explanation.

Later, Rogers and Anson [79] reported their results for the same system, Eu(III) in thiocyanate solutions. They took issue with some of the conclusions of de Kreuk et al. [76]. In particular, they feel that electron transfer is possible for both the complexed and uncomplexed species. Making this assumption gives a true rate constant of 4×10^{-5} cm/sec and $\sim 8 \times 10^{-4}$ cm/sec for the uncomplexed and complexed forms of europium, respectively. The

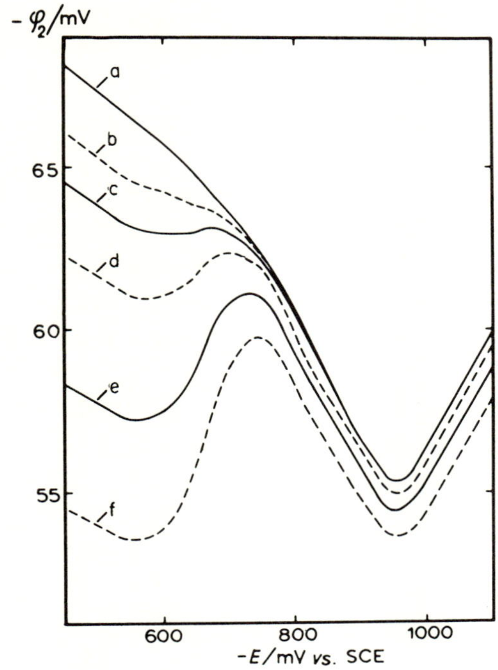

FIG. 3.2.2. Potentials ϕ_2 at the outer Helmholtz plane as a function of the electrode potential for 1 M KSCN at pH 2, with (a) 0, (b) 0.5, (c) 1, (d) 2, (e) 5, and (f) 10 mM Eu^{3+} with dc polarization (neglecting complex formation) [78]. Reproduced from the Journal of Electroanalytical Chemistry by permission of Elsevier Scientific Publishing Company.

former value is not significantly greater than that reported in chloride and iodide media, and it is no longer necessary to invoke a shifting outer Helmholtz plane. These authors feel, generally, that data analysis should not be restricted to only one reducing species when studying the electrode kinetics of labile metal systems. They also qualify the suggestion of de Kreuk et al. [76] that specific adsorption has no influence on the solution-soluble electroactive form by excluding inner sphere electron transfer pathways.

Sluyters et al. [80] also studied the europium(III)/europium(II) system in EDTA solutions. Their intention was to gain more insight in reactions accelerated by complex formations. It was necessary for them to extend an earlier theory and to prove that this was applicable to the conditions under investigation. In their study it was not always possible to maintain a large excess of EDTA. However, they were able to show in an appendix to the article that this condition was not an absolute requirement in this case. The europium(III) ion was more strongly complexed than the europium(II) ion as the standard potential shifted to more negative potentials. The electrochemical reaction can be represented by the equation

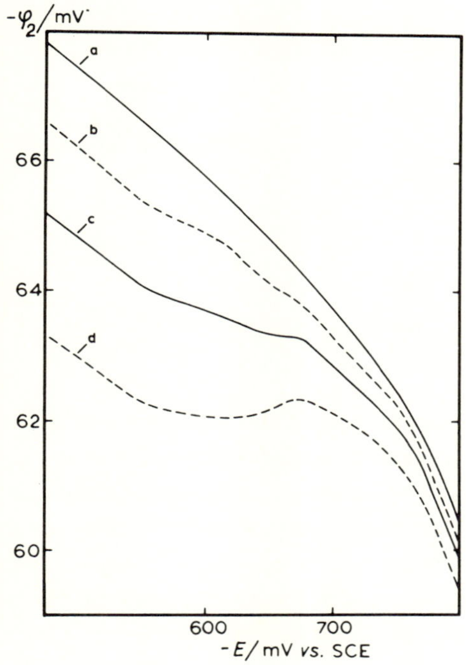

FIG. 3.2.3. Potentials ϕ_2 at the outer Helmholtz plane as a function of the electrode potential for 1 M KSCN with (a) 0, (b) 2, (c) 5, and (d) 10 mM Eu(III) with dc polarization and complex formation [78]. Reproduced from the Journal of Electroanalytical Chemistry by permission of Elsevier Scientific Publishing Company.

$$\text{Eu(III)L}_p + e = \text{Eu(II)L}_q + (p - q)\text{L}$$

They present electrochemical evidence that p - q is approximately equal to 1, and since there is no reason to suspect q is other than 1 [europium(II) is much like barium(II)], they conclude that europium(III) may be coordinated by more than one EDTA ligand. As the authors are aware, this conclusion would need further corroboration.

The final conclusions of the study were that the overall electron transfer proceeds in two steps:

$\text{Eu(III)Y}_2 = \text{Eu(III)Y} + \text{Y}$ dissociation step (fast)

$\text{Eu(III)Y} + e = \text{Eu(II)Y}$ charge transfer (fast)

where Y is EDTA. They measured k_{sh} to be greater or equal to 0.16 cm/sec. This value is larger than previously found for solutions containing chloride, iodide, or thiocyanate, and they concluded that this is strong evidence for the previously found acceleration of electron transfer by complexation.

3.3. LANTHANUM

The kinetics of the electroreduction of La^{3+} was also studied by Sancho [68] using the technique described for Ce^{3+} in Section 3.1. Values for k_0 of 1.84×10^{-4} to 1.73×10^{-5} cm/sec were found in various media.

4. ELECTROCHEMICAL STUDIES

It was found by Rairden [81] that a thin, apparently dense, oxide film is formed on yttrium by anodization in an aqueous solution containing ammonium pentaborate and ethylene glycol. The materials anodized were vacuum deposited thin films on glass substrates. Capacitance vs anodizing voltage data indicate that there are two linear regions of "oxide build": one from 0 to ~ 50 V the other from ~ 50 to ~ 400 V. Above 50 V the oxide thickness increases at a rate of 14.8 Å/V + 300 Å. The oxide film is "amorphous" and has a dielectric constant of 16.5 ± 0.5.

Goldstein and Pettit [82] determined the rate characteristics of constant voltage anodization of Y films. They found that the short-time (5 sec) current behavior is governed by a reciprocal logarithmic rate:

$$1/X = A - B \ln t$$

where X is the thickness, t is the time, $A = 1.23 \times 10^{-3}$/Å, and $B = 0.38 \times 10^{-3}$/Å. The long-time current decay was found to be proportional to $t^{-0.6}$.

The electrodeposition of the lanthanide metals has been reviewed by Hopkins [83], Holt [84], and Topp [85]. Because these metals are highly electropositive, electrodeposition must be done from nonaqueous electrolytes. They have been prepared by electrolysis of fused halides, and amalgams can be formed by reduction from selected organic solution at a Hg cathode [86].

Bivalent europium sulfate was precipitated from an acid solution of mixed rare earths by Yntema [87] using an electrolytic reduction process. The rare-earth oxides were dissolved in hydrochloric acid, the excess acid evaporated, and sulfuric acid added. This solution was diluted to 1 M and electrolyzed in a cell using a Hg cathode and a Pt anode. The europous sulfate was precipitated using a current density of 0.01 A/cm^2. A similar technique was used by Ball and Yntema [88] to selectivity precipitate $YbSO_4$.

Small quantities of high-purity cerium, lanthanum, and neodymium were prepared by Trombe [89] using a fused chloride electrolysis technique. Cerium was produced in a $CeCl_3 + KCl + CaF_2$ melt at 850°C. Nine grams of Ce were obtained in 1.5 hr from a 25-g $CeCl_3$ charge using a current density of 4 A/cm^2. A similar bath held at 960 to 980°C was used to obtain lanthanum. Neodymium was produced using a chloride bath held at 1040 to 1060°C.

Trombe's technique was used by Gray [90] to prepare cerium metal. Anhydrous $CeCl_3$ was dissolved in KCl + CeF_3 at 850°C. He found that control of current density at 3 to 4 A/cm^2 is important to avoid a loss of yield due to metal loss in the electrolyte. The yield by this process was 70% (yield vs theoretical yield calculated from Faraday's law). Gray was also successful in producing cerium electrolytically from the oxide. Five percent ceric oxide was dissolved in a mixture of CeF_3, LiF, and BaF_2.

Yttrium has been electrodeposited from halide salt melts by Butorov et al. [91] on a liquid Zn cathode at an efficiency of 94 to 96%. Polarization curves in NaCl-KCl melts were studied at 700, 750, and 800°C. On a Mo cathode, deposition of Y occurs at a voltage near to its equilibrium potential; on a Zn cathode, the deposition potential is shifted 0.6 V more positive.

Borides of the lanthanide metals (which were designated by the symbol Ln) were formed by electrolysis of Ln_2O_3 in a fused borate bath by Andrieux [92] and by von Stackelberg and Newman [93]. Alkali and alkaline earth fluorides were added to the bath which was operated at 950 to 1000°C. The cathode product was a mixture of the hexaboride and boron.

Lanthanide silicides have also been formed electrolytically by Dodero [94]. A fused silicate bath containing $CaSiO_3$, CaF_2, and $CaCl_2$ was operated at 1000°C.

Eu and Yb amalgams were electrolytically produced by McCoy [95] from a solution of potassium citrate and a rare-earth acetate made alkaline with potassium carbonate. When such solutions were electrolyzed with a Hg cathode, amalgams were formed at low voltage with good current efficiencies.

Shvedov and Stepanov [96] studied the deposition of Ce and Pr on a liquid Wood's alloy using a radioisotope technique. These metals were deposited at 15 to 20 mA/cm^2 from a citric acid solution containing 0.1 M LiOAc.

A series of articles [62, 97-99] on the electrochemical production of rare-earth amalgams has appeared recently. As noted previously, citrates, tartrates, and lacates, all complexing anions, are necessary to produce significant quantities of some of the yttric earth metals. David and Bouissieres [100] have reported that the kinetics of the formation of amalgams during the electrolysis of rare earths in citrate solutions is facilitated if a stable divalent state is formed.

5. APPLIED ELECTROCHEMISTRY

5.1. ELECTROWINNING AND ELECTROREFINING

The electrolytic reduction technique has been used extensively for producing rare-earth metals. This technology has been reviewed by Kremers [101], Love [102], and Hirschhorn [103].

The major commercial rare-earth product produced electrolytically is mischmetal which is, typically, principally a mixture of Ce, La, Nd, and Pr. The primary ore sources are monazite sand, a rare-earth phosphate, and bastnasite, a rare-earth fluocarbonate. These are converted to the anhydrous chlorides for electrolytic reduction in ceramic or graphite cells.

Electrolysis is done in a bath containing mixtures of the rare-earth chlorides and the alkali and alkaline-earth chlorides held at 800 to 900°C. The deposition is made on iron or graphite cathodes.

High-purity cerium (99.9%) has been produced electrolytically by Morrice et al. [104]. Carefully prepared CeF_3, LiF, and BaF_2 were dried at 280°C and charged to a graphite cell. After initiation of melting using an ac current, vacuum-dried CeO_2 was added to the cell and the current was changed to dc for electrolysis.

5.2. CORROSION

Love [105] investigated the rate of atmospheric corrosion of rare-earth metals at room and elevated temperature and in dry and moist air. He found that the various metals show significant differences in behavior as summarized in Table 5.2.1. The light rare earths, La and Ce, oxidize most rapidly, while the oxidation rate for heavy rare-earth metals is low at moderate temperatures, even in the presence of water.

TABLE 5.2.1. The Corrosion Rates of Rare Earth Metals in Air [105]

	mg/dm²/day						
Temperature (°C):	35		95		200	400	600
Relative humidity (%):	1	75	1	75			
Lanthanum	80	950	510	21,000	30	3,200	13,000
Cerium						200,000	
Praseodymium	8	76	900	5,500	80	38,000	130,000
Neodymium	2	7	60	2,000	70	380	4,800
Samarium	0	0	0	100	15	17	35
Gadolinium	1	2	0	35	0	210	16,000
Terbium	0	0	0		0	1,600	40,000
Dysprosium	0	0	0	43		350	6,600
Holmium	1	1	1		11	110	5,400
Erbium	1	1	0		10	90	720
Ytterbium						170	
Yttrium	1	1	2	9	4	40	1,900

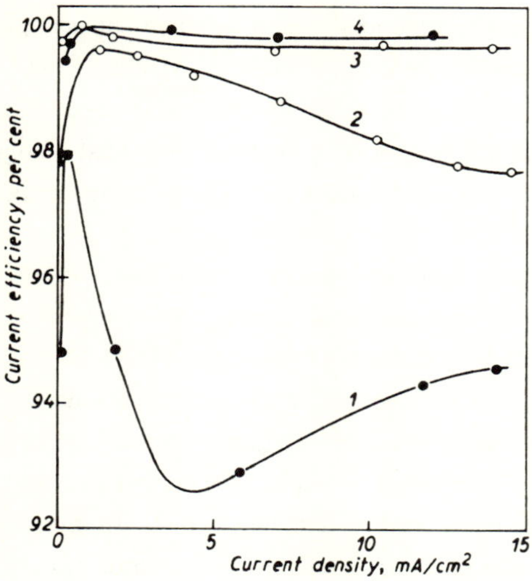

FIG. 5.3.1. Current efficiency for electrogeneration of ceric ion as a function of current density and solution composition. (1) 0.05 M cerous perchlorate in 1 M perchloric acid. (2) 0.025 M cerous sulfate in 0.9 M sulfuric acid. (3) 0.1 M cerous sulfate in 0.9 M sulfuric acid. (4) 0.1 M cerous sulfate in 3 M sulfuric acid [106]. Reproduced from Analytica Chimica Acta by permission of Elsevier Scientific Publishing Company.

5.3. COULOMETRIC TITRATION

A coulometric titration technique was reported by Lingane et al. [106] which utilizes the electrogenerated ceric ion. They found that the current efficiency for the electrogeneration of the ceric ion at a Pt anode falls considerably below 100% at both very small and at large current densities in both sulfuric and perchloric acid media. This is illustrated in Fig. 5.3.1. However, a high titration efficiency for ceric ion generation is obtained for iodine because it is oxidized to the iodate ion slightly in advance of the potential at which the cerous ion is oxidized. Since the electrogenerated iodate ion oxidizes iodide ion, only a minor fraction of the total quantity of electricity results from ceric ion generation, so the effect of its inefficient generation is greatly minimized.

Schults [107] reported on a controlled potential coulometric procedure for europium in aqueous solutions. He found it was not possible to reduce europium(III) with 100% current efficiency. However, europium(II) can be prepared electrochemically and then reoxidized back to europium(III). Coulombs measured during the reoxidation process provide the means for quantitative estimation. Another study [108] on europium described a procedure where europium(II) was produced by reaction of europium(III) with zinc metal and the europium(II) titrated by electrogenerated bromine. The end point of the titration was determined graphically with 4% precision.

A procedure for the coulometric analysis of mixtures of europium, ytterbium, and samarium in acetonitrile has been described. Cokal and Wise [109] were able to analyze these elements individually or in mixtures to a precision of about 5%. Precipitation of hydroxides and sample preparation makes their method somewhat inconvenient. They suggest some possible remedies. It was not possible to analyze for neodymium under their conditions.

In an adjacent note, these authors reported [110] observations of colored solutions in mixtures of rare earths in different oxidation states. They attributed these to be mixed valence complexes. Later Dyke and Hush [111] showed the spectral properties produced from ordinary superpositions of the spectra of the individual ions.

Interesting methods based on electrolytic generation of lanthanum(III) using a lanthanum hexaboride anode have been reported [112-113]. Normally, in constant current coulometry, the equivalents produced are calculated by Faraday's law. In this case an empirical correction had to be applied since the current efficiency was greater than 100%. Happily, in a given pH range, the current efficiency was constant to within 0.1% and the method was successfully applied to several different systems. A measured excess of EDTA was added to solutions of nickel(II), cobalt(II), and zinc(II). The excess was back-titrated with electrogenerated lanthanum(III) and an eriochrome black T end point. Precision and accuracy were reported to be in the few parts per thousand range. Fluoride ion and oxalate ion can be also titrated by electrogenerated lanthanum. End points are determined using a fluoride ion selective electrode and a turbidimetric procedure, respectively. One part per thousand precision and accuracy are claimed for the fluoride titration, and 1 to 2% for the oxalate.

REFERENCES

1. W. M. Latimer, The Oxidation States of the Elements and Their Potentials in Aqueous Solutions, 2nd ed., Prentice-Hall, Englewood Cliffs, New Jersey, 1961.
2. C. A. Hampel, Rare Metals Handbook, 2nd ed., Reinhold, London, 1961.
3. D. M. Yost, H. Russell, and C. S. Garner, The Rare Earth Elements and Their Compounds, Wiley, New York, 1947.
4. F. H. Spedding and C. F. Miller, J. Amer. Chem. Soc., 74, 4195 (1952).
5. G. Charlot, Selected Constants, Oxydo-Reduction Potentials, Pergamon, New York, 1958.
6. G. F. Smith and C. A. Getz, Ind. Eng. Chem., Anal. Ed., 10, 191 (1938).
7. A. A. Noyes and C. S. Garner, J. Amer. Chem. Soc., 58, 1265 (1936).
8. K. J. Vetter, Z. Phys. Chem., A., 196, 360 (1951).
9. A. H. Kunz, J. Amer. Chem. Soc., 50, 98 (1931).
10. L. B. Anderson and D. J. Macero, J. Phys. Chem., 67, 1942 (1963).
11. L. R. Morss and H. O. Haug, J. Chem. Thermodyn., 5, 513 (1973).

12. G. Biedermann and H. B. Silber, Acta Chem. Scand., 27, 3761 (1973).

13. G. C. Walters and D. W. Pearce, J. Amer. Chem. Soc., 62, 3331 (1940).

14. E. I. Onstott, Rare Earth Research (E. V. Kleber, ed.), Macmillan Co., New York, 1961.

15. H. A. Laitinen, J. Amer. Chem. Soc., 64, 1133 (1942).

16. L. J. Nugent, R. D. Baybarz, J. L. Burnett, and J. L. Ryan, J. Inorg. Nucl. Chem., 33, 2503 (1971).

17. L. J. Nugent, R. D. Baybarz, J. L. Burnett, and J. L. Ryan, J. Phys. Chem., 77, 1528 (1973).

18. K. E. Johnson and J. R. MacKenzie, J. Electrochem. Soc., 116, 1697 (1969).

19. T. Kuroda, Bull. Electrotech. Lab. (Tokyo), 17, 405 (1953).

20. T. Kojima and K. Kikuchi, J. Electrochem. Soc., Japan, 21, 525 (1953).

21. B. Neumann and H. Richter, Z. Electrochem., 31, 287 (1925).

22. U. K. Delimarskii and B. F. Markov, Electrochemistry of Fused Salts (trans. by A. Peiperyl and R. E. Wood), Sigma Press, Washington, D.C., 1961.

23. W. Hamer, M. Malmberg, and B. Rubin, J. Electrochem. Soc., 103, 8 (1956).

24. J. Van Muylder and N. de Zoubov, in Atlas of Electrochemical Equilibria (M. Pourbaix, ed.), (trans. by J. A. Franklin), Pergamon, New York, 1966.

25. A. A. Vlcek, Collect. Czech. Chem. Commun., 24, 181 (1959).

26. W. F. Kinard and R. H. Philp, Jr., J. Electroanal. Chem., 25, 373 (1970).

27. L. Gierst and R. Cornelissen, Collect Czech. Chem. Commun., 25, 3004 (1960).

28. E. I. Onstott, J. Amer. Chem. Soc., 74, 3773 (1952).

29. W. Noddack and A. Bruckl, Angew. Chem., 50, 362 (1937).

30. I. M. Kolthoff and J. J. Lingane, Polarography, Interscience, New York, 1946.

31. H. A. Laitinen and E. B. Mohr, private communication quoted in Ref. 42.

32. S. Misumi and Y. Ide, Bull. Chem. Soc. Japan, 33, 836 (1960).

33. H. Hamaguchi, J. Hashimoto, K. Hosohara and Y. Narusawa, ibid., 33, 562 (1960).

34. L. Treindl, Collect. Czech. Chem. Commun., 24, 3389 (1959).

35. L. Treindl, ibid., 26, 3003 (1961).

36. R. F. Large and A. Timnick, Anal. Chem., 36, 1258 (1964).

37. N. N. Sevryukov and M. Z. Hassan, Zh. Anal. Khim., 22, 1487 (1967).

38. J. Heyrovsky, Collect. Czech. Chem. Commun., 18, 756 (1953).

39. M. Heyrovsky, Advan. Polarogr., 3, 854-860 (1960).

40. A. Leveque and R. Rosset, Bull. Soc. Chim. Fr., 1971, 4205.

41. J. Grosbois, Memoire Conservatoire National des Arts et Metiers, Paris, 1967.

42. I. M. Kolthoff and J. F. Coetzee, J. Amer. Chem. Soc., 79, 1852 (1957).

43. E. J. Cokal and E. N. Wise, J. Electroanal. Chem., 11, 406 (1966).

44. J. B. Headridge and D. Pletcher, ibid., 15, 312 (1967).

45. S. H. Cohen, R. T. Iwamoto, and J. Kleinberg, J. Phys. Chem., 67, 1275 (1963).

46. L. C. Hall and D. A. Flanigan, Anal. Chem., 35, 2108 (1963).

47. S. Misumi, M. Aihara, and Y. Nanoka, Bull. Chem. Soc. Japan, 43, 774 (1970).

48. R. W. Murray and L. K. Hiller, Anal. Chem., 39, 1221 (1967).

49. V. Gutmann and G. Peychal-Heiling, Monatsh. Chem., 100, 1423 (1969).

50. V. Gutmann and E. Wychera, Inorg. Nucl. Chem. Lett., 2, 257 (1966).

51. C. K. Mann and K. K. Barnes, Electrochemical Reactions in Nonaqueous Systems, Dekker, New York, 1970, p. 453.

52. J. N. Gaur and N. K. Goswami, Proceedings of the Symposium on Electrode Processes, 1966, pp. 78-83; through Chem. Abstr., 70, 43373 (1969).

53. K. Zutshi, Rev. Polarogr., 14, 294 (1967); through Chem. Abstr., 71, 9073 (1969).

54. V. Gutmann, Monatsh. Chem., 100, 813 (1969).

55. V. Gutmann and G. Peychal-Heiling, ibid., 100, 1423 (1969).

56. G. Peychal-Heiling and V. Gutmann, Z. Anal. Chem., 248, 6 (1969).

57. J. Sancho, A. Alday, and A. Pujante, An. Real Soc. Espan. Fis. Quim., Ser. B, 66, 245 (1970).

58. J. Sancho, A. Alday, and A. Pujante, ibid., 66, 239 (1970).

59. S. Misumi, M. Aihara, and Y. Nonaka, Bull. Chem. Soc. Japan, 43, 774 (1970).

60. V. R. Chandrasekaran and A. K. Sundaram, Proc. Indian Acad. Sci., Sect. A, 74, 133 (1971); through Chem. Abstr., 76, 80167 (1972).

61. M. F. Barrett, D. Sweasey, and N. E. Topp, J. Inorg. Nucl. Chem., 25, 1273 (1963).

62. V. P. Shvedov and P. G. Antonov, Radiokhimiya, 11, 307 (1969); through Chem. Abstr., 71, 87311 (1969).

63. J. V. Petrocelli and A. A. Paolucci, J. Electrochem. Soc., 93, 291 (1951).

64. R. Greef and H. Anlich, J. Electroanal. Chem., 18, 295 (1968).

65. R. A. Bonewitz and G. M. Schmid, J. Electrochem. Soc., 117, 1367 (1970).

66. Z. Galus and R. N. Adams, J. Phys. Chem., 67, 866 (1963).

67. K. J. Vetter, Z. Phys. Chem., 196, 360 (1951).

68. J. Sancho and V. Almagro, An. Real Soc. Espan. Fis. Quim., Ser. B, 67, 129 (1971).

69. P. Delahay, Double Layer and Electrode Kinetics, Wiley-Interscience, New York, 1965.

70. L. Gierst, Transactions on the Symposium on Electrode Processes (E. Yeager, ed.), Wiley, New York, 1961, p. 109.

71. R. N. Adams, Electrochemistry at Solid Electrodes, Dekker, New York, 1969, p. 169.

72. J. Heyrovsky and J. Kita, Principles of Polarography, Academic, New York, 1966, pp. 210-211.

73. Ref. 72, pp. 232-234.

74. J. E. B. Randles and K. W. Somerton, Trans. Faraday Soc., 48, 937 (1952).

75. B. Timmer, M. Sluyters-Rehbach, and J. H. Sluyters, J. Electroanal. Chem., 14, 181 (1967).

76. B. Timmer, M. Sluyters-Rehbach, and J. H. Sluyters, ibid., 14, 169 (1967).

77. C. W. de Kreuk, M. Sluyters-Rehbach, and J. H. Sluyters, ibid., 28, 391 (1970).

78. C. W. de Kreuk, M. Sluyters-Rehbach, and J. H. Sluyters, ibid., 33, 267 (1971).

79. R. S. Rogers and F. C. Anson, ibid., 42, 381 (1973).

80. L. Kisova, M. Sluyters-Rehbach, and J. H. Sluyters, ibid., 40, 29 (1972).

81. J. R. Rairden, J. Electrochem. Soc., 114, 75 (1967).

82. R. M. Goldstein and W. J. Pettit, ibid., 115, 979 (1968).

83. B. S. Hopkins, Trans. Electrochem. Soc., 89, 295 (1946).

84. M. L. Holt, Modern Electroplating, 2nd ed. (F. A. Lowenheim, ed.), Wiley, New York, 1963.

85. N. E. Topp, The Chemistry of the Rare-Earth Elements, Elsevier, Amsterdam, 1965.

86. F. E. Ephraim, Inorganic Chemistry, 5th ed. (F. A. Lowenheim, ed.), Wiley, New York, 1963.

87. L. F. Yntema, J. Amer. Chem. Soc., 52, 2782 (1930).

88. R. W. Ball and L. F. Yntema, ibid., 52, 4264 (1930).

89. F. Trombe, Trans. Electrochem. Soc., 66, 57 (1934).

90. P. M. J. Gray, Trans. Inst. Mining Met., 61, 231 (1951).

91. V. P. Butorov, I. F. Nichkov, E. A. Novikov, S. P. Raspopin, and M. I. Solin, Izv. Vyssh. Ucheb. Zaved., Tsvet. Met., 14(3), 85 (1971).

92. L. Andrieux, C.R. Acad. Sci., Paris, 186, 1736 (1928); 194, 720 (1932).

93. M. von Stackelberg and F. Newman, Z. Phys. Chem., B19, 314 (1932).

94. M. Dodero, Bull. Soc. Chim. Fr., 1950, 545.

95. H. N. McCoy, J. Amer. Chem. Soc., 63, 1622 (1941).

96. V. P. Shvedov and V. I. Stepanov, Radiokhimiya, 13(4), 610 (1971).

97. P. G. Antonov and V. P. Shvedov, ibid., 9, 264 (1967); through Chem. Abstr., 68, 18098 (1968).

98. V. P. Shvedov and P. G. Antonov, ibid., 9, 478 (1967); through Chem. Abstr., 68, 101201 (1968).

99. P. G. Antonov and V. P. Shvedov, ibid., 11, 311 (1969); through Chem. Abstr., 71, 87310 (1969).

100. F. David and G. Bouissieries, Inorg. Nucl. Chem. Lett., 4, 153 (1968).

101. H. E. Kremers, Rare Metals Handbook, 2nd ed. (C. A. Hampel, ed.), Reinhold, New York; Chapman and Hall, London, 1961, p. 393.

102. B. Love, WADD Tech. Rep. 60-864, Wright-Patterson Air Force Base, Ohio, 1961.

103. I. S. Hirschhorn, The Encyclopedia of Electrochemistry (C. A. Hampel, ed.), Reinhold, New York; Chapman and Hall, London, 1964, p. 1000.

104. E. Morrice, J. Darrah, E. Brown, C. Wyche, W. Headrick, R. Williams, and R. G. Knickerbocker, U.S. Bur. Mines, Rep. Invest., 5549 (1960).

105. B. Love, WADC Tech. Rep. TR 57-666, Part II, Wright-Patterson Air Force Base, Ohio, 1959.

106. J. J. Lingane, C. H. Langford, and F. C. Anson, Anal. Chem. Acta, 16, 165 (1957).

107. W. D. Schults, Anal. Chem., 31, 1095 (1959).

108. M. K. Akhmadeev, Sb. Aspir. Rab., Kazan. Gos. Univ., Khim. Geol., 1967, 95; through Chem. Abstr., 70, 63877g (1969).

109. E. J. Cokal and E. N. Wise, J. Electroanal. Chem., 12, 136 (1966).

110. E. J. Cokal and E. N. Wise, ibid., 12, 170 (1966).

111. J. M. Dyke and N. S. Hush, ibid., 36, 337 (1972).

112. D. J. Curran and K. S. Fletcher III, Anal. Chem., 40, 1809 (1968).

113. D. J. Curran and K. S. Fletcher III, ibid., 41, 267 (1969).

114. L. Meites, Polarographic Techniques, 2nd ed., Wiley-Interscience, New York, 1965, p. 615.

115. J. Heyrovský and P. Zuman, Practical Polarography, Academic, New York, 1968.

116. G. Gritzner, V. Gutmann, and G. Schöber, Monatsh. Chem., 96, 1056 (1965).

Chapter VI-3

ALUMINUM

PAUL A. MALACHESKY*

Tyco Laboratories, Inc.
Chemistry Department
Waltham, Massachusetts

*Present address: Exxon Research and Engineering Co., Linden, New Jersey.

1. STANDARD AND FORMAL POTENTIALS

1.1. AQUEOUS SOLUTION

Aluminum belongs to that group of metals which have very negative electrode potentials and evolve hydrogen on contact with most aqueous electrolytes. For aluminum there is the added complication of a tendency for formation of a nonconducting oxide or amorphous hydroxide layer. For these reasons, accurate and reproducible experimental electrode potential data for aluminum in aqueous solution are not in abundance.

Table 1.1.1 presents the main features of the electrode potential data for Al as calculated from thermodynamic data by de Bethune and Loud [1]. There are minor discrepancies between the data of Table 1.1.1 and those presented at an earlier date by Latimer [2]. Very early experimental attempts to place aluminum in the electrochemical series were summarized by Heyrovsky who, ca. 1920, carried out what is perhaps the most extensive investigation of the electrode potentials of aluminum [3]. Heyrovsky used dilute aluminum amalgams (<0.1% Al) to minimize complications from hydrogen evolution and oxide formation. In more recent times, workers (Plumb and Hagyard et al.) have tried to use special techniques to generate fresh aluminum surfaces with minimal oxide coverages [4-7]. However, none of these workers have been able to confirm experimentally the thermodynamically calculated electrode potential values shown in Table 1.1.1.

Heyrovsky, using Al amalgam and working in $AlCl_3$ solutions, found a value of -1.333 V (at 25°C) for the formal potential of aluminum [3]. He attributed this electrode potential to the reaction

$$Al + 3OH^- = Al(OH)_3 + 3e^- \qquad (1.1.1)$$

Working in acidic solutions with fresh Al surfaces of minimal oxide coverages, Plumb's electrode potential data shows a value of -1.324 V for the formal potential of Al [4]. Hagyard and Earl measured a potential of -1.65 V vs NHE for oxide-free Al electrodes in 0.00033 M $AlCl_3$ (+1 M KCl) from which a value of -1.58 V can be calculated for the formal potential of Al [7]. In basic solution, Plumb and Swain observed that at pH <12.4 the electrode potential data corresponded to an Al/aluminate ion couple with [5]

$$E\ (\text{V vs SCE}) = 0.5099 + 0.0197 \log \text{aluminate ion} + 0.1182\ \text{pH} \qquad (1.1.2)$$

At pH >12.4, a change in the potential-determining reaction apparently occurred with the potential data described by the empirical relationship

$$E\ (\text{V vs SCE}) = 0.962 + 0.0788\ \text{pH} \qquad (1.1.3)$$

In Eq. (1.1.3) the potential is not dependent on the aluminum ion concentration. While Plumb and Swain invoked a reaction between OH^- and Al ions in solution to account for this nondependence on aluminum ion concentration, one must be quite skeptical of such an explanation since the measured potentials are most likely mixed potentials as pointed out by Kolthoff and Sambucetti [8].

TABLE 1.1.1.[a] Standard Electrode Potentials for Al Electrode Reactions Calculated from Thermodynamic Data (after Ref. 1)

Electrode reaction	E° (V)	dE/dT mV/K	dE/dT mV/K
$Al^{3+} + 3e = Al$	-1.662	+1.375	+0.504
$AlF_6^{3-} + 3e = Al + 6F^-$	-2.069	+0.67	-0.20
$H_2AlO_3^- + H_2O + 3e = Al + 4OH^-$	-2.33		
$Al(OH)_3 + 3e = Al + 3OH^-$	-2.30	-0.06	-0.93

[a]The thermal temperature coefficient $(dV°/dT)_{th}$ has a positive value if the hot electrode is the (+) thermal cell of two such electrodes at two different temperatures. The thermal liquid junction potential is taken as zero in KCl solutions. The first and second isothermal temperature coefficients have positive values if $d(V' - V)/dT^2$ are positive, respectively, in the isothermal cell: (E) Al|SHE||electrolyte|electrode|Al' (E'). Values given are calculated. The thermodynamic relations used are:

$$E° = \Delta G°(\text{oxidn})/nF$$

where $\Delta G°$ is that for the oxidation reaction

$$R = O + ne^-$$

and

$$(dE°/dT)_{isoth} = \Delta S°(\text{redn})/nF$$

$$(dE°/dT)_{th} = (dE°/dT)_{isoth} + 0.871 \text{ mV/°C}$$

1.2. NONAQUEOUS SOLUTION

Standard or formal potential data for aluminum in nonaqueous electrolytes are, to the best of the author's knowledge, nonexistent. This is not too surprising when one considers the paucity of data for aqueous solution. There are some reports of "decomposition potentials" and open-circuit potentials for aluminum in such systems, but these have little thermodynamic value [9, 10]. Wallace and Bruins measured a potential of 1.150 V for the cell Al|0.1 M $AlCl_3$-acetamide|AgCl, Ag at 90°C [11]. Hydrogen evolution was observed, however, due to the high acidity of the molten $AlCl_3$-acetamide electrolyte. The measurement of electrode potentials for aluminum in nonaqueous solution would seem to be an area for useful studies.

1.3. FUSED SALTS

In contrast to the paucity of experimental electrode potential data for aluminum in aqueous and nonaqueous solution, such data in fused salts are in relative abundance. In such fused salt electrolytes, oxide films are of little importance, and the aluminum electrode reactions are quite reversible. Experimental electrode potential data for Al are collected in Table 1.3.1. Delimarskii and Markov have compared a number of experimentally determined decomposition electrode potentials for Al-containing electrolytes with potential values calculated from thermodynamic data which may be of some interest [12]. Table 1.3.2 collects such data relevant to aluminum. Electrode potential data for Al in cryolite melts at elevated temperatures (~1000°C) are of importance with regard to Al electrowinning (see Section 5.1.1). Thonstad has collected and compared experimental measurements with calculated emf values for the cell [13]

$$Al_{liq}/Na_3AlF_6 + Al_2O_3/Pt,\ O_2 \qquad (1.3.1)$$

corresponding to the cell reaction

$$2Al + \frac{3}{2}O_2 = Al_2O_3 \qquad (1.3.2)$$

Most of these earlier experimental results gave values which were lower than those calculated from thermodynamic data. Since the Al electrode has been shown to behave quite reversibly in such melts, such discrepancies were attributed to the oxygen electrode. Thonstad, using potential decay curves, measured the anodic overvoltage and arrived at an extrapolated value of 2.205 V for the cell reaction at 1000°C. This value agrees quite closely with the calculated emf of 2.194 V.

TABLE 1.3.1. Experimental Electrode Potential Data for Al in Fused Salts

Solvent[a]	Electrode	Potential (V)	Reference electrode	T° (C)	Refs.
$50MgCl_2$-30NaCl-20KCl	Al/Al^{3+}	-1.990	Cl_2/Cl^-(graphite)	475	[14]
59LiCl-41KCl	Al/Al^{3+}	-2.013	Cl_2/Cl^-(graphite)	450	[15]
NaF-KF	Al/Al^{3+}	-1.500	Ni^{2+}/Ni	850	[16]
$KHSO_4$	Al/Al^{3+}	-0.63	H_2 (1 atm)/Pt	270	[17]
$70AlBr_3$-30NaBr	Al/Al^{3+}	-2.757 -2.715	Na/Na^+(glass) Na/Na^+(glass)	255 165	[18]

[a]All composition values in mole %.

TABLE 1.3.2. A Comparison of Calculated and Experimental Decomposition Potential Values for Al Containing Molten Salt Systems at 700°C (taken, in part, from Ref. 12)[a]

Solvent	E_d^{theory} (V)	E_d^{exp} (V)
$AlCl_3$	1.90	1.61
$NaCl-AlCl_3$	1.64	1.82
$NaCl-KCl-SrCl_2$	1.67	1.74
$AlBr_3$	1.52	1.20
$NaBr-AlBr_3$	1.23	1.39
NaBr-KBr	1.26	1.32
AlI_3	0.89	0.70
$NaI-AlI_3$	0.73	0.81
NaI	0.73	0.78

[a]E_d^{theory} values have been calculated from the latest JANEFF values for the pure aluminum halides [19].

1.4. EQUILIBRIUM DATA

Deltombe, Vanleugenhaghe, and Pourbaix have constructed a potential-pH diagram for the aluminum-water system at 25°C [20]. This diagram is shown in Fig. 1.4.1 and is based on hydrargillite, $Al_2O_3 . 3H_2O$ as the most stable form of aluminum oxide. The diagram of Fig. 1.4.1 is valid only in the absence of complexing anions or substances which form insoluble salts with aluminum and is based on the following equations:

(1) $Al^{3+} + 2H_2O = AlO_2^- + 4H^+ \quad \log [AlO_2^-]/[Al^{3+}] = -20.30 + 4\ pH$ (1.4.1)

The limit of the region in which Al^{3+} exists as a simple dissolved ion:

$Al^{3+}/AlO_2 \quad pH = 5.07$ (1.4.2)

(2) $Al + 3H_2O = Al_2O_3 + 6H^+ + 6e^- \quad E_0 = -1.550 - 0.591\ pH$ (1.4.3)

(3) $2Al^{3+} + 3H_2O = Al_2O_3 + 6H^+ \quad \log Al^{3+} = 5.70 - 3\ pH$ (1.4.4)

(4) $Al_2O_3 + H_2O = 2AlO_2^- + 2H^+ \quad \log AlO_2^- = -14.60 + pH$ (1.4.5)

(5) $Al^{3+} + 3e^- = Al \quad E_0 = -1.663 - 0.0197 \log Al^{3+}$ (1.4.6)

(6) $AlO_2^- = 4H^+ + 3e^- = Al + 2H_2O \quad E_0 = -1.262 + 0.0788\ pH - 0.0192 \log AlO_2$ (1.4.7)

More recently, MacDonald and Butler and also Brook have updated the data of Pourbaix et al., and have extended such calculations to elevated temperatures [21, 22]. In contrast to Pourbaix et al., MacDonald and Butler considered three ionic aluminum species: Al^{3+},

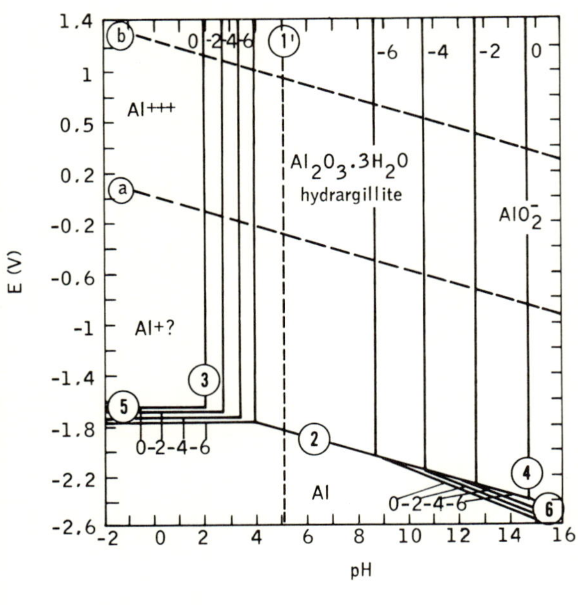

FIG. 1.4.1. Potential-pH diagram for the aluminum-water system at 25°C. (Taken from Ref. 20, p. 171, by permission of Pergamon Press.)

$Al(OH)^{2+}$, and AlO_2^-. The second species, $Al(OH)^{2+}$, represents the first hydrolysis product of Al^{3+}. In very dilute solutions (<10^{-3} M), the monomeric $Al(OH)^{2+}$ species predominates, while $Al_2(OH)_2^{4+}$, $Al_3(OH)_4^{5+}$, $Al_{13}(OH)_{32}^{7+}$, and members of the infinite series $Al[(OH)_5Al_2]n$ have been proposed in more concentrated solutions [23]. Brook based his calculations on only the Al^{3+} and AlO_2^- species.

Several hydrated aluminum oxides are known to exist, and since these hydrated oxides can exist as metastable phases, the choice of an equilibrium state is difficult. Gibbsite, $Al_2O_3 \cdot 3H_2O$, appears to be the stable form at 25°C, which dehydrates on heating to boehmite, $Al_2O_3 \cdot H_2O$ (and/or diaspore), and finally to corundum (α-Al_2O_3). There is some uncertainty, however, regarding the transition temperature of these reactions. In the range of 25 to 150°C, gibbsite was considered by MacDonald and Butler to represent the stable oxide of aluminum. At temperatures >150°C, boehmite was considered to be the most thermodynamically stable form. Figures 1.4.2 to 1.4.4 present potential-pH diagrams, after MacDonald and Butler, for Al-water at 25, 100, and 300°C based on the following reactions:

(1) $2H^+ + 2e^- = H_2$ (1.4.8)

(2) $O_2 + 4H^+ + 4e^- = 2H_2O$ (1.4.9)

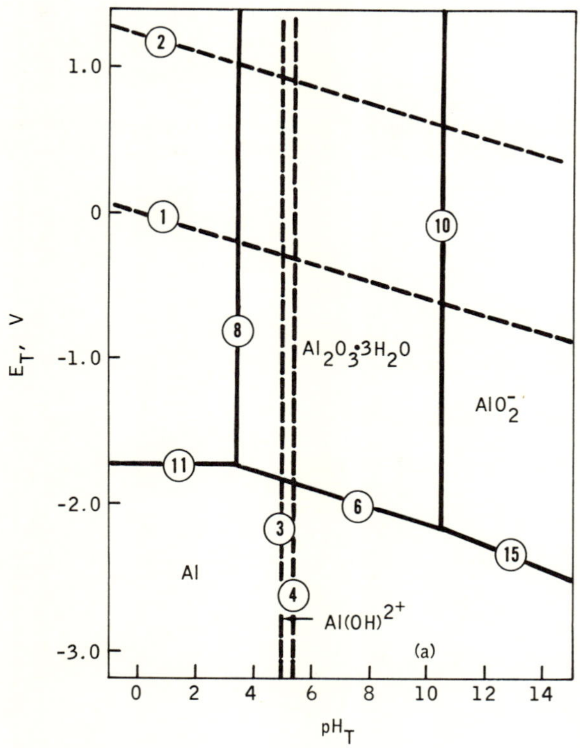

FIG. 1.4.2. Potential-pH diagram for the aluminum-water system at 25°. (a) refers to the pH_T of a 10^{-4} m OH^- solution. (Taken from Ref. 21, by permission of Pergamon Press.)

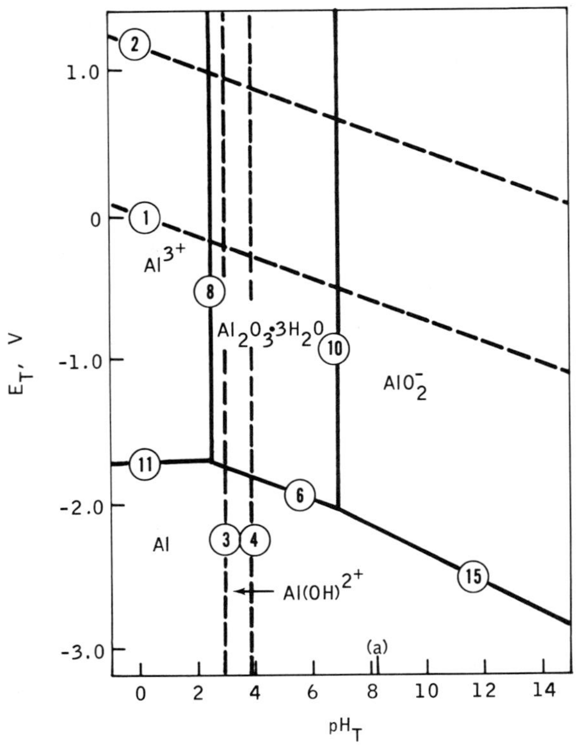

FIG. 1.4.3. Potential-pH diagram for the aluminum-water system at 100°C. (a) refers to the pH_T of a 10^{-4} m OH^- solution. (Taken from Ref. 21, by permission of Pergamon Press.)

(3) $Al(OH)^{2+} + H^+ = Al^{3+} + H_2O$ (1.4.10)

(4) $AlO_2^- + 3H^+ = Al(OH)^{2+} + H_2O$ (1.4.11)

(5) $Al_2O_3 . H_2O + 6H^+ + 6e^- = 2Al + 4H_2O$ (1.4.12)

(6) $Al_2O_3 . 3H_2O + 6H^+ + 6e^- = 2Al + 6H_2O$ (1.4.13)

(7) $Al_2O_3 . H_2O + 6H^+ = 2Al^{3+} + 4H_2O$ (1.4.14)

(8) $Al_2O_3 . 3H_2O + 6H^+ = 2Al^{3+} + 6H_2O$ (1.4.15)

(9) $2AlO_2^- + 2H^+ = Al_2O_3 . H_2O$ (1.4.16)

(10) $2AlO_2^- + 2H^+ + 2H_2O = Al_2O_3 . 3H_2O$ (1.4.17)

(11) $Al^{3+} + 3e^- = Al$ (1.4.18)

(12) $Al(OH)^{2+} + H^+ + 3e^- = Al + H_2O$ (1.4.19)

(13) $Al_2O_3 . H_2O + 4H^+ = 2Al(OH)^{2+} + 2H_2O$ (1.4.20)

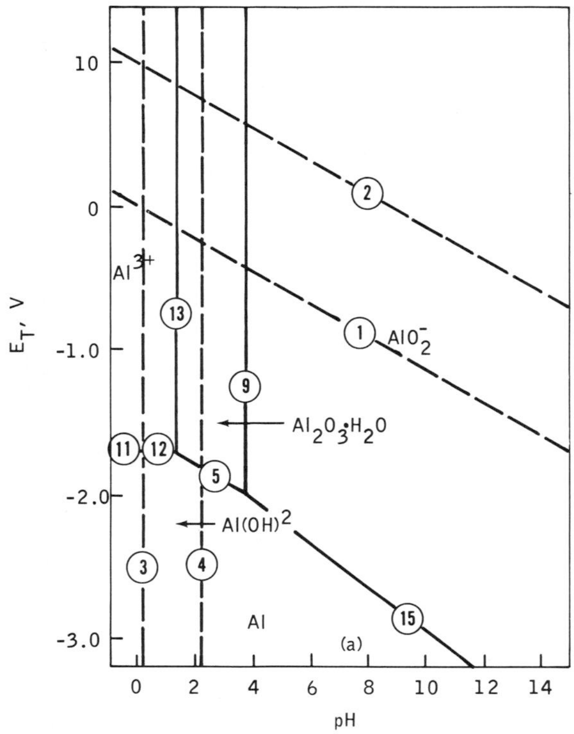

FIG. 1.4.4. Potential-pH diagram for the aluminum-water system at 300°C. (a) refers to the pH_T of a 10^{-4} m OH^- solution. (Taken from Ref. 21 by permission of Pergamon Press.)

(14) $Al_2O_3 \cdot 3H_2O + 4H^+ = 2Al(OH)^{2+} + 4H_2O$ (1.4.21)

(15) $AlO_2^- + 4H^+ + 3e^- = Al + 2H_2O$ (1.4.22)

All ionic activities and partial pressures were arbitrarily set at 10^{-4} m and 1 atm, respectively.

In Figs. 1.4.2 to 1.4.4, dashed lines 3 and 4 represent the regions of predominance of Al^{3+} and AlO_2^-. In between these limits $Al(OH)^{2+}$ is the dominant species. At temperatures >150°C, the equilibrium lines for Reactions (1.4.15) and (1.4.21) lie to the left of the line for Reaction (1.4.10) which defines the regions of predominance of Al^{3+} and $Al(OH)^{2+}$, and only Reaction (1.4.15) appears to predominate. At temperatures >150°C, Reactions (1.4.15), (1.4.20), and (1.4.21) lie to the right of Reaction (1.4.10), and $Al(OH)^{2+}$ predominates. Only Reactions (1.4.21) and (1.4.22) are included in Fig. 1.4.4. These calculations demonstrate that temperature effects on such potential-pH diagrams are not insignificant since Reaction (1.4.10) shifts from pH ~ 5 at 25°C to pH ~ 0.2 at 300°C, in addition to the shift from $Al_2O_3 \cdot 3H_2O$ to $Al_2O_3 \cdot H_2O$ as the stable oxide phase.

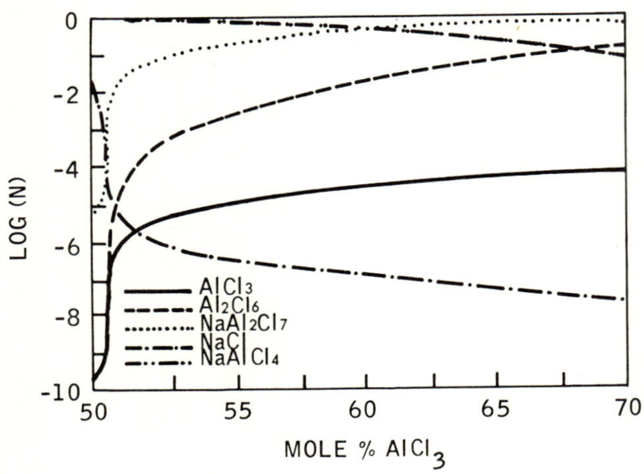

FIG. 1.4.5. Mole fraction of the species present in chloroaluminate melts at 175°C as a function of the net $AlCl_3$-NaCl ratio. (Taken from Ref. 24, by permission of The Electrochemical Society.)

In molten salt systems, some efforts have been made to describe the solvent equilibria in a manner similar to that in aqueous electrolytes. Boxall et al., Fannin et al., and others have examined solvent equilibria in $AlCl_3$-NaCl melts, while Rey has considered equilibrium diagrams in cryolite melts [24-28].

Boxall et al., using potentiometric measurements for the cell Al|$AlCl_3(C_{ref})$NaCl|fritted Pyrex|$AlCl_3$(C), NaCl|Al found that the solvent acid-base properties of $AlCl_3$-NaCl melts containing 50 to 70 mole % $AlCl_3$ could be fully described at 175°C by the three equilibrium constants [24]:

$$2AlCl_{3(l)} = Al_2Cl_{6(l)} \qquad K_0 = 2.86 \times 10^7 \tag{1.4.23}$$

$$AlCl_4^- + AlCl_3 = Al_2Cl_7^- \qquad K_2 = 2.4 \times 10^4 \tag{1.4.24}$$

$$2AlCl_4^- = Al_2Cl_7^- + Cl^- \qquad K_M = 1.06 \times 10^{-7} \tag{1.4.25}$$

Significant changes in the mole fraction of the various species occurs in the vicinity of 50 mole % $AlCl_3$ as Fig. 1.4.5 indicates. At $AlCl_3$ concentrations < 63 mole %, $AlCl_4^-$ is the principal aluminum-containing species in such melts. At higher $AlCl_3$ concentrations, $Al_2Cl_7^-$ predominates. The species $Al_3Cl_{10}^-$ has been suggested from Raman work [29],

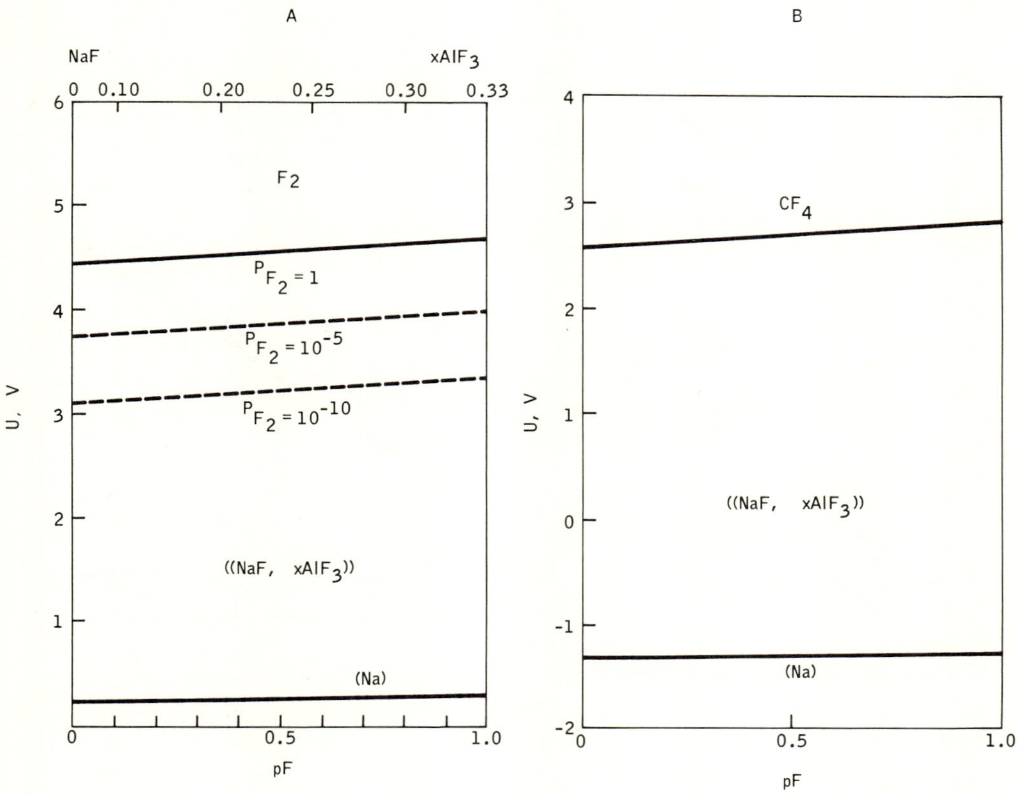

FIG. 1.4.6. Electrochemical equilibria in cryolite at 1300°K. (A) Based on inert anodes. (B) Based on carbon anodes. Potentials referred to Na/Na^+. (Taken from Ref. 28, by permission of Pergamon Press.)

and may be present at $AlCl_3$ concentrations > 70 mole %. As the temperature is increased, NaCl and Al_2Cl_6 species become more predominant.

Rey has studied the electrochemical equilibria in pure cryolite and cryolite-alumina melts using diagrams of potential versus melt basicity as expressed by pF or pO [28]. Figure 1.4.6(a) shows an equilibrium diagram for pure cryolite at 1300°K calculated using a Na/Na^+ reference electrode and an inert anode. If carbon is used as the anode, the anode reaction is the formation of CF_4, and the equilibrium diagram of Fig. 1.4.6(b) results. In the binary cryolite -Al_2O_3 system, a similar distinction can be made between inert and attackable anodes to give the equilibrium diagrams of Fig. 1.4.7. (In Fig. 1.4.7b an Al/AlO^+

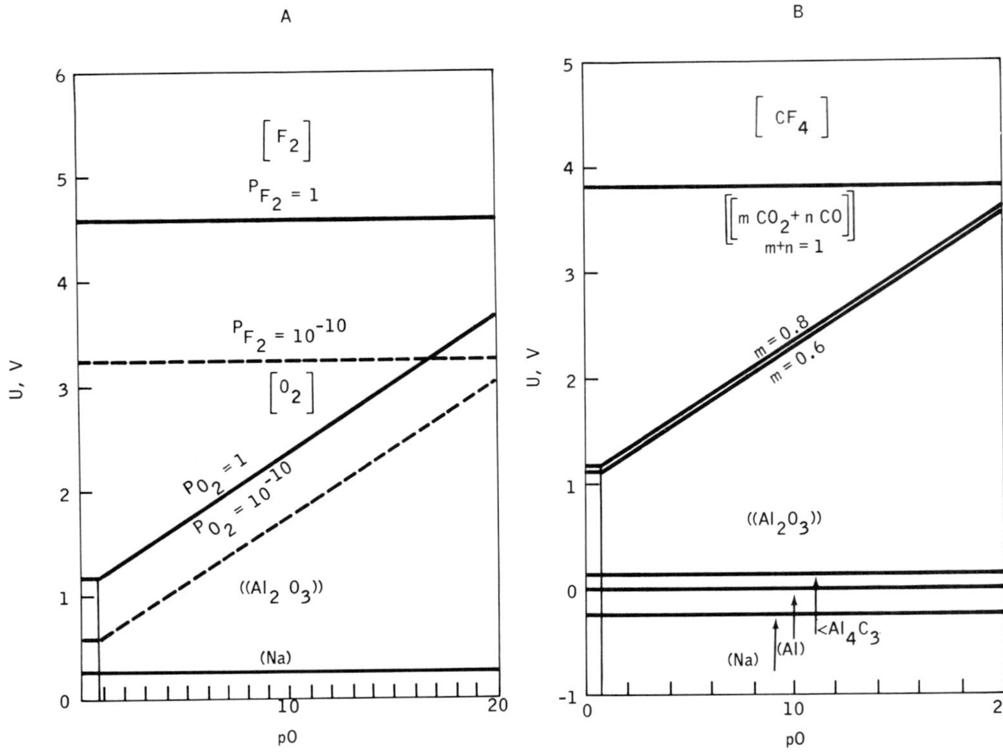

FIG. 1.4.7. Electrochemical equilibria in cryolite-Al_2O_3 melts at 1300°K. (A) Based on inert anodes and a Na/Na^+ reference. (B) Based on carbon anodes and an Al/AlO^+ reference. (Taken from Ref. 28, by permission of Pergamon Press.)

reference electrode system has been used.) Rey has also hypothesized that the net anode reaction could be of the type

$$3O^{2-} + C = CO_3^{2-} + 4e^- \quad (1.4.26)$$

rather than

$$\frac{3}{2}O_2 + yC = mCO_2 + nCO \quad (1.4.27)$$

Essentially, then, Reaction (1.4.26) corresponds to a reversible C/CO_3^{2-} electrode system. Figure 1.4.8 shows the effect of this type of anode reaction on the equilibrium diagram calculated for carbon anodes at 1300°K.

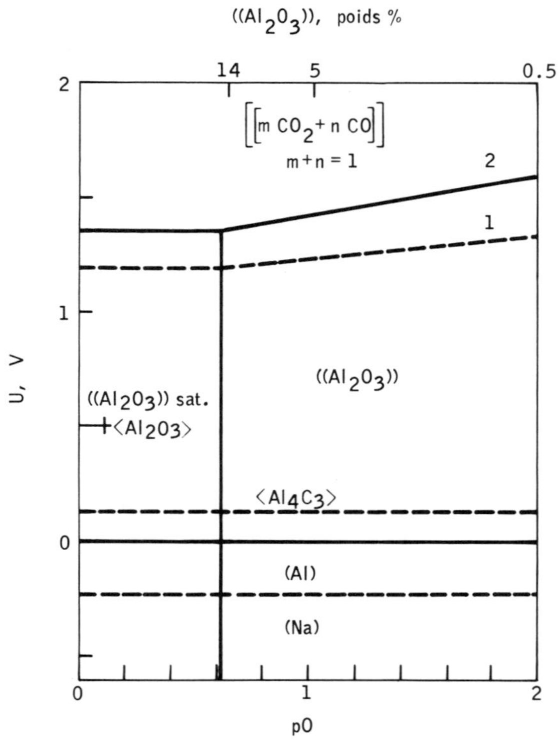

FIG. 1.4.8. Electrochemical equilibria in cryolite-Al_2O_3 melts at 1300°K. (1) is based on Reaction (1.4.27), while (2) is based on Reaction (1.4.26). (Taken from Ref. 28, by permission of Pergamon Press.)

2. POLAROGRAPHIC AND VOLTAMMETRIC STUDIES

2.1. POLAROGRAPHIC CHARACTERISTICS

2.1.1. Aqueous Solution

The direct polarographic determination of aluminum is not an oft-practiced procedure. Very early work (to ca. 1951) on the polarographic behavior of the aluminum ion has been reviewed by Kolthoff and Lingane [30]. The first reported study of the polarographic behavior of aluminum ion is attributed to Prajzler who observed a fairly well-developed reduction wave in 0.05 N BaCl at -1.75 V vs SCE [31]. Gull used this information to develop a method for

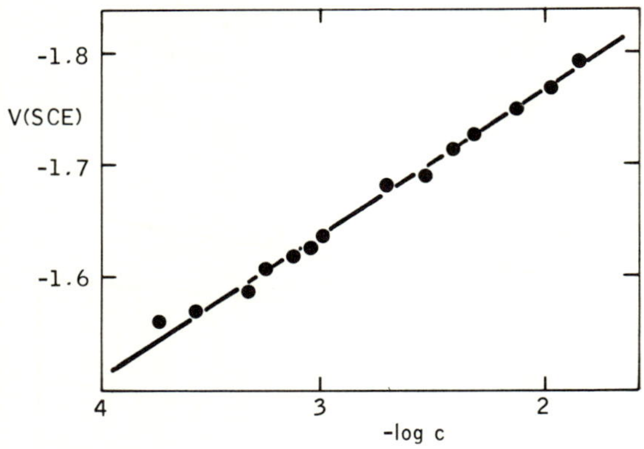

FIG. 2.1.1. Dependence of $E_{1/2}$ for aluminum(III) reduction in 0.1 N LiCl on aluminum concentration. (Taken from Ref. 35, by permission of Academic Press Inc.)

Al determination in magnesium alloys using a magnesium chloride base electrolyte at a pH of 3.8 [32]. As pointed out by Parks and Lykken, the use of this wave at -1.7 V vs SCE was subject to criticism due to its proximity to the hydrogen reduction wave [33]. These authors recommended the first indirect polarographic procedure based on an oxine precipitation technique. Willard and Dean later further developed other indirect methods for the polarographic determination of aluminum, which will be discussed later in this section [34].

M. Heyrovsky critically examined the polarographic behavior of aluminum in aqueous electrolytes, being concerned with the nature of the electrode reaction giving the polarographic waves observed in the presence of aluminum ions [35, 36]. Heyrovsky questioned the interpretation of such polarographic waves as the direct polarographic reduction of aquated Al ions and postulated that they could be due to catalytic hydrogen ion reduction. Such catalytic hydrogen ion reduction was postulated as a result of the acidic properties of the aquated aluminum ion, leading to the following electrode reaction:

$$Al(H_2O)_6^{3+} + 3e^- = Al(OH)_3 + \frac{3}{2}H_2 + 3H_2O \qquad (2.1.1)$$

Visual observation indicated formation of a white precipitate and gas evolution which are expected from Eq. (2.1.1). Stradins and Liepina had, independently, come to a similar conclusion [37]. As the concentration of Al ions is increased, the half-wave potential of Al (in 0.1 M LiCl) shifts to more negative values as can be seen in Fig. 2.1.1. Figure 2.1.2 shows that a decrease in pH (increasing acidity) causes a similar effect. The wave height and shape are dependent on the mercury column characteristics as expected for an autocatalytic process. The half-wave potential of the aquated Al cation shifts to more negative potentials with increasing concentration of the supporting electrolyte, with the exception of Li salts which

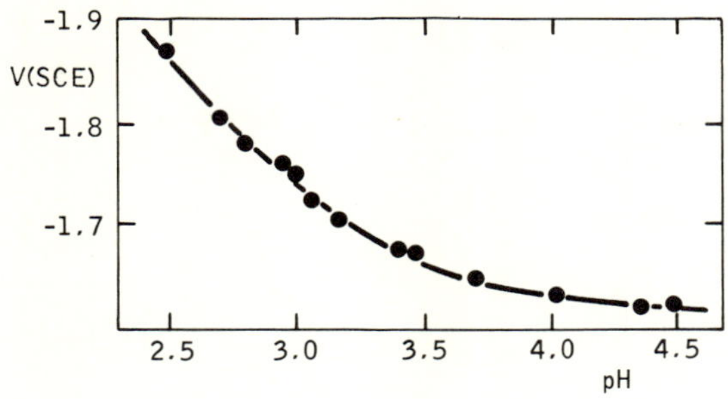

FIG. 2.1.2. Dependence of $E_{1/2}$ for aluminum(III) reduction on pH in 0.1 N LiCl + HCl. (Taken from Ref. 35, by permission of Academic Press Inc.)

cause a positive shift in $E_{1/2}$. While the $E_{1/2}$ values are not dependent on the supporting electrolyte cation at 0.1 N concentrations, in 1 N supporting electrolytes considerable differences are observed.

In the presence of complexing agents in the form of anions such as fluoride, oxalate, citrate, and tartrate, at amounts equivalent to the aluminum ion concentration, significant changes occur in the aluminum polarographic wave. Any indication of an autocatalytic process as well as polarographic maxima disappear in the presence of such complexing agents, resulting in a diffusion-controlled polarographic wave. Galova and Szmerekova examined the polarographic behavior of Al in the presence of complexing anions such as fluoride, tartrate, and citrate [38]. As the complexing anion/Al^{3+} concentration ratio is increased, the $E_{1/2}$ values shift to more negative potentials as shown in Fig. 2.1.3. The $E_{1/2}$ values are also dependent on the concentration of the complex as shown in Figs. 2.1.4 and 2.1.5. Little or no shift occurs in the $E_{1/2}$ of the citrate complex, while the shift of the $E_{1/2}$ of the 1:6 fluoride complex is quite marked. With fluoride and citrate, the $E_{1/2}$ values are independent of pH (-1.72 and -1.66 V vs SCE, respectively). It is not clear, however, what the reduction product is in the case of these aluminum complexes. In the case of fluoride and citrate, the limiting current is diffusion controlled, but tends to kinetic control with decreasing pH. With tartrate complexes, the wave height is kinetically controlled throughout the pH range.

Polarographic data for Al in aqueous electrolytes is collected in Table 2.1.1, with the reader advised to bear in mind the real nature of the polarographic wave of Al^{3+} in aqueous electrolytes.

Willard and Dean, as mentioned, developed the first successful indirect method for the polarographic determination of Al [34]. Using Solochrome Violet RS, the sodium salt of 5-sulfo-2-hydroxy benzene-azo-2-napthol (also known as Pontachrome Violet SW, Mordant Violet 5, or Eriochrome Violet B), Willard and Dean observed that, in the presence of Al^{3+}, the

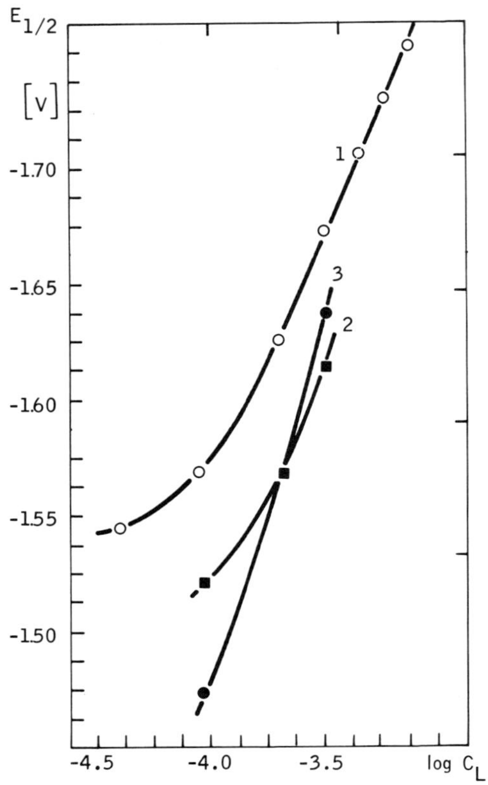

FIG. 2.1.3. $E_{1/2}$ values for aluminum(III) complexes as a function of complexing agent concentration ($Al^{3+} = 10^{-4}$ M): (1) = fluoride; (2) = tartrate; (3) = citrate. (Taken from Ref. 38, by permission of Vydavatelštvo Slovenskej Akademie Vied v Bratislave.)

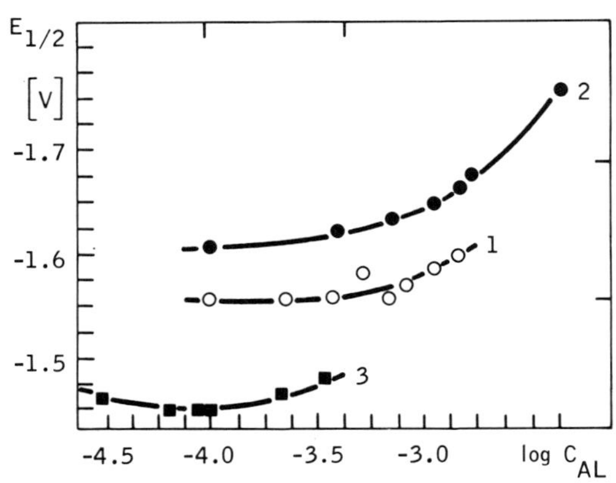

FIG. 2.1.4. $E_{1/2}$ values for aluminum(III)-fluoride complexes at various complex concentrations: (1) = 1:6 complex; (2) = 1:3 complex. (Taken from Ref. 38, by permission of Bydavatelštvo Slovenskej Akademie Vied v Bratislave.)

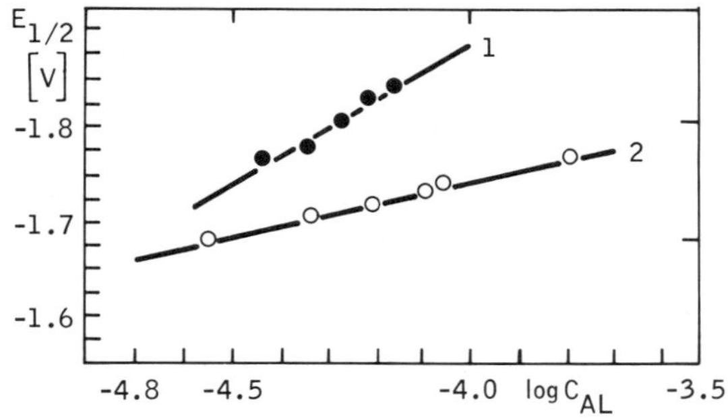

FIG. 2.1.5. $E_{1/2}$ values for aluminum(III) complexes at various complex concentrations: (1) = 1:3 tartrate complex; (2) = 1:2 tartrate complex; (3) = 1:1 citrate complex. (Taken from Ref. 38, by permission of Vydavateľstvo Slovenskej Akademie Vied v Bratislave.)

normal reduction wave of the organic dye split up into two discrete waves. The organic dye-aluminum complex was reduced at more negative potentials (~ 0.2 V) than the dye itself. The total height of the two steps was equal to that of the organic dye alone, with the height of the second wave proportional to the aluminum concentration. Latimer has reviewed the polarographic behavior of the metal chalates of o,o'-dihydroxyazo dyes [39]. Fluoride, citrate, and phosphate can interfere by preferentially complexing aluminum, while sulfate, chloride, and potassium can precipitate the dye. Metal ions which interfere by also forming complexes with the dye can be removed by electrolysis at a Hg pool. While Be does not form a polarographically reducible complex with the organic dye, it does decrease the wave height and should be removed. Some refinement of the polarographic method of Willard and Dean has occurred using linear sweep methods as reported by Florence and Florence et al. [40, 41]. In the work of Florence et al. a rather novel approach was taken [41]. Rather than using the polarographic reduction method, a technique based on the voltammetric oxidation of the Al-Solochrome Violet RS complex at a rotated pyrolytic graphite electrode was successfully used. Very recently, Lendermann et al. have used a carbon paste electrode in the determination of Al via oxidation of its Solochrome Violet RS complex in a 0.2 M sodium acetate electrolyte [42]. Little interference by Be was observed. The electrochemistry of aluminum-dye complexes in methanolic electrolytes has also been considered by Holleck et al. [43, 44]. The use of the aluminum ion catalyzed reduction of nitrate ion has also been proposed for the determination of Al [45].

Trace quantities of aluminum (25 ppb to 5 ppm) can be determined as the aluminum Solochrome Violet RS (Al.SVRS) complex by differential pulse polarography. In a 0.1 M acetate buffer at pH 4.7, the Al.SVRS complex shows a peak potential of -0.455 V vs SCE as shown in Fig. 2.1.6. The free dye shows a large differential pulse reduction peak at ~0.3 V vs SCE which can be used as a check for excess dye [352].

TABLE 2.1.1. Polarographic Characteristics of Al in Aqueous Electrolytes

Substance	Product	Conditions	$E_{1/2}$ (V vs SCE)	I	$E_{3/4} - E_{1/4}$ (mV)	Remarks	Refs.
$Al(H_2O)_6^{3+}$	$Al(OH)_3$	0.05 N $BaCl_2$	-1.75	-	-		[31]
$Al(H_2O)_6^{3+}$	$Al(OH)_3$	0.1 N KCl (LiCl)	-1.70	4.63	44	0.01% Gelatin	[46]
$Al(H_2O)_6^{3+}$	$Al(OH)_3$	LiOH (pH = 4)	-1.758	-	-		[47]
$Al(H_2O)_6^{3+}$	$Al(OH)_3$	15% $MgCl_2$	-1.69	-	-	0.05% Gelatin	[48]
$Al(H_2O)_6^{3+}$	$Al(OH)_3$	0.01 M TBAC (pH = 3.70)	-1.76	5.4-5.9	-	0.002% Triton X	[49]
$Al(H_2O)_6^{3+}$	$Al(OH)_3$	0.1 N LiCl	-1.58	-	115, 73	$KAl(SO_4)_2 = 10^{-5}$ M	[35]
$Al(H_2O)_6^{3+}$	$Al(OH)_3$	0.1 N MCl	-1.610	-	-	$AlCl_3 = 7 \times 10^{-4}$ M	[35]
$Al(H_2O)_6^{3+}$	$Al(OH)_3$	1.0 N LiCl	-1.575	-	-	$AlCl_3 = 7 \times 10^{-4}$ M	[35]
$Al(H_2O)_6^{3+}$	$Al(OH)_3$	1.0 N NaCl	-1.625	-	-	$AlCl_3 = 7 \times 10^{-4}$ M	[35]
$Al(H_2O)_6^{3+}$	$Al(OH)_3$	1.0 N KCl	-1.665	-	-	$AlCl_3 = 7 \times 10^{-4}$ M	[35]
$Al(H_2O)_6^{3+}$	$Al(OH)_3$	1.0 N TMAC	-1.710	-	-	$AlCl_3 = 7 \times 10^{-4}$ M	[35]
$Al(H_2O)_6^{3+}$	$Al(OH)_3$	1.0 N $MgCl_2$	-1.650	-	-	$AlCl_3 = 7 \times 10^{-4}$ M	[35]
$Al(H_2O)_6^{3+}$	$Al(OH)_3$	1.0 N $CaCl_2$	-1.695	-	-	$AlCl_3 = 7 \times 10^{-4}$ M	[35]
$Al(H_2O)_6^{3+}$	$Al(OH)_3$	1.0 N $BaCl_2$	1.705	-	-	$AlCl_3 = 7 \times 10^{-4}$ M	[35]

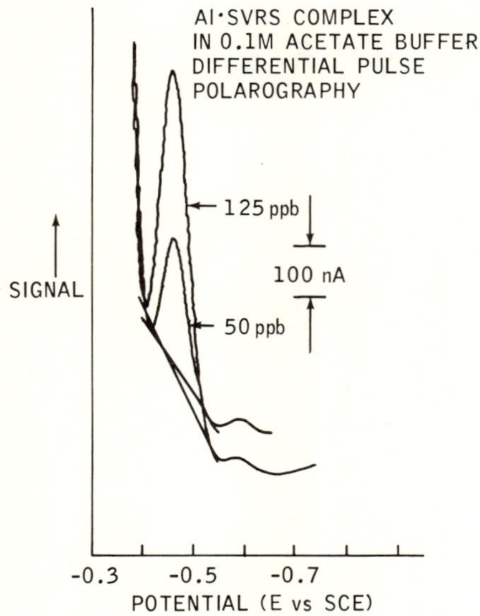

FIG. 2.1.6. Differential pulse polarography of the Al.SVRS complex in 0.1 M acetate buffer. (Taken from Ref. 352, by permission of Princeton Applied Research Corp.)

2.1.2. Nonaqueous Solution

In contrast to the complications found in the polarographic reduction of aluminum ion in aqueous electrolytes, reduction of trivalent aluminum to the metal (amalgam) does appear to occur in most nonaqueous electrolytes with the exception of pyridine. In pyridine containing LiCl, Cizak and Elving found that the observed polarographic waves actually involved the solvent [46]. Aluminum quinolinate and acetylacetonate give quite well-behaved polarographic waves in nonaqueous electrolytes [47, 48]. Coulometric studies have, in general, supported a three-electron reduction process [49, 50]. However, in DMF- and DMSO-based electrolytes, multiwave phenomena have been observed. In both DMF and DMSO, two distinct reduction waves are present. In DMF the height of the second wave is approximately twice that of the first wave. While Kumar and Pantony have broached the possibility of an initial one-electron process followed by a two-electron reduction, in DMSO the wave heights are only in a 1.4:1 ratio which would seem to belie this explanation. Adsorption phenomena or specific solvation effects could be alternative explanations.

Polarographic data for aluminum in nonaqueous electrolytes are collected in Table 2.1.2.

TABLE 2.1.2. Polarographic Characteristics of Al in Nonaqueous Solvents

Substance	Conditions	$E_{1/2}$ (V)	Reference electrode	I	$E_{3/4} - E_{1/4}$ (mV)	Remarks	Refs.
$Al(ClO_4)_3 \cdot 9H_2O$	0.1 M TEAP/MeCN	-1.42	aq SCE	6.60	irr	T = 25°C	[51]
$Al(ClO_4)_3 \cdot 6H_2O$	0.1 M TBAP/MeCN	-1.64	Ag/Ag^+ (0.1 M)	5.22	81	T = 25°C	[52]
$Al(ClO_4)_3$	0.1 M TEAP/PhCN	-1.47	aq SCE	-	48	T = 25°C	[53]
$Al(ClO_4)_3$	0.1 M $LiClO_4$/PhCN	-1.41	aq SCE	-	38	T = 25°C	[53]
$AlCl_3$	0.1 M TEAP/PC	-1.08	aq SCE	-	168	$AlCl_3 = 1.5 \times 10^{-3}$ M T = 25°C	[54]
$Al(ClO_4)_3$	0.1 M $NaClO_4$/SUL	-0.64	Ag/AgCl-TEAC(s)	-	140	T = 25°C	[55]
$Al(ClO_4)_3$	0.1 M TEAP/$PhPOCl_2$	-0.80	satd $Hg/HgSO_4$	-	irr	T = 21°C	[56]
$Al(ClO_4)_3$	0.1 M TEAP/DMF	-1.88 -2.05	aq SCE	0.44 1.53	73 30		[50]
$AlCl_3$	0.1 M TEAP/DMF	-1.92 -2.05	aq SCE	-	irr	T = 21°C	[47]
$Al(ClO_4)_3 \cdot xDMF$	0.1 M TEAP/DMF	-1.92 -2.05	aq SCE	-	-	T = 21°C	[47]
Al quinolinate	0.1 M TEAP/DMF	-1.75 -2.10 -2.17	aq SCE	-	-	T = 21°C	[47]
$Al(ClO_4)_3$	0.1 M TEAP/DMSO	-1.67 -1.95	aq SCE	0.86 1.19	144 69		[50]
$AlCl_3$	0.1 M TEAP/DMSO	-1.5	Hg pool	-	-		[48]
Al acetylacetonate	0.1 M TEAP/DMSO	-2.20	Hg pool	-	-		[48]

2.2. VOLTAMMETRIC CHARACTERISTICS

2.2.1. Aqueous Solution

Voltammetric studies of aluminum in aqueous solution have been carried out primarily to elucidate the mechanism of the polarographic reduction of aluminum. Cyclic voltammetry of Al does reveal a cathodic current peak which, under certain conditions, is proportional to aluminum concentration. No oxidation peak can be observed and there is evidence that the reduction process is associated with adsorption phenomena. Reynolds and Webber found that the Al reduction peak potential was a function of the starting potential, aluminum concentration, and composition of the solution (salt concentration, pH, and buffer concentration) [57]. The optimum conditions for Al determination were found to be 0.1 M $CaCl_2$, pH 4, and 0.8 to 1.0×10^{-2} M KHP. Heyrovsky also carried out voltammetric studies of Al using cyclic voltammetry and ac polarography [35]. The ac polarographic peaks were of an unusual shape and much smaller than expected for a three-electron process. This ill-defined voltammetric behavior of Al in aqueous solution is not unexpected since polarographic studies have indicated the complex nature of the reduction waves obtained in the presence of aluminum ions.

There are some rather specific effects in the voltammetry of aluminum in the presence of lithium ions as described by Kemula and Kublik [58-60]. It was observed that oscillographic curves gave a wave at -1.0 V vs SCE, which could not be due to Li^+ or Al^{3+} reduction. Using oscillopolarographic and ac techniques, this low potential reduction process was attributed to formation of insoluble $LiH(AlO_2).5H_2O$ (lithium dihydroaluminate).

Some mention should be made here of the work of Kolthoff et al. on the voltammetric behavior of a rotating Al wire electrode in aqueous solutions [59]. In the absence of fluoride ion, an Al electrode was found to function as a pOH electrode at pH > 9. In more acidic solutions, it can function as an indicator electrode for the amperometric determination of fluoride ion. Fluoride ion produces an anodic current at -0.75 V vs SCE which is proportional to its concentration between 0.2 and 6.0 ppm F^-.

2.2.2. Nonaqueous Solution

While the interpretation of the voltammetry of Al in nonaqueous solution is more straightforward, such studies have not been in abundance. Galova and Pantony reported that Al^{3+} gave two peaks during cyclic voltammetry in DMF, the ratio of heights varying with Al^{3+} concentration and scan rate [60]. No peaks appeared during the anodic going-portion of the cyclic sweep. With successive sweeps the more positive peak disappeared almost immediately while the more negative peak declined rapidly and vanished after a few cycles. For the more negative peak, i_p is proportional to the aluminum concentration down to 10^{-4} M Al^{3+}. The positive peak height (at -2.2 V vs SCE) was proportional to $[Al^{3+}]$ in the range of 1 to 2 X 10^{-4} M Al^{3+}, indicating the adsorptive character of this process. $i_p/v^{1/2}$ is a straight line parallel to the $(v^{1/2})$-axis for the more positive peak, while it increased at $v^{1/2} < 2\ sec^{-1}$ for the negative peak, indicating the possible presence of a chemical reaction coupled to charge transfer. Coulometric studies of Al^{3+} reduction in DMSO (0.1 M TEAP) led to the conclusion that Al could indeed be electrodeposited from this electrolyte [50], although Al could not be electrodeposited from propylene carbonate containing $AlCl_3$ [61]. Coulometry of Al^{3+} in liquid ammonia at a Hg pool has also indicated a three-electron process to form Al(Hg) [49]. Toropova et al. found that certain o,o'-disubstituted azo complexes of Al^{3+} could be reversibly reduced in DMF/0.1 M TEAP [62]. ESR signals could be measured due to formation of a stable anion radical, and a value of 1 X 10^{-2} cm/sec was estimated for the heterogenous charge transfer process using cyclic voltammetry.

Eisenberg et al. examined the voltammetric behavior of the chloroaluminate anion, $AlCl_4^-$, in aprotic electrolytes [63]. Oxidation of $AlCl_4^-$ was postulated to result in formation of Cl_2 and $AlCl_2^+$ via the reaction

$$AlCl_4^- = AlCl_2^+ + Cl_2 + 2e^- \qquad (2.2.1)$$

The corresponding reduction process involving $AlCl_2^+$ was proposed to occur via

$$AlCl_2^+ + e^- = AlCl_2\cdot \qquad \text{at } E_{(1)} \qquad (2.2.2)$$

and

$$AlCl_2\cdot + Cl_2 + e^- = AlCl_4^- \qquad \text{at } E_{(2)} \qquad (2.2.3)$$

One could also propose an ECE mechanism of the type:

$$AlCl_2^+ + e^- = AlCl_2\cdot + Cl_2 \underset{k_b}{\overset{k_f}{\rightleftharpoons}} AlCl_4\cdot + e^- = AlCl_4^- \qquad (2.2.4)$$

Fleischmann et al. examined the behavior of $NaAl(Et)_4$ and $Al(Et)_3$ at a lead electrode in 0.25 M TBAP/THF [64]. $NaAl(Et)_4$ showed a well-defined wave with $E_{1/2} = 1.25$ V vs Ag/Ag^+. $Al(Et)_3$ showed an ill-defined wave at E > -0.6 V vs Ag/Ag^+. i_l was proportional to $[NaAl(Et)_4]$, but was an order of magnitude lower than expected for a diffusion-controlled process, indicating the possible influence of a preceeding chemical reaction.

2.2.3. Fused Salts

Laitinen, Lui, and Ferguson observed a single reduction wave at about -1.8 V (vs Pt/Pt^{2+}) on tungsten and -1.7 V at Pt electrodes in LiCl-KCl eutectic at 450°C [65]. While no evidence of Al alloying was found with tungsten, evidence of surface alloying of Pt with Al was observed. Naryshkkin and Minin used a dropping Pb electrode in a LiCl-KCl melt at 450°C [66]. Al^{3+} (as $AlCl_3$) gave an $E_{1/2}$ value of -2.28 V vs Cl_2/C. A two-electron process was postulated since $n_{app} = 1.98$.

Available voltammetric characteristics of Al are collected in Table 2.2.1.

TABLE 2.2.1. Voltammetric Characteristics of Aluminum

Substance	Product	Conditions	Electrode	Technique	Potential	Rev.	Remarks	Refs.
Al^{3+}	?	0.4 M $CaCl_2$, 0.01 M KHP, pH = 4.0	Hg drp	E_{swp}	E_p = -1.75 V vs Hg	irr	-	[57]
Al^{3+}	?	0.1 M LiCl	Hg	ac pol	E_p = -1.8 V vs SCE	irr	-	[35]
Al^{3+}	Al	0.1 M TEAP/DMF	Hg drp	E_{swp}	E_p^1 = -2.2 V vs Ag/AgCl E_p^2 = -2.5 V vs Ag/AgCl	irr	1 V/sec	[60]
NaAl $(Et)_3$		0.2 M TBAP/THF	Pb dsk	E_{swp}	$E_{1/2}$ = -1.25 V vs $Ag/AgCl^+$	irr	-	[64]
$AlCl_3$	Al	LiCl/KCl (450°C)	Pt, W	E_{swp}	E = -(1.7-1.8 V) vs Pt/Pt^{2+}	fw	PtAl formation	[65]
$AlCl_3$	Al	LiCl/KCl (450°C)	Pb drp	dc pol	$E_{1/2}$ = -2.28 V vs Cl_2	fw	-	[66]

3. KINETIC PARAMETERS AND DOUBLE-LAYER PROPERTIES

3.1. KINETIC PARAMETERS

3.1.1. Aqueous Solution

As might be expected from the lack of experimental electrode potential data, kinetic studies of the aluminum electrode in aqueous electrolytes have been ignored for the most part. Hagyard and Earl measured an exchange current density of 9×10^{-4} A/cm^2 ($\alpha = 0.34$) for aluminum dissolution in 1 M KCl containing 0.00033 M $AlCl_3$, using a special technique to generate a fresh aluminum surface and minimize H_2 evolution interference [7].

3.1.2. Nonaqueous Electrolytes

Kinetic studies of the aluminum electrode in nonaqueous electrolytes are absent in the literature. Typical polarographic waves for aluminum ion reduction in nonaqueous electrolytes are quite irreversible in shape, implying a relatively low rate constant ($k° < 10^{-2}$ cm/sec) (see Section 2.1.2).

3.1.3. Fused Salt Electrolytes

In contrast to the almost nonexistent electrode kinetic studies of aluminum in aqueous and nonaqueous electrolytes, there are a number of electrode kinetic investigations of the aluminum electrode in fused salt systems. Alkali chloride-aluminum chloride and cryolite melts have been the most commonly used electrolytes.

Early studies by Piontelli found no apparent charge transfer overvoltage at aluminum electrodes in $AlCl_3$-NaCl melts at 200°C and attributed the measured overvoltage to pure ohmic polarization [67]. King et al. found no measurable charge transfer or diffusion overvoltage at current densities as high as 1 to 2 A/cm^2 in 60 mole % $AlCl_3$-40 mole % NaCl melts [68]. Del Duca attempted a comprehensive study of the electrode kinetics of the aluminum electrode in $AlCl_3$-NaCl and $AlCl_3$-$(LiCl\text{-}KCl)_{eut}$ electrolytes over a temperature range of 175 to 313°C [69]. The apparent exchange current density varied from 1 to 56 mA/cm^2. However, as Piontelli and Ducati have commented [70], the experimental techniques followed by Del Duca leave serious doubts about the validity of electrode kinetic data obtained under such conditions.

Holleck and Giner, and Schulze and Hoff, have carried out more reliable kinetic studies of the aluminum electrode in aluminum chloride-alkali halide melts [71-74]. Using the galvanostatic method, an apparent exchange current density of 0.27 A/cm^2 was measured in an $AlCl_3$-KCl-NaCl (57.5-12.5-30 mole %) melt at 130°C by Holleck and Giner [71]. Rotating disk studies were also carried out and indicated passivation due to formation of a solid salt layer at the electrode surface caused by local concentration changes during current flow. Dendrite formation was also observed during cathodic polarization.

Schulze and Hoff carried out the most detailed studies of the kinetics of the aluminum electrode in melts of approximate composition MCl:$AlCl_3$ = 48.5:51.5 mole % (M = Li, Na, K, Rb, Cs), in addition to the ternary melt NaCl:KCl:$AlCl_3$ = 20:14:66 mole % [72-74]. Schulze and Hoff used a galvanostatic double pulse method to determine the activation overvoltage. One may write four possible change transfer sequences for the deposition or dissolution of Al if parallel steps are excluded:

$$Al^{3+} + 3e^- = Al \tag{3.1.1}$$

$$Al^{3+} + e^- = Al^{2+} + e^- = Al^+ + e^- = Al \tag{3.1.2}$$

$$Al^{3+} + e^- = Al^{2+} + 2e^- = Al \tag{3.1.3}$$

$$Al^{3+} + 2e^- = Al^+ + e^- = Al \tag{3.1.4}$$

For reaction sequences (3.1.1-3.1.4), six different rate-determining steps are possible, with the appropriate α-values determined by an Allen-Hickling plot:

$$\log\left\{i/[1 - \exp(-3\phi\eta)]\right\} = \log(i_0 n/z) + (z\alpha - 1)\phi\eta/2.3 \tag{3.1.5}$$

where

- i = current density
- ϕ = F/RT
- η = overpotential
- n = total number of electrons
- z = number of electrons involved in the rate-determining step
- l = number of electrons which must be removed (added) to permit the rate-determining step to occur once

The mean value of the experimental slope was 1.66 ± 0.08. Within the range of the experimental measured overvoltages, it was impossible to distinguish between

$$Al^{2+} + e^- = Al^+ \tag{3.1.6}$$

$$3Al^{3+} + 3e^- = Al \tag{3.1.7}$$

with α-values of 0.66 ± 0.03 (Reaction 3.1.6) and 0.55 ± 0.03 (Reaction 3.1.7). Exchange current density values were calculated on the assumption of n/z = 1.

In the binary $AlCl_3$-alkali halide melts, on changing from LiCl to NaCl, the i_0 values (at 720°K) increase by about an order of magnitude, and decrease on going from Na^+ to Cs^+-containing melts. The i_0 values in the NaCl-$AlCl_3$ and NaCl-KCl-$AlCl_3$ (20-14-66 mole %)

melts were approximately equal. A later study by Schulze and Hoff found a very strong dependence of the exchange current density for the Al electrode on the mole fraction of aluminum chloride, $\underline{n}$-$AlCl_3$, in alkali halide-aluminum chloride binary melts. In fact, calculated standard rate constant (k°) values based on the fraction of free Al^{3+} ions in these melts show almost the same k° values ($\sim 10^{-2}$ cm/sec), indicating that the observed difference in i_0 values merely reflects variations in free aluminum ion concentration.

Electrode kinetic data for the aluminum electrode in aluminum chloride-alkali halide melts is collected in Table 3.1.1.

TABLE 3.1.1. Kinetic Parameters for Al in Fused Alkali Chloride-Aluminum Chloride Melts

Reaction	Conditions[a]	T (°K)	Technique	i_0 (A/cm^2)		Refs.
Al/Al^{3+}	52.4$AlCl_3$-47.6LiCl	720	i stp	1.56	0.54[b] (0.63)[c]	[73]
Al/Al^{3+}	52.8$AlCl_3$-47.2NaCl	720	i stp	18.0	0.55 (0.66)	[73]
Al/Al^{3+}	51.5$AlCl_3$-48.5KCl	720	i stp	8.4	0.54 (0.62)	[73]
Al/Al^{3+}	52.5$AlCl_3$-47.8RbCl	720	i stp	0.76	0.55 (0.67)	[73]
Al/Al^{3+}	51.2$AlCl_3$-48.8CsCl	720	i stp	0.38	0.55 (0.65)	[73]
Al/Al^{3+}	66$AlCl_3$-20NaCl-14KCl	720	i stp	20.6	0.64	[72]
Al/Al^{3+}	57.5$AlCl_3$-12.5KCl-30NaCl	403	i stp	0.27	-	[71]
Al/Al^{3+}	66AlC_3-20NaCl-14KCl	490	i stp	0.86	-	[72]

[a]All compositions values in mole percent.

[b]Based on: $Al^{3+} + 3e^- = Al$ as the rate-determining step.

[c]Based on: $Al^{2+} + e^- = Al^+$ as the rate-determining step.

Vetyukov et al. used a faradaic impedance method to determine exchange current density values for the aluminum electrode in an equimolar KCl-NaCl melt containing 1 to 3 wt% cryolite [75]. A working electrode composed of a 50% Al-50% Cu alloy was used to eliminate difficulties with wetting of the boron carbonitride working electrode capillary by liquid Al. Table 3.1.2 shows the dependence of the measured exchange current densities on the cryolite concentration. Based on a volume concentration, a k° value of 8.9×10^{-4} cm/sec and an α-value of 0.66 were calculated at 734°C. The temperature dependence of k° and the measured diffusion coefficient, D°, led to activation energies of 13 and 6 kcal/mole, respectively, when calculated from the data of Table 3.1.3.

TABLE 3.1.2. Dependence of the Exchange Current Density Values for Aluminum in a 1:1 KCl:NaCl Melt as a Function of the Cryolite Concentration [75]

Na_3AlF_6 (w/o)	i_0 (A/cm^2)
1.0	1.88
2.0	2.10
3.0	2.64

TABLE 3.1.3. Dependence of the Exchange Current Density Values for Aluminum in a 1:1 KCl:NaCl Melt (1 w/o Na_3AlF_6) as a Function of Temperature [75]

T (°K)	i_0 (A/cm^2)
989	2.03
1021	2.21
1057	2.37
1090	2.60

3.2. DOUBLE-LAYER PROPERTIES

Studies of the double-layer properties of aluminum in aqueous electrolytes are complicated by the omnipresent oxide film. Antropov, according to Vijh [76], estimated the zero charge of aluminum in aqueous solution to be -0.4 V (vs NHE). Jakuszewski and Kozlowski determined the zero charge potential of aluminum to be -0.523 ± 0.005 V vs NHE in 0.01 N KCl using an immersion method [77]. One should treat these p.z.c. values with some caution, recognizing the difficulties associated with such measurements on solid electrodes. In this context, the observations of Andersen et al. should be mentioned [78]. Using a scraping technique to generate a fresh surface, Andersen et al. found that the measured potentials (termed "scrape" potentials) fell into two categories: (1) those metals which yield "scrape" potentials corresponding to the p.z.c., or (2) metals whose "scrape" potential was shifted significantly from the p.z.c. due to interfering faradaic processes. Al, in 1 N $NaClO_4$ (or K_2SO_4) solution, showed "scrape" potentials which varied from -1.24 V vs NHE (at pH = 1) to -1.58 V vs NHE (pH = 11) which were not interpreted as p.z.c. values. These "scrape" potential values are significantly more negative than those p.z.c. values reported above, and it would be expected that they would be more typical of an "oxide-free" Al surface than an Al surface used in the immersion technique of Jakuszewski and Kozlowski.

Jakuszewski and Kozlowski have also used the immersion method to determine the zero charge potential of aluminum in anhydrous methanol and ethanol containing 0.01 N LiCl [79]. Zero charge potential values of -0.632 and -0.751 V (vs NHE) were reported in methanol and ethanol, respectively. These same workers have reported values of -0.810 and -0.770 V (vs NHE) for Al in 2-butanol and isopropanol [80]. No mention was made of the presence of an electrolyte.

Table 3.2.1 summarizes the p.z.c. charge values for Al in aqueous and nonaqueous electrolytes.

TABLE 3.2.1. Zero Charge Potentials for Aluminum in Aqueous and Nonaqueous Electrolytes as Measured by the Immersion Method

Electrolyte	E_Z (V vs NHE)	Refs.
0.01 N KCl/H_2O	-0.523	[77]
0.01 N LiCl/MeOH	-0.632	[79]
0.01 N LiCl/EtOH	-0.751	[79]
2-Butanol	-0.810	[80]
Isopropanol	-0.770	[80]

TABLE 3.2.2. Zero Charge Potentials for Aluminum in Binary Alkali Halide Melts [83]

Melt	T (°C)	E_Z (V vs Pt/Pt^{2+})	Technique
LiCl-KCl	500	-2.15	dl capacity min
LiCl-KCl	700	-2.23	dl capacity min
NaCl-KCl	700	-2.23	dl capacity min
LiCl-KCl	750	-1.61	electrocapillary curve max

Table 3.2.2 collects the reported measurements of the zero charge potential of Al in binary alkali halide melts. The data given is that of Russian workers [81, 82] recalculated by Graves et al. to refer to a Pt/Pt(II) reference electrode [83]. In the kinetic studies in molten salts referred to earlier in this section, double-layer capacitance measurements at the equilibrium potentials showed values $< 100\ \mu F/cm^2$ indicating the absence of any specific adsorption phenomena [71-74]. Little effect of the alkali metal cation was observed on the double-layer capacitance values in these binary alkali halide-aluminum halide melts.

4. ELECTROCHEMICAL STUDIES

In this section we will discuss the general electrochemical behavior of aluminum. First we will consider the reduction of dissolved Al ions in various electrolyte systems and then consider the electrochemical oxidation of Al. This latter subject, the electrochemical oxidation of Al, is of considerable importance in the applied chemistry of aluminum as discussed in Section 5 with regard to corrosion, anodizing, electrochemical power sources, and capacitors. In particular, the anodic formation of oxide layers on Al has been, without a doubt, the primary focus of electrochemical studies of aluminum, and this section will concentrate heavily on such processes.

Before considering the electrochemical behavior of aluminum in detail, it is pertinent here to temporarily digress somewhat and consider the question of the existence of subvalent Al species (Al^+, Al^{2+}).

4.1. SUBVALENT ALUMINUM SPECIES

As early as 1934, Bhaduri and Fowler established the existence of Al^+ in the vapor phase equilibrium reaction [84]

$$2Al(l) + AlCl_3(g) = 2AlCl(g) \qquad (4.1.1)$$

The +1 state appears to be well established in the gas phase for AlF, AlCl, and AlBr [85]. Evidence for Al^{+2} in the gas phase is less positive. Since Al^{+2} should have a single s-electron, one might expect it to occur as a dimer, viz., Hg_2^{2+}. Chai et al. have found indications for $(AlCl)_2$ dimer in the gas phase [86]. While the preparation of solid AlI has been reported [87], its existence is dubious [88], and no serious claims have been made for the existence of solid aluminum monohalides. Similarly, gaseous Al_2O and AlO have been shown to exist above 1000°C, but, again, no solid oxides containing lower valent aluminum have been defined.

In the liquid phase (including molten salts), the existence of subvalent Al species has been a controversial subject. As early as 1857, Wohler and Buff found that Al anodes dissolved in aqueous NaCl with current efficiencies greater than 100% based on n = 3 [89]. Turrentine later observed H_2 evolution at Al anodes and proposed that hydrogen formation was due to reduction of water by Al^+ or Al^{2+} [90]. Similar postulations have been made for the generation of lower valent Al species in anhydrous acetic acid [91, 92], liquid ammonia [93, 94], acetone [95], ethanol [92], and pyridine [96]. The work of Raijola and Davidson appeared to be most convincing with respect to the formation of Al(I) during oxidation of Al in aqueous electrolytes [97]. A flowing anolyte was used and was passed over the Al anode and flowed into a solution of an oxidant (Ag^+ or MnO_4^-). Since reduction of the oxidant occurred, lower valent aluminum was postulated.

However, Straumanis questioned the necessity for postulating lower valent aluminum, particularly since a dark grey deposit was observed in the electrolyte [98]. This dark grey deposit had been presumed to be metallic Al formed by the disproportionation reaction

$$3Al^+ = 2Al^0 + Al^{3+} \qquad (4.1.2)$$

Straumanis felt that Reaction (4.1.2) was highly improbable since it assumed growth of Al particles in an aqueous solution from nuclei free from an oxide layer. He suggested that the reducing properties of the anolyte were due to finely divided Al particles and not subvalent Al. This led to the use of anodic disintegration theory to explain the deviations from Faraday's law as shown by Straumanis and Poush [99]. They showed that, during anodic dissolution of Al, forcible breakdown of the oxide film caused Al to be removed from the bulk metal. No evidence was found for the presence of subvalent Al in their study. It seems highly improbable that Al^+ exists in aqueous solutions.

The existence of subvalent Al species in nonaqueous electrolytes appears to be more probable. As mentioned above, subvalent Al (primarily Al^+) was postulated in anhydrous acetic acid, acetone, ethanol, and pyridine. In liquid ammonia containing nitrate and halide ions, a minimum valence of 1.5 was found for the dissolution of Al [94]. The formation of N_2 by reduction of nitrate was presumed to occur via Al^+ and the reaction

$$5Al^+ + 2NO_3^- + 12NH_4^+ = 5Al^{3+} + N_2 + 12NH_3 + 6H_2O \quad (4.1.3)$$

Kiss et al. recently reported on a rotating ring-disk electrode study of the anodic dissolution of Al in anhydrous acetic acid [100]. Their data showed that Al^+ could be detected indirectly by the reduction of ClO_4^- in $LiClO_4$-containing solution.

As discussed in Section 3, Al^+ and Al^{2+} have been postulated in kinetic studies of the Al electrode in $AlCl_3$-alkali halide melts. Delimarskii et al. and Storozhenko also found current efficiencies greater than 100% (based on n = 3) for the dissolution of Al in such melts, particularly at low current densities [101, 102]. This dependence on current density would be consistent with the formation of Al^+ at low overpotentials. Mass transfer of Al metal has been observed in $AlCl_3$-NaCl eutectic at 100 to 400°C by Midorkawa and also by Notoya, presumably by Al^+ [103, 104]. Recently Gale and Osteryoung carefully measured Al anode current efficiencies in 50 mole % NaCl-50 mole % Cl_3 and 33 NaCl-67 $AlCl_3$ melts [105]. In the current density range of 0.03 to 6.3 mA/cm^2, only the faradaic three-electron oxidation appeared to occur. No subvalent ion effects could be positively established within the inherent limitations placed on the coulometric determinations by impurity effects. Belyaev has discussed the importance of Al^+ in the electrowinning of Al, as will be discussed in Section 5.1 [106]. In these high-temperature cryolite melts, the existence of Al^+ appears to be well-established.

Summarizing, subvalent Al species do appear to exist as definite species in the electrochemical behavior of Al in molten cryolite melts. In organic electrolytes and lower temperature melts their existence is open to further experimentation, since they appear to be ephemeral. There does not appear to be any convincing evidence for the existence of Al^+ in aqueous electrolytes.

4.2. ELECTROREDUCTION OF ALUMINUM IONS

The studies of the polarographic and voltammetric reduction of aluminum ions in aqueous electrolytes discussed in Sections 2.1 and 2.2 do not show any convincing evidence for the successful reduction of aluminum ions to the metal. Only in aprotic electrolytes, including both organic solvents and molten salts, does the reduction of dissolved aluminum ions to aluminum metal occur. The reduction of complex oxyfluoride aluminum ions is the basis for the commercially important Hall-Heroult process for the electrowinning of Al in cryolite melts (Section 5.1). Aluminum can be electroplated from aluminum salt-containing organic electrolytes and from molten salt electrolytes as considered in Section 5.2 which deals with the applied electrochemistry of aluminum.

4.2.1. Organic Electrolytes

Numerous organic electrolyte baths for the electrodeposition of aluminum have been investigated over the past 50 years. In Section 5.2 we briefly review the historical development of Al plating from organic electrolytes, which has been comprehensively reviewed by Brenner [107]. Basically, such plating baths consist of an aluminum halide or aluminum organometallic compound in a solvent such as an alkyl halide, ether, or aromatic hydrocarbon. The reported deposition of Al from $Al(ClO_4)_3$ in DMSO appears to be the only successful use of the more common organic electrolyte systems for Al deposition [50]. There has not been any indication of Al deposition from organic electrolytes containing symmetrical tetrahaloaluminum ions ($AlCl_4^-$). Al deposition does appear to occur from organic electrolytes when unsymmetrical aluminum-containing complex ions can be formed such as $Al_2X_7^-$ or $AlX_3X'^-$.

Perhaps the most widely studied electrolyte for Al deposition has been the so-called NBS (United States National Bureau of Standards) bath of Couch and Brenner [108, 109]. The original electrolyte of Couch and Brenner consisted of 2 to 3 $\underline{M}$ $AlCl_3$ and 0.5 to 1.0 $\underline{M}$ LiH in diethyl ether. Later, the use of $LiAlH_4$ became common and the usual NBS bath now contains 3 $\underline{M}$ $AlCl_3$ and 0.4 $\underline{M}$ $LiAlH_4$. While the original bath of Brenner did not appear to be capable of plating more Al than was present in the original bath as $AlCl_3$, later workers showed that such baths, when well-filtered and chemically maintained, have almost unlimited life [110]. The electrochemistry of this aluminum plating electrolyte has been studied, but it is by no means well-defined.

$LiAlH_4$ is a well-known reducing agent in both inorganic and organic syntheses. Combinations of $LiAlH_4$ with $AlCl_3$ have been shown to be a more selective "mixed hydride" reagent, typically with $LiAlH_4/AlCl_3$ ratios of about 1:4, although ratios of 3:1, 1:1, and 1:3 have been used. Ashby and Prather studied the reaction of lithium aluminum hydride with aluminum chloride in diethyl ether to form the intermediate hydridoaluminum halides via the reactions [111]

$$LiAlH_4 + 3AlCl_3 \overset{Et_2O}{\rightleftharpoons} 4HAlCl_2 + LiCl \qquad (4.2.1)$$

$$LiAlH_4 + AlCl_3 \overset{Et_2O}{\rightleftharpoons} 2H_2AlCl + LiCl \qquad (4.2.2)$$

$$3LiAlH_4 + AlCl_3 \overset{Et_2O}{\rightleftharpoons} 4H_3Al + 3LiCl \qquad (4.2.3)$$

$HAlCl_2$, H_2AlCl, and AlH_3 could be identified by their IR spectra and isolation as their trialkylamine complexes. At high $AlCl_3/LiAlH_4$ ratios (4:1) as commonly found in aluminum plating baths, the following reaction might occur:

$$LiAlH_4 + 4AlCl_3 \overset{Et_2O}{\rightleftharpoons} 3HAlCl_2 + LiAlCl_3H + AlCl_3 \qquad (4.2.4)$$

where the $LiAlCl_3H$ could actually be ionized as

$$LiAlCl_3H = Li^+ + AlCl_3H^- \qquad (4.2.5)$$

The ionization of $LiAlCl_3H$ per Eq. (4.2.5) was also postulated by Clay et al. [110]. These authors considered the $LiAlH_4 . AlCl_3$/diethyl ether bath to contain these species:

$AlCl_3.(Et_2O)_n$ $\quad$ $Al^{3+}.(Et_2O)_n$

$AlCl_2H.(Et_2O)_n$ $\quad$ $AlCl_3H^-$

$Li^+.(Et_2O)_n$ $\quad$ $AlCl_4^-$

The Li^+.etherate was presumed to be the principal current-carrying species. Couch and Brenner found, from transference studies, that an anion containing hydride existed, which could be $AlCl_3H^-$.

Later work by Ishibashi and Yoshio showed that the volatile diethyl ether could be replaced by tetrahydrofuran or tetrahydrofuran-benzene mixtures [112, 113]. Conductometric studies on the THF-Bz bath first indicated that the ionic species present were different from those in the NBS-type bath. More thorough studies in anhydrous THF led to the conclusion that $AlHCl_2$ and AlH_2Cl are formed along with the ionizable complexes $[Li^+AlHCl_3^-]$ and $[Li^+AlH_2Cl_2^-]$ [114]. Since LiCl is soluble in THF and insoluble in diethyl ether, an $AlCl_3/LiAlH_4$ molar ratio of 1 can lead to the following reactions:

$$LiAlH_4 + AlCl_3 = 2AlH_2Cl + LiCl \qquad (4.2.6)$$

$$AlH_2Cl + LiCl = LiAlH_2Cl_2 = Li^+ + AlH_2Cl_2^- \qquad (4.2.7)$$

At an $AlCl_3/LiAlH_4$ ratio of 3, the following reactions can occur:

$$LiAlH_4 + 3AlCl_3 = 4AlHCl_2 + LiCl \qquad (4.2.8)$$

$$AlHCl_2 + LiCl = LiAlHCl_3 = Li^+ + AlHCl_3^- \qquad (4.2.9)$$

Up to a molar ratio of 1, the main conductive species were proposed to be Li^+ and $AlH_2Cl_2^-$. From a molar ratio of 1 to 3, the main conducting ions were presumed to be Li^+ and $AlHCl_3^-$. Beyond a molar ratio of 3, Li^+ and $AlCl_4^-$ are formed. In contrast, due to the insolubility of LiCl in diethyl ether, $Li^+AlH_2Cl_2^-$ should not exist at lower $AlCl_3/LiAlH_4$ ratios than 3, while at molar ratios greater than 3, only $AlCl_3$, $AlHCl_2$, and $Li^+AlHCl_3^-$ may exist. Accordingly, Yoshio and Ishibashi attributed the difference in the plating behavior of THF and diethyl ether to the absence of ionizable $LiAlH_2Cl_2$ and LiCl in diethyl ether. From this work on the plating of Al from ether type baths, it would appear that $AlCl_3H^-$ may be the species from which electrodeposition of Al occurs.

Aluminum halides dissolved in ethyl bromide, nitrogen-containing aromatics, or aromatic hydrocarbon solvents comprise another class of electrolytes for Al deposition. Capuano and Davenport have attempted to rationalize the choice of solvent for this type of Al plating bath in terms of Lewis acid (salt)-Lewis base (solvent) interaction [115]. While some Lewis acid-base interaction between solute and solvent is required for dissolution or ionization of the solute, strong coordination of the dissolved aluminum with nitrogen or carbonyl groups was assumed to yield an aluminum complex which was too stable for reduction of Al from the complex. Solvents such as ethers or aromatic hydrocarbons, which possess relatively weak coordination centers, were felt to be suitable for deposition of aluminum. This rationale led Capuano and Davenport to investigate alkyl benzenes as solvents for aluminum halide electrolytes. Using $AlBr_3$ as the solute, alkylated aromatic solvents such as ethylbenzene and diethylbenzene were successfully used for Al deposition. Some interesting chemistry exists in such $AlBr_3$-alkylbenzene electrolytes.

The use of an aluminum halide-aromatic hydrocarbon mixture is an integral part of Friedel-Crafts reactions, in which highly colored, hydrocarbon insoluble oils are formed. These oils were first demonstrated by Gustavson to have marked catalytic properties [116]. While a number of workers formulated the existence of such complexes as $[AlBr_3.ArH]$ or the dimer $[Al_2Br_6.2ArH]$, Brown et al. showed that in the presence of hydrogen halides a different type of complex could be formed such as $[ArH_2{}^+Al_2Br_7{}^-]$ or $[ArH_2{}^+AlBr_4{}^-]$ [117, 118]. The complexes were termed "σ-complexes" by Brown and Wallace and were equated with the colored oils of Friedel-Crafts reactions. These σ-complexes could be associated with additional aromatic hydrocarbon molecules. The complexes formed in the absence of acid halide were postulated to yield π-complexes such as [119]

$$Al_2Br_6.ArH = Al_2Br_5.ArH^+ + Br^- \quad (\pi\text{-complex}) \qquad (4.2.10)$$

Electrodeposition of Al was assumed to be via such π-complexes by Simanavicius and Levinskiene [120].

Capuano and Davenport examined the effects of moisture on the deposition of aluminum from $AlBr_3$-alkylbenzene electrolytes [115]. In a nominally anhydrous electrolyte, Al could not be plated, in contrast to successful plating of Al from baths in which moisture was not rigidly excluded. (It should be mentioned here that although a certain level of water appeared to be desirable for Al deposition, excess water precipitated Al as the hydroxide or hydroxychloride.) The effect of water was postulated by Capuano and Davenport to be via displacement of bromine by OH^- as

$$2(Al_2Br_6.ArH) = H_2O = \underset{\text{(hydroxy σ-complex)}}{[Al_2Br_4OH.ArH^+ + Br]} + \underset{\text{(σ-complex)}}{[ArH^+ + Al_2Br_7{}^-]} \qquad (4.2.11)$$

The σ-complex thus formed would be equivalent to that formed by the reaction of HBr with $ArH.AlBr_3$ as proposed by Brown and Wallace. The deposition of Al was assumed to be by the following electrode process:

$$ArH_2{}^+ + Al_2Br_4OH.ArH^+ + 6e^- = 2Al + 4Br^- + H_2O + 2ArH \qquad (4.2.12)$$

It would seem possible to this author that an alternative to the hydroxy σ-complex mechanism exists. The presence of water should lead to HBr formation by the reaction

$$AlBr_3 + H_2O = AlBr_2OH + HBr \tag{4.2.13}$$

The HBr thus formed could lead to formation of the σ-complex of Brown and Wallace, $[ArH^+Al_2Br_7^-]$. It seems clear that further work is required on the nature of this type of Al plating bath.

Another type of electrolyte which seems similar to the $AlBr_3$/aromatic solvent electrolyte is the $AlBr_3$/N, N-dimethylaniline (DMA) electrolyte of Hisano et al. [121]. In such electrolytes the molecular complexes $AlBr_3.C_6H_5N(CH_3)_2$ and $AlBr_3.2C_6H_5N(CH_3)_2$ were indicated from a freezing point-composition diagram and the former complex was isolated. The current efficiency for Al deposition was a maximum at $AlBr_3$ concentrations of 35 and 65 mole %, paralleling the observed maximum in specific conductivity. At the equimolar composition, both cathode current efficiency and specific conductivity passed through a minimum. Above 55 mole % $AlBr_3$, the dissolution of an Al anode was 100% based on n = 3. While no explanation was put forth by Hisano et al. for the conductivity of such mixtures, it was observed that irreversible changes occurred in the conductivity on heating $AlBr_3$ solutions of less than 50 mole %. A 33 mole % $AlBr_3$ solution had an initial conductivity of 3×10^{-4} ohm^{-1}/cm at 80°C. After heating to 150°C, the conductivity at 80°C increased to 4×10^{-3} ohm^{-1}/cm. Such $AlBr_3$/DMA solutions were quite sensitive to oxygen and temperature. Below 100°C and at any concentration, such solutions were colorless. In vacuo or in nitrogen, such solutions developed a light yellow to reddish-orange color, except for the 50 mole % composition which was colorless. In the presence of air, solutions with <50 mole % $AlBr_3$ turned blue. The same blue color could be caused by sunlight or UV radiation even in the absence of oxygen. This adsorption maximum at 605 μm appeared to be close to that of crystal violet, postulated as the reaction product in DMA with $AlCl_3$ in vacuo. The author feels that a Lewis acid-base reaction could occur between $AlBr_3$ and DMA to form a charge transfer complex:

$$DMA + AlBr_3 = [DMA.AlBr_3]_{CT} \tag{4.2.14}$$

Reaction of this charge-transfer complex with O_2 could lead to formation of $DMA^{+\cdot}$ which should dimerize to form N, N-tetramethylbenzene (TMB). This reaction of DMA with the strong Lewis acid, $AlBr_3$, is analogous to the formation of aromatic hydrocarbon cation radicals by reaction with H_2SO_4 or other strong acids as postulated by Aalbersbery et al. [122]. Since the DMA cation radical is unstable, it readily dimerizes. Since TMB is easily oxidized, the final product could be the TMB dication. Thus the ionic conducting species could be the TMB dication and $AlBr_4^-$. Further electrochemical studies in this type of electrolyte would be quite informative.

4.2.2. Molten Salts

The next general class of electrolytes from which Al deposition occurs is the molten salts. The binary or ternary aluminum halide-alkali halide melts have been quite well studied, while Al deposition from cryolite-based melts is the basis of commercial Al production. Other melts for Al deposition are based on organic salts. Keyes et al. used a 50 mole % $AlBr_3$-50 mole % tetraethylammonium bromide melt at ~ 100°C [123], while later Hurley and Weir used an $AlCl_3$-ethylpyridium bromide melt [124]. One may also consider as an organic salt melt the aluminum plating bath of Ziegler and Lehmkuhl, molten $NaF.2Al(Et)_3$, which has a conductivity of 0.08 ohm^{-1}/cm at 160°C [125, 126].

In the previous section we considered the kinetics of the Al electrode in the binary and ternary $AlCl_3$-alkali chloride melts. Al deposition appears to be quite straightforward from such melts as $AlCl_3$-NaCl, $AlCl_3$-KCl, and ternary $AlCl_3$-NaCl-KCl mixtures. In contrast, Senderoff claimed deposition of LiAl from 50 mole % $AlCl_3$-50 mole % LiCl ($LiAlCl_4$) [127]. He rationalized the formation of LiAl alloy by the existence of a stable intermetallic and mutual solubility of Li and Al. In contrast, the Na-Al and K-Al phase diagrams show no mutual solubility or intermetallic phases.

Deposition of Al from the aluminum chloride-alkali halide melts is often accompanied by dendrite formation. Early work by Midorikawa, carried out with regard to electrorefining of Al, examined such factors as the nature of the substrate, current density, use of superimposed ac, electrolyte composition (and temperature), and effects of foreign ions [128, 129]. Pb was found to be the best cathode compared to Au, Cu, Fe, Al, and C. Using $AlCl_3$-NaCl at 130 to 160°C, current densities up to 1 to 3 mA/cm^2 led to satisfactory deposits. Superimposed ac (60 to 500 Hz) was found to prevent the occurrence of rough crystals of Al. While iron impurity $>10^{-3}\%$ in the bath led to low current efficiency and dendrite formation, 10^{-3} to $10^{-2}\%$ $PbCl_2$ prevented dendrite formation at a Pb cathode and gave a more adherent deposit. Howie and MacMillan, more recently, carried out a very informative study of dendrite formation during electrodeposition of Al on steel from a 64 mole % $AlCl_3$-36 mole % NaCl melt at 175°C [130]. In a highly purified melt, dendrite formation could only be avoided at current densities ≤ 1 mA/cm^2. It was thought that ferric chloride and/or some of the products of the hydrolysis of $AlCl_3$ (HCl and Al_2O_3) were most likely to affect the electrodeposition process. Ferric chloride alone added to the bath did not lead to an improvement in the deposit. Addition of hydrolysis products did lead to improved deposits, and this effect was traced to HCl rather than Al_2O_3. The action of HCl (~ 0.07%) was postulated to occur by 1) removal of Al from high spots where dendrite growth could originate or 2) by providing an alternative reaction (H^+ reduction) at the tip of the dendrite rather than further deposition of Al. Both mechanisms could yield a smooth compact plate. Under these conditions, the dendrite growth threshold level was the region of 20 to 25 mA/cm^2.

Sato, Ishikawa, and Midorikawa carried out a study of Al deposition from $AlCl_3$-NaCl melts of different compositions [131]. On the basis of the composition dependence of the

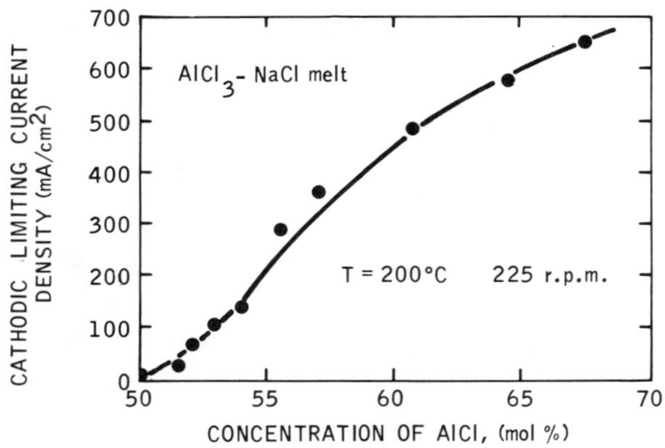

FIG. 4.2.1. Limiting current for Al deposition from NaCl-$AlCl_3$ melts at 200°C as a function of melt composition. (Taken from Ref. 131, by permission of The Electrochemical Society, Japan.)

cathodic limiting current on a rotating W cylindrical electrode, as shown in Fig. 4.2.1, the deposition of Al was postulated to occur from $Al_2Cl_7^-$ rather than $AlCl_4^-$. This would be in agreement with the earlier conclusion that Al deposition occurs primarily from unsymmetrical Al complex anions. Codeposition of alkali metal was also found at potentials less than predicted for the alkali metals.

Hurley and Wier extended the early work of Keyes et al. on the deposition of aluminum from aluminum halide/quaternary ammonium halide molten salt mixtures. While the $AlBr_3$-tetraethylammonium bromide bath of Keyers et al. required cell voltages of ~ 20 V, $AlCl_3$-ethylpyridinium bromide (EtPyBr) mixtures appeared to be much more conductive, allowing cell voltages on the order of 2 to 3 V. The phase diagram of the $AlCl_3$-EtPyBr system shows two eutectic mixtures at 1:2 and 2:1 $AlCl_3$/EtPyBr mole ratios and compound formation at the 1:1 mole ratio. The 2:1 eutectic proved to be the most suitable for the Al deposition and allowed deposition of uniform, coherent deposits at current densities up to 10 mA/cm^2. Based on transference studies on the compound $AlCl_3$.EtPyBr in benzene, the dissociation mechanism was considered to be

$$AlCl_3.EtPyBr = EtPy^+ + AlCl_3Br^- \qquad (4.2.15)$$

To eliminate the need for elevated temperatures, Hurley and Wier also used the 2:1 $AlCl_3$/EtPyBr complex in an aromatic solvent such as benzene or toluene to form a room-temperature electrolyte [132].

The electrolyte of Ziegler and Lehmkuhl, molten sodium fluoride-bistriethylaluminum complex (NaF.2AlEt) has a conductance of 2 X 10^{-2} ohm^{-1}/cm at 63°C and is a liquid at room temperature. In contrast, the 1:1 complex (NaF.$AlEt_3$) melts at 64 to 66°C and has a conductance of only 10^{-3} ohm^{-1}/cm. As mentioned earlier, the conductivity of the 2:1 complex is 0.08 ohm^{-1}/cm at 160°C. Electrolysis of molten $NaF.2Al(Et)_3$ has been postulated to proceed as follows [126]:

$$(Et_3Al)_2F^- = Et^0 + Et_3Al + Et_2AlF + e^- \quad \text{(anode)} \tag{4.2.16}$$

$$(Et_3Al)_2{}^{2-} + 3e^- = Al + Et_3AlF^- + 3Et^- \quad \text{(cathode)} \tag{4.2.17}$$

It is difficult to rationalize the production of $CH_3CH_2{}^-$ in the above cathode reaction. It was also observed that if the 1:1 $NaF.Al(Et)_3$ complex was electrolyzed, Na and Al were codeposited. Some of the applications of these organoaluminum complexes in syntheses are discussed in Section 5.3. However, the nature of this type of electrolyte is not well-defined and the mechanism of Al deposition is not clear.

In Section 5.1 we will discuss the reduction of Al from cryolite melts in some detail. The basic electrochemistry in such melts is quite complex and will be presented in respect to the actual process of Al electrowinning.

4.3. ELECTROOXIDATION OF ALUMINUM

Studies of the anodic oxidation of aluminum occupy a voluminous portion of the literature of the electrochemical behavior of aluminum. In general, these anodic reactions of aluminum fall into three different categories:

$$\text{Type I} \quad Al = (Al^{3+})_{solv} + 3e^- \tag{4.3.1}$$

$$\text{Type II} \quad mAl + nX^{x-} = Al_mX_n + nxe^- \tag{4.3.2}$$

$$\text{Type III} \quad 2Al + 3H_2O = Al_2O_3 + 6H^+ + 6e^- \tag{4.3.3}$$

Type I reactions of aluminum are best exemplified by the behavior of aluminum in molten salts and certain organic electrolytes. Type II reactions are typified by the behavior of Al in aqueous electrolytes. As will be discussed in Section 5.4, the corrosion behavior of Al is governed by the properties of Al_mX_n, with corrosion of Al occurring in electrolytes where Al_mX_n is soluble. Type III reactions, in which an oxide film is formed on Al by anodic oxidation, are of great commercial interest. The formation and properties of these aluminum "oxide" films are the reasons for the corrosion resistance of anodized aluminum surfaces as discussed in Section 5.6. Their dielectric properties allow their application in electrolytic capacitors as discussed in Section 5.7. While it is not within the scope of this chapter to comprehensively review the formation of these oxide films on aluminum, we will consider here the generally accepted nature of these technologically important and scientifically intriguing processes. This subject has been the object of some very recent reviews by Taijima, Wood, and Diggle et al. which should be consulted for more detailed exposes [133-135]. The work of Vijh may serve as a general introduction to the anodic oxidation of metals [136]. In this subsection we will consider: 1) the nature and properties of the oxide films formed on aluminum, and 2) the various mechanisms which have been proposed for the formation and growth of such films.

4.3.1. Nature and Properties of Anodic Al Oxides

Stoichiometrically speaking, there is only one aluminum oxide, Al_2O_3, at normal temperatures and pressures. However, a number of polymorphs, hydrates, etc., exist, of which six forms are most often considered. Carborundum, α-Al_2O_3, is found in nature and is quite refractory. Gamma alumina, γ-Al_2O_3, comprises a number of anhydrous aluminas with ill-defined structures and is the main constituent of "activated alumina." Two well-defined hydrates of α-Al_2O_3 exist, gibbsite and boehmite. Gibbsite, α-$Al_2O_3.3H_2O$ is the prominent constituent of bauxite, the ore from which aluminum is produced. Boehmite (α-$Al_2O_3.H_2O$ or AlO.OH) can be obtained by the hydrothermal decomposition of gibbsite and has been proposed by many workers to be the main constituent of the surface oxide formed on Al in water at elevated temperatures (>80°C). However, more recently, it has been demonstrated that such oxide films, usually formed by immersion of Al in boiling water, are really pseudoboehmite, a poorly crystalline form of boehmite, $Al_2O_3.xH_2O$ (x = 1.8 to 2.5). Bayerite (β-$Al_2O_3.H_2O$) does not occur in nature, although it is the usual film which forms on Al in water at temperatures below 80°C and under certain conditions of corrosion. Diaspore (β-$Al_2O_3.H_2O$) is a relatively rare aluminum oxide which occurs primarily in some bauxite ores.

There is general agreement that the loss of aluminum from the metal surface during anodic oxidation is strictly a faradaic three-electron process. Al_2O_3, thus produced, can dissolve according to the solvent action of the electrolyte, which is usually controlled by the nature of the anion. Moreover, the nature of the usual anodic film on Al is such that electrolyte anions, their polymers, or solvent and electrolyte anions are usually incorporated. It is generally recognized that the nature of the electrolyte is the most important single factor in the type of aluminum oxide film. In fact, the classification of oxide films on aluminum is usually done in terms of the solubility of the oxide in the electrolyte.

Electrolytes in which the aluminum oxide film is completely insoluble lead to barrier-type films. Those electrolytes in which the oxide film is slightly soluble, produce porous-type films. These porous-type anodic films on Al are more appropriately called duplex films, since they have been shown to consist of an inner, thin, compact barrier-type layer and an outer, relatively thick porous layer. Barrier-type films are thin, compact, and non-porous. Their thickness is controlled solely by the voltage applied, apart from temperature. The maximum film thickness of these barrier-type films is restricted to a voltage below the oxide breakdown voltage. For aluminum oxide films these breakdown voltage values are 500 to 700 V, which corresponds to thicknesses of 7,000 to 10,000 Å. For duplex films, although the inner layer behaves as a barrier-film layer, the thickness of the outer layer is current density, time, and temperature dependent. At low temperatures (0 to 5°C), the

TABLE 4.3.1. Properties of Barrier and Duplex Anodic Films on Aluminum [135]

Property	Barrier type	Duplex type
Structure	Thin, compact, nonporous	Inner layer, compact barrier-type; outer layer, thick and porous
Thickness	Voltage dependent (~14 Å/V)	Inner layer as for barrier-type; outer layer, voltage independent, current density, time, and temperature dependent
Typical electrolytes	Boric acid-borax; citric acid-citrate; ammonium tartrate (all aqueous)	Aqueous sulfuric, oxalic, and chromic acids
Uses	Electrolytic capacitors	Corrosion resistant surfaces

outer porous film is thick, compact, and hard, while at high temperatures (60 to 75°C), the outer porous film is thin, soft, and nonprotective. Table 4.3.1 summarizes the properties of these two types of films which can be produced on aluminum.

Based on the faradaic conversion of aluminum to Al_2O_3, 1.889 g of Al_2O_3 are formed from 1.0 g of Al. This ratio, the weight of oxide formed per gram of aluminum reacted, is commonly known as the "coating ratio." The "coating ratio" is a measure of the overall anode efficiency with respect to the coating efficiency. For many duplex films, coating ratios of 1.35 to 1.46 are found. Lower temperature, high current densities, and low electrolyte concentrations favor high "coating ratios."

Al_2O_3 in these anodic films is present most often as the amorphous form. In barrier films, some γ-Al_2O_3 and γ'-Al_2O_3 were shown to exist by Verwey [137]. Crystalline γ-Al_2O_3 and γ'-Al_2O_3 are distinguished by a different arrangement of cations in the structure, with the oxygen lattice remaining the same. γ'-Al_2O_3 has been regarded as an intermediate, poorly crystalline form between amorphous Al_2O_3 and γ-Al_2O_3. This coexistence of amorphous and crystalline γ-Al_2O_3 is an important part of most theories of oxide formation on aluminum as we will see later. Table 4.3.2, after Taijima [133], summarizes the thicknesses and structure of the oxide films produced on aluminum under various conditions.

While there is general agreement that both barrier and duplex oxide films on aluminum contain water when formed in aqueous electrolytes, the nature or state of this water is not definitively established. Ginsberg and Wefers postulated that the water-anodic oxide interaction is similar to the adsorption of water by zeolites with the water existing in the molecular state in a fairly strong bond [138]. Dorsey, based on IR reflectance measurements on anodic films formed in boric acid, indicated the barrier layers could be distinguished from

TABLE 4.3.2. Characteristics of Oxide Films Formed on Al under Various Conditions [133]

Conditions	T (°C)	Barrier layer thickness (μ)	Total film thickness (μ)	Structure and composition
Dry air	20	0.0010-0.0020	0.0010-0.0020	Amorphous Al_2O_3
Humid air	20	0.0004-0.0010	0.05-0.10	Boehmite and hydragillite
Chemical oxidation	70-100	0.0002-0.0008	1.0-5.0	Boehmite + electrolyte anions
Chemical polishing	50-100	~0.0005	0.01-0.10	Boehmite + electrolyte anions
Anodic oxidation	18-25	0.0100-0.0150	5.0-30.0	Amorphous Al_2O_3 + electrolyte anions
Hard anodizing	+6 to -3	0.0150-0.0300	150-200	Amorphous Al_2O_3 + electrolyte anions
Electropolishing	50-60	0.0050-0.0100	0.1-0.2	Unknown form of Al_2O_3 + electrolyte anions
Anodic barrier layer formation	50-100	0.0300-0.0400	0.0-3.0	Crystalline and amorphous Al_2O_3 + electrolyte anions

porous layers even in duplex films [139]. Dorsey postulated that barrier layers are in fact alumina trihydrates which are subject to structural changes in the formation of duplex films. This structural change was interpreted as a transformation of a cyclic aluminic acid trihydrate into a decyclized form of lower polymer weight. Kormany also reported the presence of alumina trihydrate in barrier layers [140]. Some workers have regarded the presence of water in these anodic films as a necessity for the stabilization of the spinel structure of alumina [141, 142]. It would appear that the water which is present in anodic oxides on Al is not in the free form, but is present in the form of a hydroxide, hydrated oxide, or both.

Anion incorporation into anodic aluminum oxides was demonstrated quite early by Norden who showed that anodic oxidation of Al in dilute H_2SO_4 led to an anodic layer containing 13% SO_3 [143]. Incorporation of oxalate, borate, and phosphate anions appears to be well-documented [144-146]. This incorporation of anions appears to be closely related to the nature of the anodic film. For pore-forming electrolytes, i.e., in duplex films, the incorporation of anions is greatest, while it is minimal in barrier-type layer electrolytes. Two methods of anion incorporation have been proposed. The first relies on incorporation of "bound" anions during the conversion of the barrier layer to the porous layer. "Free" electrolyte anions could be present by accommodation in the pores of the oxide. Ginsberg and Wefers showed that in H_2SO_4 an oxide film was formed which possessed a total sulfur content of 13% SO_3 [138]. Eight percent of this SO_3 was considered to exist in the "bound" form, while 5% was considered to be present in a "free" form which could eventually be removed by prolonged washing. The anion has been considered by many workers to be present as the basic salt of the anion.

In nonaqueous media such as fused salts or organic electrolytes, some of these same concepts regarding incorporation of electrolyte and solvent are also applicable, although studies in this area are not as extensive as those in aqueous media. For example, anodization of aluminum in boric acid-formamide electrolyte leads to films which incorporate solvent (as $HCONH^-$ ion) and also boric acid (as B_2O_3) [147].

4.3.2. Kinetics and Mechanism of Anodic Oxide Formation

4.3.2.1. Barrier Films. In the ideal case a barrier film is a nonporous, thin oxide layer possessing electronic and ionic conductivity at high electric field strength. While this field strength can be calculated by dividing the voltage drop across the oxide film by its thickness, it should be recognized that this voltage drop is not always equal to the applied voltage. The electronic conductivity of aluminum oxide barrier films is quite low. Therefore, ionic conduction is the predominant mode of charge transport, a phenomenon shared by aluminum and the so-called valve metals (Ta, Nb). This ionic conduction can be due to metal cations, oxygen anions, or other ionic species (protons, impurities). During the formation of the barrier film, the metal substrate is consumed and the anodic film grows. There has been some controversy over the reaction site for aluminum oxide barrier films, i.e., does film growth occur predominantly at the oxide-electrolyte interface, at the metal-oxide interface, or somewhere in-between these two interfaces? This question, of course, is related to whether the predominant mode of ionic conduction is via aluminum ion transport, oxygen anion transport, or both cation and anion transport.

Historically, early workers favored oxygen ion transport with reaction occurring at the metal-oxide interface [148-150]. However, the preponderance of later research evidence indicated that cations, rather than anions, were the mobile species. Thus a number of detailed theories were proposed for the growth of barrier oxide layers on aluminum and other valve metals assuming a cation mobile system. Basically, these theories assume high field ionic conduction since in such oxide layers field strengths of 10^6 to 10^7 V/cm exist and mitigate against ionic motion against the field direction. Early work by Guntherschultze and Betz led to the experimentally determined exponential relationship between the ionic current density, I_i^+, and the electric field strength, F, as [151]

$$I_i^+ = A^+ \exp(B^+ F) \tag{4.3.4}$$

where A^+ and B^+ are temperature-dependent constants. While cation transport was considered to predominate, this does not, in itself, define the reaction environment. Three possible rate-determining steps could be postulated: 1) ionic transport across the metal-oxide interface, 2) ionic transport through the bulk oxide, and 3) ionic transport across the oxide-electrolyte interface. The oft-quoted Cabrera-Mott, Verwey, and Dewald theories are based, respectively, on 1), 2), and a combination of 1) and 2) being the rate-determining step (r.d.s.). Ionic transport at the oxide-electrolyte interface does not appear to be a significant kinetic factor under the conditions of barrier layer formation. The three theories mentioned above are related to some extent. Cabrera and Mott postulated that the r.d.s. would be the rate at which cations entered into the oxide from the metal with the consequence that the number of cations in transit through the oxide film during any time period would be quite small. This mitigates against any accumulation of any ionic space charges within the film. Cabrera and Mott derived the following equation with these assumptions [152]:

$$I_i^+ = nvz \exp\left[-\frac{(W - zaE)}{kT}\right] \qquad (4.3.5)$$

where n is the surface density of mobile ions at the metal, v is a frequency factor, W is the activation energy, z is the effective charge on the mobile ion, and a is the half jump distance. Equation (4.3.5) is similar to Eq. (4.3.4) in that A^+ is now expressed in terms of a jump frequency and thus

$$A^+ = nvz \exp(-W/kT) \qquad (4.3.6)$$

$$B^+ = za/kT \qquad (4.3.7)$$

Verwey assumed that ion movement through an electrically neutral bulk was the rate-determining step [153]. Equations similar to those of Cabrera and Mott could be derived whose parameters are now related to the properties of the bulk oxide rather than those of the metal-oxide interface. Essentially then, the Cabrera-Mott and Verwey theories are equivalent, except that the Cabrera-Mott case would hold for thin films and that of Verwey would hold for thick oxide films which are electrically neutral. Both of these theories would predict a dependence of B^+ on temperature. However, experimentally this expected dependence was not always found [153, 154], indicating a dependence of the activation distance, a, on temperature. Young also concluded that a might be dependent on the field [155]. This behavior led Dewald to consider that the change in the activation distance could be a result of a change in the rate-determining energy barrier from the metal-oxide interface to one within the bulk oxide [156, 157]. Dewald also concluded from this dual barrier model that accumulation of space charges within the oxide could occur and lead to a near-independence of B from temperature. According to Dewald, the average field, $\overline{E}$, through the oxide is given by

$$\overline{E} = \left[\frac{kT}{za} \ln \frac{I_i^+}{nvz}\right] + \frac{W}{za} + \frac{kT}{za'}\left[\left(1 + \frac{1}{\delta}\right)\ln(1 + \delta) - 1\right] \qquad (4.3.8)$$

where δ is a dimensionless parameter which includes the thickness of the oxide film, the activation energy, W', for diffusion through the bulk oxide, the dielectric constant of the oxide, and the vibrational frequency of an ion in an interstitial position. a and a' are the activation distances for the metal-oxide interface and bulk oxide, respectively. W in Eq. (4.3.8) refers to the activation energy for diffusion at the metal-oxide interface. An inspection of Eq. (4.3.8) shows that the first term is identical to the Cabrera-Mott formulation. The second term, which takes into account space charge effects, becomes negligible when $\delta \ll 1$, and Dewald's theory reduces to that of Cabrera and Mott when space charge effects are absent.

The experimental evidence, as pointed out by Diggle et al. [135], indicates that the Cabrera-Mott theory is valid for anodic growth of thin films on Al and that the high field conduction expressions of Cabrera-Mott and Verwey do account satisfactorily for some of the observed experimental behavior. However, they do have some serious deficiencies. For example, the theoretically predicted dependence of B on temperature is often not observed, and plots of ln i versus field are not always strictly linear. While the theory of Dewald can theoretically account for some of these observations, there is no direct and conclusive evidence for generation of space charges within the oxide film. The major deficiency of all these theories is that they fail to account adequately for anodic transient behavior. This transient behavior cannot be accounted for by a simple, electrically uncompensated, single charge carrier, be it cations, anions, or vacancies.

Another significant problem with these earlier theories is that they assume cationic conduction. There appears to be a significant amount of relatively recent work which indicates that both cations and anions are mobile [158-162]. For example, Davies et al. reported a cation transference number of 0.58 ± 0.04 for the anodization of Al in a sodium borate-ethylene glycol electrolyte. In an aqueous ammonium citrate, the cation transport number varied from 0.37 to 0.72 as the experimental current density increased from 0.1 to 10 mA/cm^2. While there is some controversy regarding the validity of the isotopic tracer technique of Davies et al., both cation and anion motion appear to be necessary to account adequately for oxide growth on Al.

The original expression of Guntherschulze and Betz and those of Cabrera-Mott, Verwey, and Dewald are basically Frenkel-type expressions wherein the mobile cation overcomes the energy barriers against migration by thermal excitation [163]. Young indicated that the linear relationship between log I_i^+ and field strength was not strictly linear [155, 164]. Such deviations from linearity could be explained, according to Young, by assuming that the activation distance, a, is a function of the field so that

$$a = \alpha - \beta E \qquad (4.3.9)$$

where α and β are positive constants. I_i^+ can then be expressed as a quadratic function of the field strength:

$$I_i^+ = nvz[\exp -(W - z\alpha E + z\beta E^2/kT)] \qquad (4.3.10)$$

Equation (4.3.10) is more of the type of equation ascribed to Shottky-type behavior over a short range of electric field strengths. Shottky behavior implies ion movement which is only hindered when high oppositely charged centers exist where the moving ion is trapped by coulombic attraction. Dignam and co-workers verified the results of Young in that the simple exponential relationship between the ionic current density and electric field strength should be replaced by [165-167]

$$I_i^+ = \alpha' \exp\left[q\frac{(\beta E - \gamma E^2)}{kT}\right] \qquad (4.3.11)$$

where α', β, and γ are all positive constants. While Eq. (4.3.10) or (4.3.11) does appear to apply to steady-state experimental data, it again does not account for transient behavior.

It should be obvious that all of these proposals conceived with bulk oxide control assume a defined lattice model of an oxide film which is apparently amorphous. Dignam then proposed a new conduction model which was deemed more appropriate for amorphous oxides [168, 169]. Dignam drew an analogy between the arrangement of dipoles in a polarizable liquid and an oxide which was composed of a mosaic structure consisting of very small crystallites or polymeric units. Ionic migration would take place within or between polymeric units. As in polarizable liquids, the rate-controlling step in the conduction process could be: 1) between individual units or 2) within individual units. This model considers polarization of the dielectric medium, and good agreement between theory and kinetic data has been obtained for the anodization of Al (and other valve metals) with regard to the following points [170, 171]:

1. The observed transport number phenomena are adequately accounted for.
2. The experimental and theoretical correlation of the electrical field dependence of the inverse Tafel slopes (dI_i^+/dE) are obtained.
3. Transient behavior can be better accounted for.

Dignam has also developed a more detailed model of the anodic oxide film which he regards as a vitreous structure (a random, three-dimensional network) in which ionic transport occurs via defects [172]. Transient growth kinetics have been successfully explained in terms of dielectric relaxation processes within the film.

It is apparent that mechanistic studies of barrier oxide films on Al are still proceeding vigorously. Vijh has summarized the important points which have been brought out by these theoretical and experimental studies of anodic barrier layers [136]:

1. Oxide film growth occurs simultaneously at both the oxide-electrolyte and metal-oxide interfaces. Thus anion (electrolyte) incorporation into the oxide can occur and, at the same time, a more defined oxide layer can be formed. A gradual change in composition can occur across the oxide layer, leading to concentration gradients and density differences within the oxide.

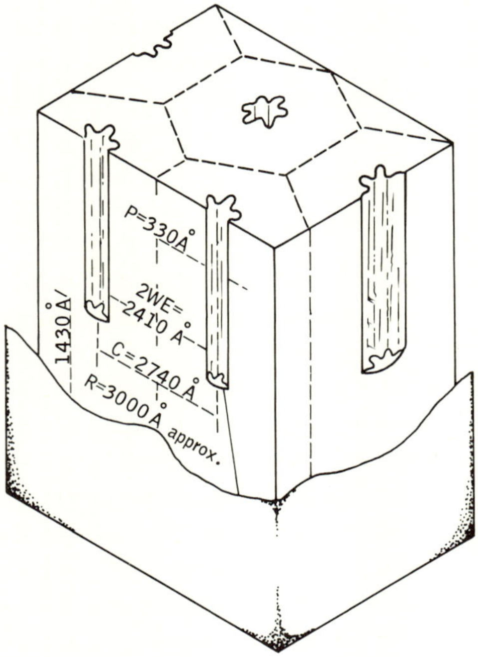

FIG. 4.3.1. Alcoa model of the unit cell formed on Al in H_3PO_4 at 120 V. (Taken from Ref. 174, by permission of The Electrochemical Society.)

2. Oxide growth occurs at the metal-oxide interface by metal ion transport while oxygen ion transport governs growth at the oxide-electrolyte interface, i.e., cation and anions, possess comparable transference numbers.

3. While the steady-state growth kinetics of barrier layers can be understood in terms of kinetic expressions such as Eq. (4.3.11), the complex transient phenomena are not yet definitively interpreted and remain an area of active interest.

4.3.2.2. Duplex Films. These distinct duplex-type films formed on Al are not found during anodization of other "valve metals." Setoh and Miyata historically were the first to propose that the anodic oxide films on aluminum consist of two regions: an outer region of thick, porous-type oxide and a thin, compact inner region adjacent to the metal [173]. While this duplex nature is well-recognized, structural details have been the subject of considerable controversy.

The generally accepted structural model of the porous anodic film on aluminum is the well-known cylindrical pore model reported by Keller, Hunter, and Robinson in 1953 and often called the Alcoa model [174]. As shown in Fig. 4.3.1 for a unit cell formed in phosphoric acid at 120 V, the oxide film is composed of hexagonal columnar cells with a

TABLE 4.3.3. Pore Diameter and Wall Thickness of Duplex Films Formed on Al under Various Conditions [174]

Electrolyte	T (°C)	Pore diameter (Å)	Wall thickness (Å/V)
15% Sulfuric acid	10	120	8.0
2% Oxalic acid	24	170	9.7
3% Chromic acid	38	240	10.9
4% Phosphoric acid	24	330	10.0

TABLE 4.3.4. Pore Density Values for Al Anodized in 15% H_2SO_4 at 10°C at Various Applied Voltages [174]

Anodizing voltage (V)	Pore density ($\times 10^9$ cm^{-2})
15	83
20	56
30	30

cylindrical pore of star-shaped section in each center. The following general formulas were proposed for this model:

$$R = \frac{H}{2} + \frac{W^2}{8H} \tag{4.3.12}$$

$$C = 2WE + P \tag{4.3.13}$$

$$V = \frac{78.5P^2}{C^2} \tag{4.3.14}$$

Here R is the radius of curvature of the cell bottom, H is the height of the scallop, C is the cell size, W is the wall thickness (Å/V), E is the forming voltage, P is the pore diameter, and V is the pore volume. Table 4.3.3 gives pore diameter and wall thickness values of some typical duplex films as originally reported by Keller et al. [174]. Since the width of the oxide cell (C) increases as the anodizing voltage increases, the pore density should decrease as the anodizing voltage increases. Pore density values as a function of anodizing voltage are shown in Table 4.3.4 for Al in 15% H_2SO_4 at 10°C.

Both electron microscopy and B.E.T. gas adsorption studies have supported the model of Keller, Hunter, and Robinson [175-180]. The work of Wood, in particular, has provided strong evidence for the geometrical pore model of Keller et al. [180]. By using a sectioning technique, vertical to the aluminum surface, Wood et al. clearly showed the pores, barrier layer, and the scallops, above which the pores are located. In 4% phosphoric acid at 25°C the dimensions of the porous structure reported by Wood et al. were somewhat different from those of Keller et al., although these variations did not detract seriously from this confirming evidence.

Observations by Wood et al. and others have led to modifications of the Alcoa model which involve nonuniform pore geometries. Paolini et al. proposed that the true pore geometry is that of a truncated cone whose basal diameter is that reported by Keller et al., and whose mouth diameter is dependent on current density and time (actually film thickness) [179]. Table 4.3.5 shows data tabulated from the work of Paolini et al. for the film formed on Al in 20% sulfuric acid. Actually the angle of the pore walls is quite slight since the diameter changes only from 120 to 159 Å over a distance of 9 Å. The maximum pore mouth diameter of 246 Å is to be compared with the reported oxide cell width of 316 Å. It would be expected that the pore mouth diameter could approach 316 Å and thus define the maximum thickness of the film. Some evidence in support of this concept has been obtained by Liechti and Treadwell [144]. Franklin and Stirland have also reported that in these types of films the pore axis is not vertically inclined to the metal surface, the degree of axis slant being dependent on the grain orientation of the Al substrate [181].

Csokan reported irregularities in the pore geometries of oxide films formed on Al in dilute H_2SO_4 at 0 to 5°C, with the pores twisted or bent [182-184]. A high density of pore openings at the surface produced a fibrous structure of variable size, shape, and orientation. Under polarized light, this laminated structure appeared to be parallel to the aluminum surface. This irregular fibrous structure proposed by Csokan is similar to that of Ginsberg and Wefers. According to Ginsberg and Wefers, the porous aluminum oxide structure should not be considered an array of regularly distributed, densely packed oxide cells, but rather a regular fibrous structure with hollow fibers standing vertically to the oxide surface [138]. The oxide was considered to be amorphous with the outer wall consisting of amorphous alumina and the inner wall consisting of aluminum hydroxide containing the electrolyte anion. This discussion of the models of Csokan and Ginsberg and Wefers prefaces a discussion of the Murphy-Michelson model of porous anodic films on aluminum, which is a radical departure from the model of Keller et al.

Murphy and Michelson originally criticized the physical pore model of the porous oxide layer, arguing that such pores are formed during preparation of the film for study [185]. (This original criticism does not seem justified by a great amount of evidence obtained by direct transmission, sectioning, and other methods.) Murphy and Michelson also pointed out that the oxide base pattern was determined by the most recent anodizing conditions, and that the change from one oxide base pattern to another occurred readily regardless of the

TABLE 4.3.5. Pore Mouth Dimensions for Al Anodized in 20% H_2SO_4 as a Function of Current Density and Time [179]

Current density (mA/cm^2)	Time (min)	Pore base d (Å)	Pore mouth d (Å)
10	30	120	159
15	30	120	182
15	60	120	246
25	30	120	208

direction of change. The simple physical pore model would require, then, that some of the pores present in a low voltage coating would cease to function on increasing the voltage, and conversely, for voltage decrease, new pores would have to be produced. The mechanism of such events is difficult to visualize by the physical pore model. It is also difficult for the physical pore model to provide a mechanism for oxide growth in the regions between pores if the pore wall is of a quality similar to that of the barrier layer at the pore base.

Using a colloid chemistry viewpoint, Murphy and Michelson proposed a triple layer or colloidal gel model, with the three layers consisting of:

1. An inner, relatively compact, anhydrous form of alumina of nonuniform composition existing as submicrocrystallites.
2. An intermediate transition region where the transition from the anhydrous inner region to an outer region of high hydration occurs. Here hydrated regions of oxide exist where electrolytic conduction is assumed to occur.
3. An outer region of hydrated oxide of variable hydration and density. This region, where the minimum number of submicrocrystallites exist, is suggested to be the precursor of the pores observed via electron microscopy.

Essentially, then, the porous anodic film could be considered an agglomeration of submicrocrystallites of anhydrous Al_2O_3 surrounded by, and held together by, submicroscopic regions containing electrolyte in a relatively complex hydrogen-bonded system. According to the Murphy and Michelson model, as shown in Fig. 4.3.2, regions of lower density which appear as pores in electron microscopy result from dissolution effects promoted by localized heating. This heating increases the rate of conversion of anhydrous oxide to hydrous oxide and finally to dissolution of the oxide, forming a pore. Selective dissolution of the barrier layer proceeds to interpore material which consists of particles of anhydrous material separated by a hydrous, anion-containing region of high electrolytic conductivity. Oxygen transport to the barrier layer occurs by OH^- or H_2O via electrolytic conduction through or along the internal surfaces of the porous layer above the barrier layer.

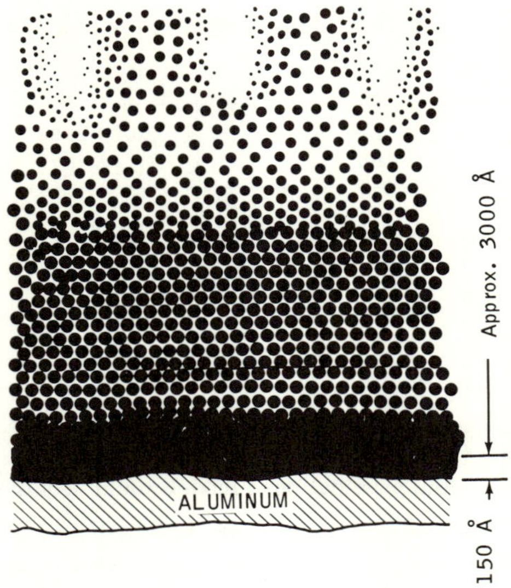

FIG. 4.3.2. Murphy-Michelson model of the porous oxide film on aluminum. (Taken from Ref. 185, by permission of The Aluminum Federation.)

This oxygen transport modifies the previously formed barrier layer into a more protonated and hydrous oxide. At the same time, protons are transported outwards and dehydration of the oxide is promoted. The association of only certain acids with duplex film formation was assumed to be due to the special hydrogen bonding properties of such acid anions as sulfate, oxalate, and phosphate which are known to minimize the rate of rearrangement and recrystallization of aluminum hydroxide gels. Thus films of such oxides appear "amorphous" by X-ray or DTA analysis since their particle size precludes any distinguishable diffraction pattern.

The original proposal of Murphy and Michelson was further developed by Murphy, taking into consideration proton space charge and colloid and electrokinetic effects [186, 187]. The first consideration was that, under the influence of the high anode potential, anions and water penetrate the coating, and proton space charge accumulates via reactions of the following type:

$$2Al^{3+} + 3OH^- = Al_2O_3 + 3H^+ \quad (4.3.15)$$

$$2Al^{3+} + 3H_2O = Al_2O_3 + 6H^+ \quad (4.3.16)$$

$$2Al^{3+} + 3HSO_4^- = Al_2(SO_4)_3 + 3H^+ \quad (4.3.17)$$

Neutralization of these protons could occur by the field-assisted migration of anions or by diffusion of these protons out of the film into the bulk electrolyte. The second consideration by

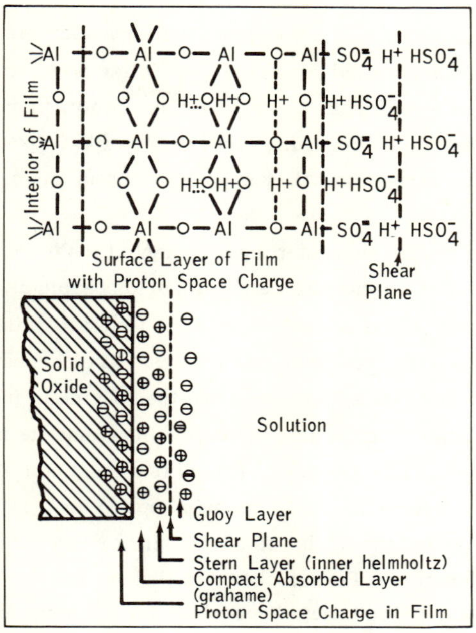

FIG. 4.3.3. Possible ionic arrangement at the aluminum oxide film-electrolyte interface as depicted by Murphy. (Taken from Ref. 187, by permission of The American Electroplater's Society, Inc.)

Murphy dealt with the presence of electrokinetic and adsorption processes. The normal isoelectric point (point of zero charge) of various aluminas is at pH values greater than 9, and therefore a high hydroxyl ion concentration is required to neutralize this adsorbed hydrogen ion capacity. This point of zero charge can be shifted to more acidic pH values by specific adsorption of anions such as sulfate and phosphate. Since duplex films seemed to form only in dibasic acid electrolytes, Murphy correlated this fact with the stronger specific adsorption of divalent ions compared to monovalent ions. The influence of these specifically adsorbed anions rather than hydroxyl ion could permit the build-up of proton space charge. In barrier layer forming electrolytes, OH^- adsorption predominates and the contained and adsorbed protons can be neutralized, resulting in a hydrated alumina with low foreign ion content. In the presence of high anion concentrations and low hydroxyl ion concentration, anion adsorption predominates and the properties of the barrier layer are modified and its conversion into a porous coating can occur. Figure 4.3.3 shows one of the two possible ionic arrangements at the anode-electrolyte interface according to Murphy, illustrating proton space charge in the film, balanced by anions in the compact adsorbed layer and in the solution. Bogayavlenskii and Schreider also proposed somewhat similar models to that of Murphy and Michelson based on a gel-like anodic layer [188, 189].

The Murphy-Michelson model is, of course, still speculative, but appears to be a viable alternative to the simple physical pore model of Keller et al. and its variations. Before either model can be accepted with any degree of certainty, more research must be carried out. Keeping this uncertainty regarding the validity of the physical pore model, some of the details of duplex film formation will now be considered, primarily from the viewpoint of the physical pore model.

Figure 4.3.4 shows typical current-time transients for the formation of barrier-type and duplex films on Al. Initially, the curves are superimposable as the barrier layer is formed and the ionic current decreases as the electric field strength decreases. At point a, the curve diverges as pore initiation apparently occurs. The current for duplex film formation still continues to decrease for a short time and then begins to increase. The final current in the case of the barrier layer (which is mainly electronic) is much smaller than those observed in the case of duplex film formation (which is almost completely ionic). The time to point a is typically 25 sec for a 15-V voltage step applied to Al in 15% H_2SO_4 [190]. It is implied from Fig. 4.3.4 that the thickness of the barrier layer must decrease as pore formation is initiated. Hunter and Fowler observed that the barrier layer thickness does indeed show such a decrease. The exact shape of the I-t transient for duplex layer formation is dependent on the voltage applied and the electrolyte, with the current minimum occurring at earlier times with higher voltages and more acidic electrolytes. Lower pH's and higher temperatures result in larger values of the current density minimum.

It is clear that the electrolyte plays a major role in duplex film formation. While Kissin has suggested that di- or trivalent anions are required for porous film formation on Al [191], sulfamic acid (NH_2SO_3H) and phosphoric acid solutions where the predominant anion is $H_2PO_4^-$ belie this generalization. According to Tajima [133], the following electrolyte properties, again with some exception, seem to be prerequisites for duplex film formation:

1. High electrolyte specific conductivity.
2. Two easily ionizable hydrogens.
3. Absence of ionizable chlorine (since chloride causes pitting).

Identification of these necessary electrolyte properties has been far ahead of a suitable mechanism to explain their effects.

Two main growth mechanisms have been considered for porous oxide films on aluminum. The first involves migration of cations across the film, together with oxygen-bearing anions (OH^-, O^{-2}, SO_4^{-2}) to form a porous oxide at the film-electrolyte interface. The second mechanism involves migration of cations and anions across the barrier layer to form compact layers at the oxide-electrolyte and metal-oxide interfaces, respectively. By some process, as yet undefined, this layer is transformed into a porous layer at a rate equal to that of barrier layer formation. The thickness of the barrier layer therefore remains

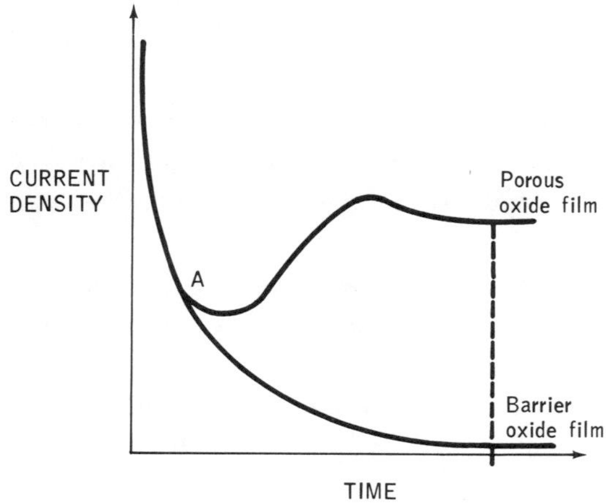

FIG. 4.3.4. Typical current-time transients for anodic film formation on aluminum. (Taken from Ref. 135, by permission of The American Chemical Society.)

constant. This second mechanism is generally accepted, although the manner in which the conversion of the barrier layer into a porous layer occurs is quite controversial.

Keller et al. proposed a purely chemical process for this transformation of the barrier layer due to local Joule heating at the pore base, causing accelerated chemical dissolution [174]. However, Nagayama and Tamura have shown that such temperature rises at the pore base are almost always negligible in practice [192]. Moreover, it is well-known that slight changes in temperature and electrolyte concentration greatly affect duplex film formation.

Hoar and Mott were the first to suggest that dissolution at the pore base was a field-assisted process [193]. They assumed that OH^- migration led to barrier-layer formation at the metal-oxide interface. The protons released in this reaction could migrate under the influence of the electric field toward the oxide-solution interface, favoring dissolution at the pore base. Hoar himself and Yaholom later rejected this model for which no experimental evidence had been obtained [194]. They then proposed that proton entry into the preformed layer, at preferred sites, was important. Pore initiation was assumed to occur at these preferred sites which become points of high local current density and at which high dissolution rates ensue. These preferred sites were not identified, yet their existence has been suggested by observations that the pores of a duplex film appear to occur at surface defects, subgrain boundaries, and preferred orientations. This is not surprising, considering the demonstrated importance of the metal substrate dislocation density in metal dissolution and deposition [195]. At this point any further discussion of porous oxide layer formation seems unwarranted, simply because of the lack of agreement on a physical model of the film. There is every indication that the initiation and growth of the porous part of such films is something more than a pure chemical dissolution process and will no doubt continue to be an area of active interest.

Before closing this discussion of the anodic oxidation of aluminum, three minor topics of some practical interest will be considered briefly; 1) electropolishing, 2) the anodic oxidation of an aluminum surface already covered with a layer of hydrated oxide, and 3) oxide formation on Al in fused salts.

Electropolished surfaces involving oxide formation can arise naturally from anodic oxide films provided that the outer surface dissolution process is continued. Grilikhes and Fedot'ev have concluded that electropolished and anodized aluminum surfaces differ only in the type of control during the electrochemical operation [196]. Electropolishing is controlled by a slow electrochemical reaction and chemical dissolution is rapid, as contrasted to anodic film formation where the electrochemical reaction is rapid compared to chemical dissolution. Neufield and Ali have interpreted electropolishing as a form of porous-type anodizing where the angle, ϕ, between the pore wall and substrate approaches 90° [197].

During the manufacture of high voltage aluminum capacitors (see Section 5.7), the etched Al anode foil is immersed in boiling water for several minutes prior to anodizing. This pretreatment, which results in a film of pseudoboehmite as shown by Vedder and Vermilyea [198], leads to a substantial savings of electricity during subsequent anodization. In the presence of the hydrous oxide, the electrochemical oxide forms with 100% current efficiency in contrast to the usually inefficient anodic oxide formation due to aluminum dissolution and oxygen evolution [199]. Alwitt et al. have recently presented a model for the conversion of this hydrous oxide to a barrier oxide layer [200]. Alwitt et al. proposed that conversion of the hydrous oxide into barrier oxide occurs via reaction of aluminum cations with OH^- and H_2O present in the hydrous oxide rather than in the electrolyte species. During this reaction, which occurs at the barrier/hydrous oxide interface, AlOOH is converted to Al_2O_3. Typical reactions might be:

$$Al^{3+} + 3AlOOH = 2Al_2O_3 + 3H^+ \tag{4.3.18}$$

$$Al^{3+} + AlOOH + H_2O = Al_2O_3 + 3H^+ \tag{4.3.19}$$

This mechanism adequately accounts for these experimental observations: 1) less anodic oxide is needed to support a particular voltage, 2) the rate of voltage rise is faster at a given current density, 3) the water content of the composite oxide decreases, and 4) the absence of side reactions during anodization.

Oxide films can be formed on Al by anodization in oxygen-anion-containing melts such as nitrates, nitrites, and bisulfates. In the nitrate melts, α-, η-, and β-Al_2O_3 are formed with α- and γ-Al_2O_3 prevailing [201]. These films are very hard and compact, with a nonporous, partially crystalline structure. Coating ratios approaching the theoretical value of 1.8 have been observed [202]. Growth of such films appears to follow the usual high field expression as in Eq. (4.3.4). Anodization of Al in the low temperature (100 to 160°C) bisulfate melts such as $NaHSO_4$, $KHSO_4$, and NH_4HSO_4 at current densities of 1 A/dm^2 leads to formation of a porcelain-like corundum (α-alumina) film of thicknesses $\geq 100\ \mu$ [203]. Such oxide films appear to be formed by conversion of an initial amorphous γ-alumina to α-alumina. Tajima et al. have proposed that an amorphous oxide is formed first which is immediately transformed to γ-alumina and finally to α-alumina by local heating due to electrical arcing within the film.

5. APPLIED ELECTROCHEMISTRY

5.1. ELECTROWINNING AND ELECTROREFINING

5.1.1. Electrowinning

Second only to the chlorine-caustic soda industry, the electrowinning of aluminum is one of the major industrial applications of electrochemistry. The estimated 1973 international consumption of Al was in the vicinity of 14 million tons [204], all produced by the electrochemical reduction of alumina in a cryolite melt by the Hall-Heroult process.

The electrolytic process for commercial production of aluminum was discovered almost simultaneously in 1866 by the American, Charles M. Hall, and the Frenchman, P. L. T. Heroult [205]. Both established that molten cryolite was an effective solvent for the electrolysis of alumina. Cryolite, in addition to dissolving alumina, had other attractive features such as a relatively low melting point (~1000°C), a low operating cell voltage (~6 V), and a specific gravity low enough to allow Al to sink to the bottom of the cell where it could easily be drawn off. The process has basically been unchanged in principle since its inception, although it has undergone much improvement and some modification. Since Karl Bayer received the basic patents on the production of alumina from bauxite ore, Heroult's name is often omitted, and thus Bayer-Hall process technology not only refers to the electrolysis procedure, but also includes the alumina production step. Some recent publicity has been given to aluminum production processes which are not based on cryolite electrolysis, and installations using this new technology are apparently on the drawing board [206, 207].

All aluminum originates in bauxite ore, which is a mixture of aluminum oxides, iron oxides, silica, and titania. It is a major export of Australia, Jamaica, Guyana, and Surinam. In the basic Bayer process, bauxite ore is first treated with NaOH solution which dissolves hydrated aluminum oxide but does not dissolve iron oxide. After removal of the iron oxide, the solution of aluminate ions is diluted and $Al(OH)_3$ is precipitated. The precipitated $Al(OH)_3$ is then calcined to Al_2O_3. Various alumina production processes have been discussed in detail by Mantell [208].

The electrolytic cell for aluminum production is essentially a carbon-lined vessel containing a liquid aluminum cathode which has a slightly greater density than the cryolite melt (at 1000°C, $d_{Na_3AlF_6} = 2.095$ g/cc; $d_{Al} = 2.29$ g/cc). Consumable carbon anodes are supported within the cell. Some of the basic electrochemistry of the aluminum production process and practical cell design and operation will now be discussed.

The electrolytic production of aluminum is based on the following cell reaction:

$$\frac{2}{3}Al_2O_3 + C = \frac{4}{3}Al = CO_2 \tag{5.1.1}$$

The equilibrium potential of Reaction (5.1.1) is 1.163 V at 1010°C as compared to 1.024 V for the reaction:

$$\frac{1}{3}Al_2O_3 + C = \frac{2}{3}Al + CO \tag{5.1.2}$$

However, at the current densities used in practical cells (>0.05 A/cm^2), CO_2 is the normal anode reaction product [209, 210].

As discussed by Piontelli [211], the primary cathode reaction is indeed the deposition of aluminum, and sodium deposition occurs only under unusual circumstances. The charge-transfer overvoltage in cryolite melts at such temperatures is very small (<10 mV), and any cathode polarization which is observed has usually been attributed to concentration polarization. This concentration overvoltage is a result of the low concentration of reducible species present in the cryolite-alumina melt. Piontelli has estimated that, while such concentration polarization effects may be on the order of 0.05 to 0.2 V, they are largely attenuated in practical cells due to the presence of electromagnetic and thermal convection.

At 100% current efficiency, one would expect the anode reaction to consist of the production of CO_2 only. In fact, some formation of CO does occur, and an equivalent redissolution of Al occurs via the universally accepted overall reaction as shown by Pearson and Thonstad [210, 212]:

$$2Al + 3CO_2 = Al_2O_3 + 3CO \tag{5.1.3}$$

There is some controversy over the nature of the actual reductant. While the rate-determining step appears to be the mass transport of the reductant from the Al/cryolite interface to the gas/cryolite interface [213, 214], Al (dissolved in the melt), Na, Na_2F, and AlF have all been postulated as the reductant. Such species can arise as a result of the following equilibria [215]:

$$Al = 3NaF = AlF_3 + 3Na \tag{5.1.4}$$

$$2Al + AlF_3 = 3AlF \tag{5.1.5}$$

Therefore, the overall Reaction (5.1.3) could be a combination of the following, or similar reactions:

$$3AlF + 3CO_2 = Al_2O_3 + 3CO + AlF_3 \tag{5.1.6}$$

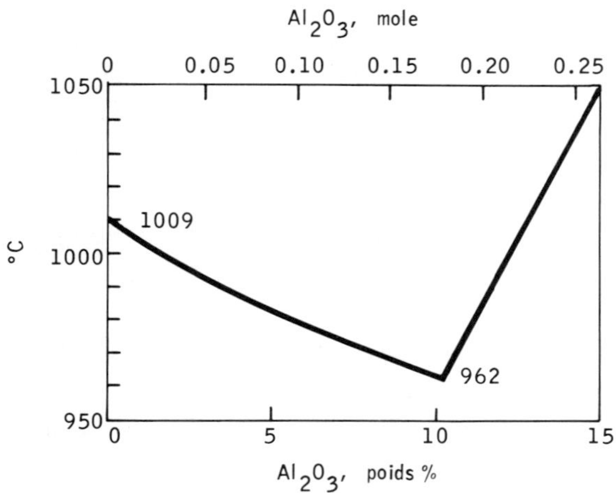

FIG. 5.1.1. Liquidus diagram for cryolite-alumina. (Taken from Ref. 28, by permission of Pergamon Press.)

$$6AlF + 3CO_2 = 2Al_2O_3 + 3C + 2AlF_3 \tag{5.1.7}$$

$$6Na + 2AlF_3 + 3CO_2 = Al_2O_3 + 6NaF + 3CO \tag{5.1.8}$$

$$12Na + 4AlF_3 + 3CO_2 = 2Al_2O_3 + 12NaF + 3C \tag{5.1.9}$$

As will be discussed later, the gaseous composition above an operating Al production cell can be used to measure the coulombic efficiency of the process based, in general, on Reaction (5.1.3).

It can be recognized that the nature of the cryolite-alumina electrolyte is quite complex. While there is a significant amount of interest in the actual nature of the species which is actually discharged at the carbon anode, the properties of the electrolyte are also of considerable importance.

The basic aluminum cell electrolyte consists of sodium cryolite, Na_3AlF_6, as the major (~ 80 wt%) constituent, and of course alumina, Al_2O_3, at 2 to 8 wt%. Other components (CaF_2, AlF_3, LiF) are added to the alumina-cryolite mixture to modify: 1) the melting point of the electrolyte, 2) the density of the electrolyte, 3) the electrical conductivity of the electrolyte, and 4) the electrode processes or chemical reactions occurring in the melt.

Figure 5.1.1 shows the liquidus diagram for cryolite-alumina. Pure cryolite has a melting point of 1009°C, with the cryolite-alumina eutectic at 962°C and 10 wt% Al_3O_3 [216]. There are extensive reports of liquidus curves for various ternary systems for cryolite-alumina with the addition of an alkali or alkaline earth halides or AlF_3.

Grjotheim et al. have reviewed the influences of NaCl and LiF and other additives on the aluminum cell electrolyte and their conclusions are of some interest [217]. Basically, any substance to be added to the Al cell electrolyte should: 1) lower the melting point, 2) decrease melt density, and 3) increase the melt conductivity of cryolite. It was reported that LiF and NaCl lower the melting point and density while improving the melt conductivity. While NaF is also an effective additive, it is not recommended because it increases the cryolite (NaF/AlF_3) ratio of the electrolyte, and by displacing the equilibria in Reaction (5.1.4) would lower current efficiency as demonstrated in actual cells. In the ternary systems the influence of such additives is quite complex. For example, LiF is a more effective additive even at low (5 wt%) concentration than NaCl due to its more significant lowering of the freezing point and increase of the conductivity, while its effects on alumina solubility and decomposition potential are minimal.

The nature of the species present in cryolite-alumina melts has not been unequivocally determined. While Na^+ has been shown to be the main current-carrying species [218], the nature of the anionic species is still the subject of some controversy, particularly from the viewpoint of the anode reaction. While one may write the anode process as [219]

$$\frac{1}{2}C + 2O^{2-} = \frac{1}{2}CO_2 + 2e^- \tag{5.1.10}$$

the use of an oxide ion symbolism is a simplification, since in such melts complex oxy- and oxyfluoride ions may exist. Foster and Frank selected AlO_2^- and $AlOF_2^-$ as the most likely oxygen-containing anions based on cryoscopic data [220]. $Al_2O_4^{2-}$, a dimeric species, was suggested at Al_2O_3 contents >5 mole %. Brynestad et al. presented evidence from quench studies and X-ray investigations that complex oxyfluoride anions existed in which the Al/O ratio was unity, e.g., $Al_2O_2F_4^{2-}$, $AlOF_3^{2-}$, and $AlOF_5^{4-}$ [221].

In contrast to the cathode reaction, which appears to be quite straightforward, the anode reaction in aluminum production cells has received considerable attention. Since the behavior of carbon electrodes in such electrolytes is covered in another chapter in this treatise, we will only briefly mention the chief considerations of the anodic behavior of carbon in Al production cells. While a considerable amount of effort has been devoted to the "anode effect" (as will be discussed later in this section) in practical Al production cells, which occurs under abnormal operating conditions, considerable anode polarization exists under normal conditions. Piontelli reported anodic overvoltage values on the order of 0.3 to 0.5 V at current densities up to 100 mA/cm^2 in early studies [211], while Richards and Welch reported a rate-determining two-electron transfer reaction [222]. Later works by Thonstad, in particular, have considered the anode reaction in some depth and the reader is referred to these papers for a more detailed expose [223-225].

The so-called "anode effect" in aluminum production cells refers to the sudden rise in voltage which sometimes occurs in such cells. Normally, this voltage rise is accompanied by [223]: 1) nonwetting at the anode-melt interface, 2) arcing at the anode melt interface, and 3) evolution of appreciable amounts of CF_4. The anode effect occurs when the current

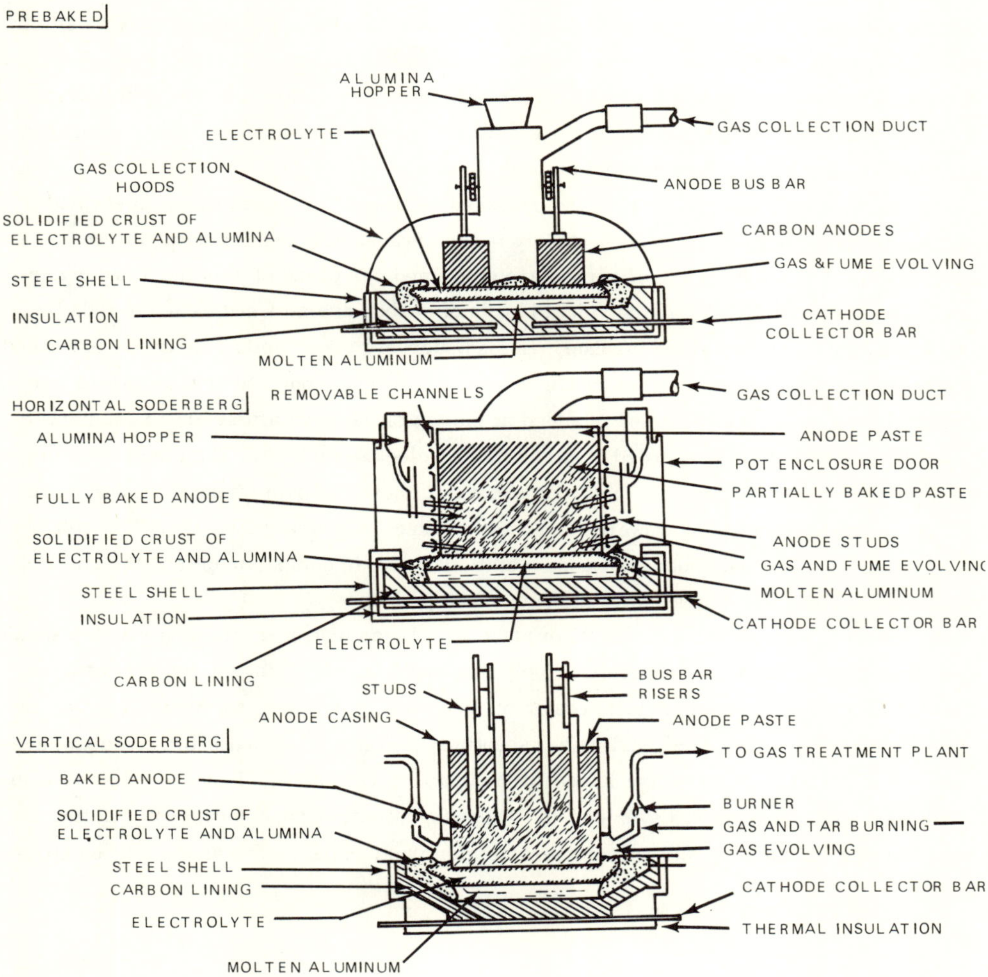

FIG. 5.1.2. Details of aluminum production cells. (Taken from Ref. 353, by permission of The AIME.)

exceeds a critical current density which is primarily a function of the melt composition. Apparently, depletion of dischargable oxygen-containing ions in the vicinity of the anode leads to fluorine discharge once the limiting current density for alumina discharge is exceeded [219]. We will now consider some of the more practical aspects of Al cell operation.

Normally, a number of cells (50 to 200) are connected in series to form what is called a potline. Typical cell dimensions are ~ 10 ft x 30 ft x 3 ft. Figure 5.1.2 shows details of the three basic types of Al production cells which are distinguished by the type of anode.

Prebaked (PB) cells use prebaked carbon anodes and account for ~ 59% of the United States aluminum industry. The two types of Soderberg cells use anodes which are formed in situ by baking of a "green" carbon paste formed from coke bound with pitch. The VSS cells (Vertical Stud Soderberg), which account for ~ 13% of United States production, and the HSS (Horizontal Stud Soderberg) cells, which account for ~ 25% of United States production, are distinguished by the position of the electrical connection to the anode.

As can be seen from Fig. 5.1.2, the prebaked anode cell process differs in several ways from the Soderberg cell process, mainly in the arrangement for replacing the spent anodes. Typically, 0.6 to 0.8 lb of carbon are consumed per pound of aluminum under normal cell operation. In PB cells, individual blocks of carbon are attached to upper rods by baking in a separate anode baking furnace. In the Soderberg anode cells, the anode is formed continuously by baking the coke-pitch mixture until it is fully carbonized at the working surface of the bath. This anode "paste" is added at the top of the rectangular anode compartment in HSS cells which is suspended over the molten electrolyte bath. The electrode connections or studs are inserted into the anode about 3 ft above the bath where the paste is fairly soft. As the anode is moved downward the paste is baked and grips the stud. A VSS cell also uses a continuously formed anode, but uses a stationary anode compartment and the studs always remain in electrical contact with the semi-fluid paste.

At 100% current efficiency in an Al production cell, an oxygen-containing species should be discharged at the anode to produce CO_2 only. In practice, the principal current efficiency loss in an aluminum production is a result of oxidation of reduced aluminum species by the CO_2 produced as discussed earlier. Essentially this is a chemical short circuit wherein there is a reaction between the anode product and a reductant. Based on this reasoning, Pearson and Waddington developed the following equation as a means of relating the cathode current efficiency, ε_c, in commercial aluminum cells to the anode gas composition [209]:

$$\varepsilon_c(\%) = 0.5(\%CO_2) + 50 \qquad (5.1.11)$$

Equation (5.1.11) assumes 100% aluminum reduction efficiency at the cathode and an equivalent formation of CO_2 at the anode. Thus, if no CO is present in the anode gas, $\%CO_2 = 100\%$, and the cathode current efficiency is 100%. If all the CO_2 reacts with reduced Al species, the anode gas is 100% CO and the current efficiency is 50%. Practical cell anode gases contain 60 to 80% CO_2, equivalent to current efficiencies of 80 to 90%. Other investigators have concluded that Eq. (5.1.11) was valid, but somewhat inaccurate, since it did not take other reactions into account. For example, Beck showed that more CO was generated (in an experimental aluminum cell) than Eq. (5.1.3) could account for, and that this excess CO was a result of the Boudouard reaction [226]:

$$CO_2 + C = 2CO \qquad (5.1.12)$$

At 950 to 1000°C, the equilibrium is displaced to the right in Eq. (5.1.12). Beck then introduced a g term which is the ratio of the carbon in the gas to carbon theoretically oxidized to CO_2. Beck's final equation is [227]

$$\epsilon_c(\%) = \left| g[50 + 0.5\%(\%CO_2)] - Z \right| \left| 1 - Z'/100 \right| \qquad (5.1.13)$$

where Z and Z' are additional correction terms. Z accounts for reactions other than those involving CO_2, while Z' accounts for losses due to electronic conduction.

Cammarota and Schlain used a small laboratory cell and developed the following equation, which was shown to give reliable estimates of aluminum production cell efficiency [228]:

$$\epsilon_c(\%) = 0.45(\%CO_2) + 47.6 \qquad (5.1.14)$$

Equation (5.1.14) again accounts for the Boudouard reaction, which varies with the anode type and operating conditions. These authors also developed an alternative equation, Eq. (5.1.15), which depends on measuring the anode current efficiency (ACE) by the oxygen present in the anode gases, and the metal efficiency loss (EL) resulting from solubility and vaporization:

$$\epsilon_c(\%) = \text{ACE} - \text{EL} \qquad (5.1.15)$$

Very recent work by Haupin has tried to corroborate these predictions of industrial cell efficiency by laboratory measurements [229]. According to Haupin, one must consider quite carefully the differences between laboratory and industrial Al cells, including effects of moisture and changes in the cell lining.

Of particular interest in practical cell operation is the effect of additives and conditions on Al production cell power and current efficiency. In order to understand the contribution of various cell components and processes to the total cell voltage, the data of Table 5.1.1 are quite informative. As a point of reference, the theoretical potential of the alumina reduction reaction is ~ 1.2 V at 1010°C. Therefore, 0.6 V polarization exists due to both the anode and cathode. Resistive losses in the electrolyte are also considerable.

TABLE 5.1.1. Voltage Losses in Al Production Cells [230]

Component	Voltage loss (V)
Electrolysis polarization	1.8
Anode effect	0.1
Pot lining drop	0.4
Carbon electrode iR drop	0.35
Terminal iR drop	0.15
Electrolyte iR drop	1.9

Common additives to the alumina-cryolite electrolyte include NaF, AlF_3, MgF_2, CaF_2, and LiF as mentioned. As might be expected from the equilibrium and reactions shown earlier, the influence of any additive is a quite complicated phenomenon. The exact method of assessing current efficiency has led to some difficulties in correlating the effects of various electrolyte additives. As reported by Richards, many studies on current efficiency show agreement only in regard to the optimum NaF/AlF_3 molar ratio (2.7) and in the fact that 5 mole % additions of MgF_2 and LiF lead to 1 to 5% increases in current efficiency, all other conditions being equal [231]. The beneficial effect of MgF_2 and LiF_2 are apparently similar to CaF_2 in that such additives improve melt conductivity and lower the liquidus temperature. While Szeker's data show a minimum in current efficiency at 5.5 wt% Al_2O_3 [232], Schmitt showed that higher alumina concentrations led to increased efficiency [233].

In general, current efficiency in laboratory and industrial cells decreases with decreasing electrode separation, which can be simply interpreted as more favorable mass transport conditions for reactant crossover, i.e., Al/ClO_2 recombination. Obviously, one is confronted here with a trade-off in efficiency due to ohmic drop in the electrolyte and current inefficiency due to Al/CO_2 recombination. No complete parametric study has apparently been carried out which examines all Al production cell variables on current efficiency. Schmitt has made an attempt at such a study and his results serve as an utilitarian example [233].

Figure 5.1.3, after Schmitt, shows the effects of interelectrode distance, temperature, Al_2O_3 content, and the free or excess AlF_3. From Fig. 5.1.3 the optimum Al production cell efficiency was predicted to occur at 940 to 960°C, an NaF/AlF_3 mole ratio of 2.5 to 3.0, and an Al_2O_3 content of 5 to 8 wt%. An Al production cell using these improved operating conditions showed a current efficiency of 91 to 93% at 4.35 V, compared to previous operation at 4.70 V. As commented on by Richards, more studies such as that of Schmitt most likely exist, but are naturally held proprietary by aluminum producers [231].

Al production cells operate at currents up to 130,000 A, with cathode current densities of 0.5 A/cm^2 (anode current densities are ~ 1 A/cm^2) [208]. Typical alumina requirements are on the order of 2 lb alumina/lb aluminum. Cryolite consumption or melt consumption may be between 0.04 and 0.07 lb/lb of Al. Power requirements are $\sim$ 8 kW-hr/lb of Al ($\sim$ 9 kW-hr ac/lb of Al). Half a pound of carbon is normally consumed per pound of aluminum.

In practical cells the anode effect is of considerable concern since the cells are connected in series and the associated rise in cell voltage to 30 to 60 V can have a significant impact on the required applied voltage. In practice, alumina addition to the molten salt is a practical solution since the anode effect is encountered at low Al_2O_3 concentrations. Since a crust of electrolyte tends to form in production cells, crust breakers are often utilized to facilitate alumina addition.

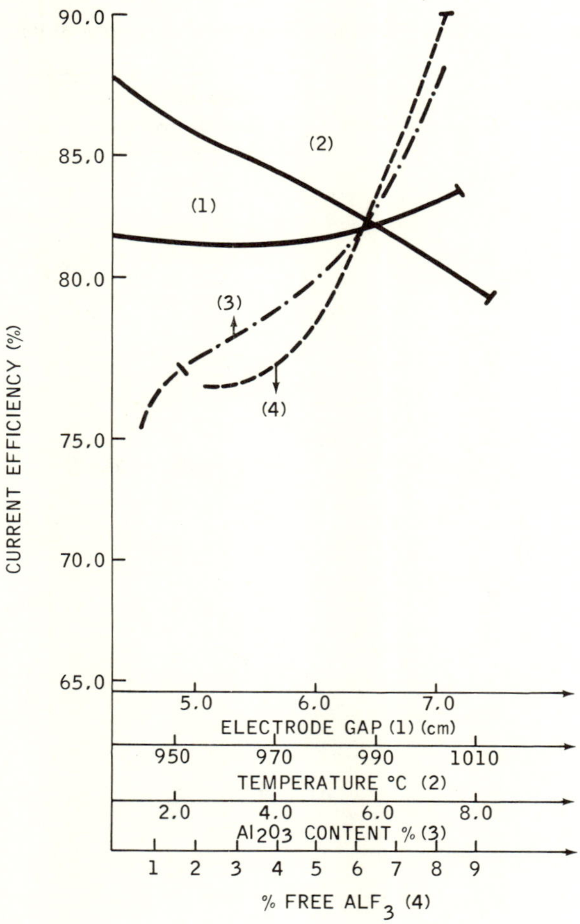

FIG. 5.1.3. Effect of various parameters on Al reduction cell efficiency. (Taken from Ref. 233, by permission of The AIME.)

For more details of practical Al production procedures, the reviews of Mantell [208], Richards [231], and Schmitt-Hatting [234], are duly noted. Very recent reports indicate the interest of aluminum producers in more economic Al production processes [206, 207]. Alcoa has publicized a new process wherein conventionally processed alumina is chlorinated and $AlCl_3$ is electrolyzed to yield Al and recyclable chlorine [206]. This process is based on the use of a 3 to 10 wt% $AlCl_3$/NaCl-LiCl electrolyte at ~ 700°C [235]. Operation of such cells at voltages of 3.3 V at 1.3 A/cm^2 led to a power consumption of 5.2 kW-hr/lb Al. This is 65% of a normal Hall process cell.

TABLE 5.1.2. Essential Details of Aluminum Electrorefining Processes [208]

Parameter	Hoopes process	Gadeau process	AC Neuhausen process
Electrolyte composition, w/o			
AlF_3	30-38	23	48
NaF	25-30	17	18
BaF_2	30-38	-	18
$BaCl_2$	-	60	-
CaF_2	-	-	16
Electrolyte resistivity, ohm/cm^3	0.3 at 1000°C	0.75-0.85 at 740°C	1.1 at 740°C
Anode composition, w/o			
Al	75	67	70
Cu	25	33	30
Cell conditions:			
Temperature, °C	900-1000	740	740
Voltage, V	5-7	6-7.5	5.5-5.8
Current density, mA/cm^2	-	400	360
Metal purity, % Al	99.90-99.98	99.99+	99.99+

5.1.2. Electrorefining

In addition to its universal production by the electrolytic Hall-Heroult process, aluminum is also refined industrially by molten salt electrochemical processes. Normal purities for aluminum as produced from a Hall-Heroult cell are in the range of 99.5 to 99.98% aluminum. Since Al is so reactive and has such a high boiling point (~2500°C), it cannot be refined by such methods as aqueous solution electrolysis or distillation, and is refined commercially only by molten salt electrolysis.

Basically, Al refining consists of a "three-layer" process. Table 5.1.2, after Mantell [208], summarizes the essential details of the three practical industrial Al refining processes: 1) the Hoopes process, 2) the Gadeau process, and 3) the Aluminum Industrie (AI) A-G (Neuhausen) process. All these processes essentially rely on adjusting the densities of the molten impure aluminum and of the electrolyte so that pure molten aluminum will float on the electrolyte.

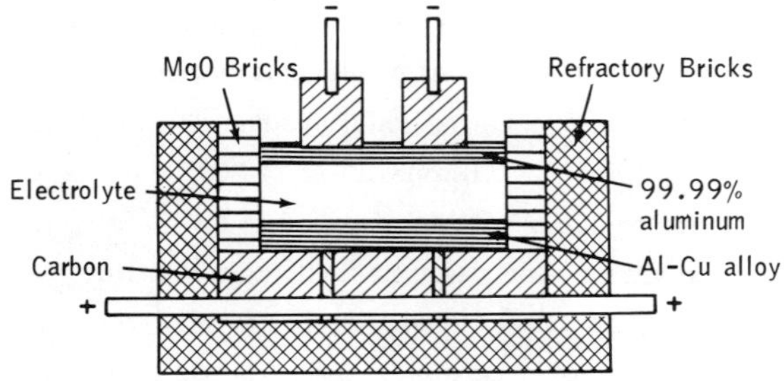

FIG. 5.1.4. A practical Al refining cell of the Hoopes type. (Taken from Ref. 208, by permission of McGraw-Hill.)

In the Hoopes process [236], patented in 1925, the anode consists of impure Al alloyed with 25% copper to give an anode with a density of 2.8 g/cc at 1000°C. The electrolyte typically consists of a cryolite-BaF_2 molten salt mixture which has a density of 2.5 g/cm^3 at 1000°C. At 1000°C, pure aluminum has a density of 2.29 g/cm^3. As current is passed, aluminum dissolves from the Al-Cu anode, diffuses through the electrolyte to the cathode (utilizing graphite contact electrodes), and forms a liquid layer of purified aluminum. This liquid Al layer can then be drawn off. Figure 5.1.4 shows a typical Al refining cell of the Hoopes type.

The Gadeau process is quite similar to the Hoopes process but utilizes a lower melting and less reactive electrolyte, using a cryolite-aluminum fluoride-barium chloride mixture with a liquidus temperature of about 720°C [237, 238]. While Hoopes process cells use a frozen alumina-rich lining to protect the cell wall, Gadeau cells, because of the less reactive electrolyte, utilize a magnesite (MgO) brick lining. However, the chloride-fluoride electrolyte is somewhat unstable, and its composition must be monitored and adjusted.

The Aluminum Industrie (AI) A-G of Neuhausen (Switzerland) Process (patented in 1937) uses an all-fluoride electrolyte at a temperature similar to that of the Gadeau process [239]. The AI process electrolyte is a 48% AlF_3-18% NaF-18% BaF_2-16% CaF_2 mixture. Here the NaF-AlF_3 ratio is in excess of the cryolite composition. Again magnesite brick can be used for the cell lining, with the all-fluoride electrolyte more stable than the Gadeau process electrolyte. However, the AI process electrolyte has a lower conductivity and requires larger cross-section cell areas or shallower baths. Graphite consumption is also higher in the all-fluoride electrolyte.

In all three processes, all operations must be carried out with a minimum of disturbance to the fluid layers. For example, if the depth of the molten cathode is reduced too far when tapping the cell, highly turbulent conditions, due to local electromagnetic forces, can result in distortion of the electrolyte layer, and the cathode may become contaminated by contact with the anode. Cell temperature fluctuations and changes in anode composition can also result in contamination of the pure Al metal layer by the Al-Cu alloy anode.

Hoopes process aluminum typically is 99.90% Al with 99.98% Al produced under certain circumstances, while both the Gadeau and AI processes lead to 99.99+% Al. Akerman et al. suggest that a double electrolysis procedure could produce 99.999% pure Al [240].

In addition to these relatively high temperature (>700°C) baths for the refining of Al, some efforts have been made to use other electrolytes at lower temperatures. Midorikowa has examined the use of the various low melting $AlCl_3$-alkali halide melts to refine Al by deposition onto a Pb cathode [128, 129]. There do not appear to be any reports of the practical application of this approach to Al refining, although this type of electrolyte seems to have some application to Al electroplating (see Section 5.2). Ziegler and Lehmkuhl claimed that aluminum of 99.99% purity could be produced by electrolysis of molten sodium triethylaluminum fluoride [$NaF.2Al(C_2H_5)_3$] using commercial anodes [125, 126]. Gaugin and Nury, however, have not found any advantage to the use of this type of bath [241].

5.2. ELECTROPLATING AND ELECTROPOLISHING

5.2.1. Electroplating

There is no evidence for the successful deposition of aluminum from aqueous electrolytes. Aluminum has been successfully deposited from both molten salt and organic electrolyte baths. The organic electrolyte plating systems, in particular, have received a considerable amount of attention.

Plotnikov first deposited aluminum from an aluminum bromide-ethyl bromide bath [242]. Brenner has reviewed in detail the work in the area of electrodeposition of aluminum from organic electrolytes to 1968 [107], as will now be succinctly presented. Blue and Mathers, in the mid-thirties, developed the first aluminum-plating process to arouse any interest [243, 244]. The bath, similar to a Friedel-Crafts reagent, was an $AlCl_3/AlBr_3$ mixture (1:1) in a mixed ethyl bromide-benzene-xylene solvent. Hurley and Weir later (1948) developed a successful aluminum plating bath based on a $AlCl_3$-ethylpyridinium bromide-benzene (or toluene) electrolyte [124, 245], which was improved by Safranek et al. by use of additives and good engineering practice [246]. However, this bath was quite sensitive to moisture and electrolytic decomposition.

Couch and Brenner, in 1952, reported on the development of an $AlCl_3$-LiAlH-ethyl ether Al plating bath which later became known as the NBS bath [108, 247]. Such a bath yields pure ductile deposits of aluminum. The $AlCl_3$ concentration was 3.0 M, while the optimum lithium aluminum hydride composition was 0.4 M. Methyl borate and β,β'-dichloromethyl ether have been used as addition agents and apparently promote grain refinement.

Some features of the NBS bath have been undesirable, primarily its limited life. The principle cause of aging is apparently electrolyte decomposition which is evidenced by H_2

evolution at the anode. While the aluminum anodes used do dissolve with theoretical efficiencies, the aluminum which goes into solution cannot be redeposited at the cathode. Therefore, aluminum chloride must be added at intervals to the bath. While the bath is stable in the absence of air, it is decomposed by moisture and CO_2, while oxygen has no effect. Despite these problems, the NBS bath has been used successfully in various applications which include electroforming of wave guides, mirrors, and cladding of uranium fuel elements [248]. Also, Rothlein reported the electrodeposition of high surface area aluminum deposits from a NBS-type bath by the use of high current densities (~ 0.5 A/cm^2) which seemed applicable to electrolytic capacitors [249].

More recent workers have substituted THF or THF-Bz solutions for diethyl ether in the NBS bath with significant improvements. Ishibashi and Yoshio used THF/Bz with $AlCl_3$/$LiAlH_4$ ratios of 3:1 and obtained excellent deposits at 0.1 A/cm^2 with stirring [112]. The operating life of such a modified NBS bath was significantly longer than the original diethyl-ether-based bath. Yashio et al. concluded that there are major differences in the ionic species present in THF-based $AlCl_3$-$LiAlH_4$ electrolytes which allow the use of aluminum dissolved from the anode to extend the lifetime of the bath. Yashio and Ishibashi also examined the use of a pure THF solvent bath for high rate Al plating which is desirable from an industrial point of view [113]. Smooth, coherent Al deposits were obtained at 18 A/dm^2 (0.18 A/cm^2) without stirring, while the conventional NBS bath allowed good plating up to only 5 A/dm^2 (0.05 A/cm^2) under similar conditions.

Other organic-solvent-based electrolytes have also been proposed for Al deposition, although they have been less extensively investigated than the NBS-type baths. Holland developed an $AlCl_3$-containing formamide-nitrobenzene-benzoyl chloride electrolyte which was capable of operation at >0.3 A/cm^2 [250], although others found it extremely difficult to operate [251]. Beach and co-workers developed the "Etmac" bath which is composed of $AlCl_3$-$AlHCl_2$-2-ethoxy ethyl trimethylammonium chloride salt mixture in a diethyl ether solvent [252]. The "Etmac" bath gives good Al deposits at current densities up to 0.16 A/cm^2, while Hisano et al. used an $AlBr_3$-N, N-dimethylaniline bath which allowed operation at 0.3 A/cm^2 [121]. In a recent study of the aluminum bromide/alkyl benzene-type bath first studied by Blue and Mathers, Capuano and Davenport recommended electrolytes of $AlBr_3$ in ethylbenzene type solvents [115]. They stressed the applicability of technical grade solvents in the procedure which was apparently insensitive to moisture.

The sodium aluminum-ethyl fluoride [$NaF/Al(C_2H_5)_3$] bath of Ziegler and Lehmkuhl has also been considered to be quite successful for aluminum deposition (see Section 5.1) [125, 126]. It is better classified as a molten salt bath since it melts at 72°C and readily conducts current at temperatures less than 180°C, where the compound decomposes. At 160°C the conductivity is 0.08 ohm^{-1}/cm. It can be dissolved in various solvents to yield a room temperature bath.

There is also extensive work reported in the deposition of aluminum from binary molten aluminum chloride/sodium chloride melts. Collins reported on the use of an 80 wt%

$AlCl_3$-20 wt% NaCl electrolyte for aluminum plating on steel at 180°C [253]. Midorikawa also made significant contributions to the understanding of Al deposition from $AlCl_3/NaCl$ melts, but mainly from the viewpoint of electrorefining Al as mentioned (see Section 5.1.2) [128, 129]. A lead cathode was recommended while iron contamination (>0.001%) of the bath lowered current efficiency and led to rough deposits. Howie and MacMillan tried to establish the optimum conditions for electroplating Al onto steel and recommended the use of purified $AlCl_3$ (free of iron) with HCl (~0.07 wt%) also present [130]. The HCl is formed by small amounts of moisture and appears to be effective in preventing dendrite formation. At the recommended current density of 16 mA/cm^2, the observed current efficiency was about 80%. Matiasovsky and Paucirova reviewed the electroplating of Al from molten bromides and chlorides, including the effects of various additives to the bath, e.g., $SnCl_2$ and $PbCl_2$ [254]. Sound Al electroplates could be deposited at 160 to 200°C at 0.02 A/cm^2 from an 80 wt% $AlCl_3$-10 wt% NaCl-10 wt% KCl melt.

5.2.2. Electropolishing

The electropolishing of aluminum and its alloys is practiced commercially using three principal processes, all carried out in aqueous electrolytes. Such processes are often the precursor to an anodizing process and may, in fact, leave a thin oxide film themselves. These processes are often referred to as electrobrightening since very little metal is actually removed and the term electropolishing is perhaps better reserved for those processes which involve a high metal removal. Table 5.2.1 lists typical conditions for the three most popular commercial Al electropolishing processes: 1) the Alzak Aluminum Company of America (Alcoa) process, 2) the Battelle process, and 3) the Brytal process (British Aluminum Co.).

The Alzak process is based on the original work of Mason [255]. Actually, the Alzak process in toto includes all steps (buffing, degreasing) to the final anodizing and sealing, and the Mason method is merely a step in the process. The Alzak process does leave an oxide smudge which is usually removed by a hot acid dip. The Battelle process is based on phosphoric acid-sulfuric acid mixtures, frequently with chromic acid additions, and requires mechanical agitation in contrast to the Alzak and Brytal processes [256]. Sulfuric acid serves to increase the bath conductivity and allows the use of higher current densities. Sometimes nitric acid is also added instead of chromic acid. Alteration of the phosphoric/sulfuric acid ratio can be used to achieve brightening or smoothing as desired.

In contrast to these acid electrolyte processes, the Brytal process is based on an alkaline electrolyte and was patented by Pullen [257]. The process produces a thin oxide film which can be dyed or anodized by an accompanying Brytal anodizing process. Aluminum phosphate and alkali metal sulfates are often added to improve the reflectivity of aluminum alloys treated by this process.

TABLE 5.2.1. Typical Conditions for Widely Used Commercial Al Electropolishing Processes

Processes	Electrolyte	T (°C)	I (mA/cm^2)	t (min)
Alzak	HBF_4 (2.5%)	30	18-55	5-10
Battelle	H_2SO_4 (4-14%) H_3PO_4 (57-75%) Chromic acid (6.5-9%)	80	100-150	15
Brytal	Na_2CO_3 (15%) Na_3PO_4 (5%)	80	40-50	8-10

Aside from these three major processes, a number of other procedures exist for the electropolishing of Al [256]. A phosphoric acid/n-butyl (or amyl) alcohol bath forms the basis of the propriety British Alkubrite process. Jacquet has developed baths based on perchloric acid (34.5 vol%)/acetic anhydride (66.5 vol%) mixtures which are used mainly in metallographic procedures and can be used at low temperatures (20°C). A nonaqueous electrolyte (1.2 M NH_4ClO_4 in formamide) has also been claimed to give good results. A sulfuric acid (80 wt%)-chromic acid (20 wt%) bath forms the basis of the German "Aluflex" process which is capable of rather high metal removal rates of 0.2 μm/sec at current densities of 200 to 300 mA/cm^2.

As might be expected, alloy composition or impurities affect electropolishing depending on their corrosion resistance under the specified conditions [258]. Generally, superpure Al or its alloys are best suited for electropolishing. In general, high percentages of Si, Fe, or Cu are deleterious, while Al-Mg alloys respond quite well, particularly those high in Al. Cast Al alloys are usually unsuitable for electropolishing in contrast to fine-grained, wrought alloys.

The alkaline electropolishing process is less expensive compared with the concentrated acid electrolyte process, due to lower chemical costs, equipment costs, and power requirements. However, the concentrated acid processes require less critical control, particularly with regard to alloy composition, and require less bath maintenance than the fluoroboric acid or alkaline processes. Chemical electropolishing of Al is often preferred to electropolishing for such reasons in many commercial applications.

For the interested reader, more detailed reviews of Al electropolishing can be found in the works of Andrus and Pettit, George and Vandenberg, and Wernick and Pinner [256, 258, 259].

5.3. ELECTROSYNTHESIS

There is no significant use of aluminum in industrial electrosynthesis at the present time. I. G. Farben (Ludwigshafen), as reported by Jackson and Kuhn [260], has electrolyzed $AlCl_3$ in a diaphragm cell at graphite electrodes to yield Cl_2, H_2, and any of three compounds: $Al(OH)_3Cl$, $Al_2(OH)_3Cl_3$, and $Al_2(OH)_5Cl$. $Al(OH)_2Cl$ and $Al_2(OH)_3Cl_3$ are tanning agents, while $Al_2(OH)_5Cl$ is a hydrophobicizing agent for textiles. Some efforts, however, have been directed toward the use of aluminum electrodes in organic synthesis and these may be of some interest. Basically, these processes at Al electrodes may be of two types: 1) electrochemical reactions wherein Al is not incorporated into the product of the electrode reaction, and 2) electrochemical reactions wherein Al becomes incorporated into the products of the electrode reaction.

Variation in the products of an electrode reaction due to changes in the electrode material are well-recognized. Finkelstein, Petersen, and Ross have contrasted the use of Al cathodes with Hg cathodes in the electrolytic reduction of various quaternary ammonium salts in different electrolytes [261-263]. While mercury electrodes result in organic-mercury compounds due to attack on Hg, Al electrodes result in cleavage products. For example, electrolysis of benzyltriethylammonium ion in DMF at Al gave both toluene and bibenzyl while the electrolysis of benzyldimethylanilinium ion gave bibenzyl, toluene, and benzene. Brown and Gonzalez later postulated that the formation of bibenzyl in the reduction of benzyltrialkylammonium salts in an aprotic solvent was due to the peculiar nature of the Al cathode which could allow extremely high local concentrations of benzyl radicals [264]. Since Al electrodes are known to be subject to local passivation, Brown and Gonzalez attributed bibenzyl formation to the presence of local active centers which resulted in high local concentrations of benzyl free radicals. On Hg or other electrodes free from insulating films, such radicals may be further reduced to toluene. Brown and Gonzalez also compared the reduction of α, ω-dibromoalkanes on Al and Hg electrodes [265]. Reduction of 1,3-dibromopropane and 1,4-dibromobutane on Hg proceeds through adsorbed radicals and leads to solid organomercury polymer and dibutylmercury from dibromobutane, while cyclopropane and propane are formed from dibromopropane. On Al cathodes, higher bromides and dibromides are formed, i.e., 1-bromohexane from 1,4-dibromobutane and 1,6-dibromohexane from 1,3-dibromopropane.

The work of Khomutov and Tsodikov is a complimentary example of the variation in reduction products afforded by the use of an aluminum cathode in a reduction reaction carried out in an aqueous electrolyte [266]. Reduction of quinoline is a possible path to 1,2,3,4-tetrahydroquinoline (THQ), a precursor to the production of N-acetyl-1,2,3,4-THQ, an insect repellant. Table 5.3.1 summarizes the results of the nature of the cathode material on the production of THQ and DHQ (trimers and dimers of dihydroquinoline). The use of Al electrodes leads to significant yields of DHQ, in contrast to the high yields of THQ at electrodes of lead, thallium, and indium, which have low hydrogen overvoltages.

TABLE 5.3.1. Effect of Cathode Materials on the Yield of the Electrolytic Reduction Products of Quinoline [266]

Cathode material	% THQ	% DHQ
Hg	33	34
Pb	73	12
Al	15	30
C(graphite)	22	38

The other type of use of Al electrodes in electrosynthesis involves electrode reactions wherein Al is incorporated into the products of the electrode reaction. Such reactions were first recognized during the electrolysis of Grignard reagents by French and Drane [267]. Evans and Lee later showed that such reactions were of the type [268]:

$$6C_2H_5MgBr + 2Al = 3Mg + 3MgBr_2 + 2Al(C_2H_5)_3 \quad (5.3.1)$$

Ziegler and Lehmkuhl then showed that electrolysis of molten $NaF.2Al(C_2H_5)_3$ at attackable anodes could produce metal alkyls, notably tetraethyllead [126]. Marlett has comprehensively reviewed the work of Ziegler and co-workers in the electrosynthesis of organometallics [269]. According to Tomilov and Brago, such reactions first involve an initial oxidation of complex fluoroalkylate ions, viz. [270]:

$$AlF(C_2H_5)_3^- = AlF(C_2H_5)_2 + C_2H_5\cdot + e^-$$

followed by attack of the alkyl radical on the anode:

$$M + 4C_2H_5\cdot = M(C_2H_5)_4$$

Table 5.3.2 summarizes the extent of such reactions which occur at attackable Al electrodes or which occur in Al-containing electrolytes, taken primarily from Tomilov and Brago [270].

More recently, Mottus and Ort have dealt with the use of consumable Al anodes in the preparation of organometallic compounds [271]. Electrolysis of dichloromethane solutions at an Al anode and a passive cathode in a divided cell led to formation of methylene (bis) aluminum dichloride ($Cl_2AlCH_2AlCl_2$) when using $AlCl_3$, $HOAlCl_2$, or as electrolytes. Further work by Mottus and Ort has dealt with formation of olefin polymerization catalysts by the electrolysis of methylene (bis) aluminum dichloride electrolytes at Al alloy anodes to yield complex Al alloy complexes of Al-Mg, Al-V, Al-Mn, and Al-Ti [272]. Such electrochemical procedures may allow the preparation of catalyst compositions which allow better control of rate formation, yield, and molecular weight of polyethylene polymers. In a similar vein, the polymerization of ethylene has been observed during the production of an initiator formed by the electrolysis of 1,2-dichloroethane containing TBAC at an Al anode [273].

TABLE 5.3.2. Organometallic Electrochemical Syntheses Involving Al Electrodes or Compounds [270]

Anode	Cathode	Compound	Solvent	T (°C)	Product	Yield (%)	Refs.
Al	Pt	C_2H_5MgI	Et_2O	-	$Al(C_2H_5)_3$	-	[274]
Al	Cu	$NaAl(C_2H_5)_4$	-	140	$Al(C_2H_5)_3$	99	[275]
Al	Cu	$NaAl(C_3H_7)_4$	-	100	$Al(C_3H_7)_3$	94	[275]
Al	Cu	$NaAl(C_4H_9)_4$	-	-	$Al(C_4H_9)_3$	-	[275]
Pb	Pt	$Al(C_2H_5)_3$	Et_2O	25	$Pb(C_2H_5)_4$	99	[276]
Pb	Hg	$NaF.2Al(C_2H_5)_3$ $NaAl(C_2H_5)_4$	-	112-117	$Pb(C_2H_5)_4$	100	[277]
Pb	Hg	$NaAl(CH_3)_4$	THF	90	$Pb(CH_3)_4$	90	[278]
Pb	Hg	$NaAl(CH_3)_4$	DIGLYME	100	$Pb(CH_3)_4$	96	[279]
Pb	Hg	$NaAl(CH_3)_3(OC_4H_9)$	-	90	$Pb(CH_3)_4$	94	[279]
Pb	Cu	$NaAl(C_2H_5)_4$ $NaAl(C_3H_7)_4$	-	-	$Pb(C_3H_7)_4$	97	[280]
Mn	Al	Fluorenyl aluminum	DEGDE	-	Bis(fluorenyl) manganese	-	[281]
Pb	Cu	$NaF.2Al(i\text{-}C_3H_7)_3$	-	50	$Pb(i\text{-}C_3H_7)_4$	-	[281]
Pb	Cu	$NaAl(C_2H_5)_4$ $NaAl(C_4H_9)_4$	-	-	$Pb(C_4H_9)_4$	-	[280]
Pb	Cu	$NaF.2Al(C_6H_5)_3$	Bz	-	$Pb(C_6H_5)_4$	-	[282]
Mg	Cu	$NaAl(C_2H_5)_4$	-	100	$Mg(C_2H_5)_2$	95	[275]
Mg	Cu	$NaF.2Al(C_2H_5)_3$	-	-	$Mg(C_2H_5)_2$	-	[283]
Zn	Cu	$NaF.2Al(C_2H_5)_3$	-	-	$Zn(C_2H_5)_2$	-	[283]
Zn	Cu	$NaAl(CH_3)_4$ $NaAl(C_2H_5)_4$	-	105	$Zn(C_2H_5)_2$	80	[284]

(continued)

TABLE 5.3.2 (continued)

Anode	Cathode	Compound	Solvent	T (°C)	Product	Yield (%)	Refs.
Sn	Hg	$NaAl(CH_3)_4$ $NaAl(C_2H_5)_4$	-	100	$Sn(C_2H_5)_4$	80	[280], [285]
Sn	Cu	$NaF.2Al(C_2H_5)_3$	-	-	$Sn(C_2H_5)_4$	-	[283], [286]
Sb	Cu	$NaF.2Al(C_2H_5)_3$	-	70-100	$Sb(C_2H_5)_3$	80-90	[283]
Hg	Hg	$NaF.2Al(C_2H_5)_3$ $NaAl(C_2H_5)_3(OC_4H_9)$	-	-	$Hg(C_2H_5)_2$	100	[277]
Hg	Cu	$NaAl(C_2H_5)_4$	-	100	$Hg(C_2H_5)_2$	82	[280]
Cd	Cu	$NaAl(C_2H_5)_4$	-	100	$Cd(C_2H_5)_2$	81	[280]
Bi	Cu	$NaAl(C_2H_5)_4$	-	100	$Bi(C_2H_5)_2$	95	[280]
In	Cu	$NaF.2Al(C_2H_5)_3$	-	-	$In(C_2H_5)_3$	-	[283], [286]
Al	Pt	AlI_3	MeI	-	CH_3AlI_2	-	[287]

5.4. CORROSION

The corrosion behavior of aluminum and its alloys is intimately connected with the behavior of its surface oxide film. A coherent oxide film effectively passivates the Al surface under a variety of environmental conditions. However, the complex chemical and crystallographic nature of these oxide films precludes reliable predictions of the corrosion resistance of Al or its alloys based solely on the solubility of the various aluminum oxides in the media in question. Generally speaking, the corrosion behavior of Al is in accord with the amphoteric character of the oxide film, and dissolution of aluminum occurs readily in strong (nonoxidizing) acids and bases. In neutral or weakly acidic solution, Al is quite resistant to corrosion as the data of Fig. 5.4.1 indicate [288].

The main practical interest in the corrosion of Al is not in the corrosion behavior of commercially pure (99.2 to 99.7%) Al but in the performance of the many Al-based alloys. Therefore, general statements concerning the corrosion behavior of Al are difficult to make. Since it is not really possible to definitively cover the voluminous literature on the corrosion

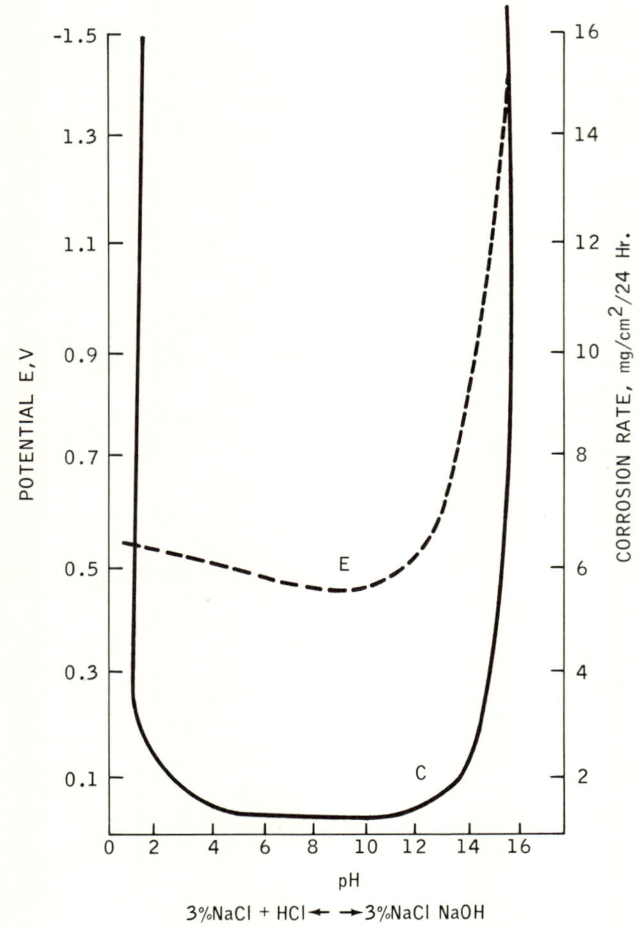

FIG. 5.4.1. Dependence of the corrosion rate (C) and electrode potential (E) of aluminum on pH. (Taken from Ref. 288, by permission of The MacMillan Company.)

of Al and its alloys in this brief expose, the reader is referred to the excellent monograph of Godard et al. for more comprehensive information [289]. Briefer summaries of the corrosion behavior of Al can be found in the works of Uhlig [290], Bailey and Porter [291], and Tomashov [288].

The following subsections will be considered:

1. Specific types of corrosion experienced by Al and its alloys.
2. Effects of alloying elements on Al corrosion.
3. Specific media effects in the corrosion of Al and its alloys.

5.4.1. Specific Types of Al Corrosion

Al and its alloys are subject to five major types of corrosion: 1) general dissolution, 2) galvanic corrosion, 3) pitting corrosion, 4) intergranular corrosion, and 5) stress corrosion. There are references in the literature to more specialized types of corrosion such as crevice corrosion, corrosion fatigue, and cavitation and impingement corrosion [289].

General dissolution is, of course, the gross corrosion of Al which occurs in highly acidic and alkaline environments with few exceptions. As shown earlier in Fig. 5.4.1, in neutral or mildly acid pH media, the oxide film is an effective corrosion inhibitor.

Galvanic corrosion of Al occurs when Al is in contact with a dissimilar metal other than Mg, Zn, Cd, or Cr. Under such circumstances, Al corrodes more rapidly than it would if it alone were exposed to the same medium. Commonly, metals are classified in the corrosion literature in terms of their position in a galvanic series. Such galvanic series are a function of the specific medium and are distinctly different from standard or formal electrode potentials since they are in most cases mixed potentials. In addition to the difference in galvanic potential between Al and the metal in question, the severity of such galvanic corrosion is dependent on the effective corrosion current density at the anode surface. This corrosion current density is a function of: 1) the electronic resistance of the metal-metal path, 2) the electrolyte conductivity, 3) the cathode/anode surface area ratio, and 4) electrode couple polarization.

A high electronic resistance between the two metals obviously leads to lower corrosion current densities. High electrolyte conductivity decreases the electrical resistance in the solution path and thus enhances galvanic corrosion. The cathode surface area controls the amount of corrosion current, while the anode current density determines the rate of material dissolution. Thus large cathode/anode area ratios tend to increase the severity of galvanic corrosion. Electrode polarization factors usually prevent predictions of the severity of galvanic corrosion based on galvanic series potential data. High polarization leads to lower corrosion currents, while low electrode polarization (high exchange current densities) allows significant corrosion currents to be maintained.

A special case of galvanic corrosion, deposition corrosion, occurs when Al is in contact with heavy metal ions in solution. Such metals deposit on the Al surface and set up local cells. Cu, Hg, Pb, and Co ions, when present in solution, can cause deposition corrosion of Al.

Pitting is perhaps the most commonly encountered form of aluminum corrosion. It occurs in near neutral aqueous media where local cell action is sufficiently vigorous to disturb the homogeneity of the solution and local pH decreases occur. As the dissolved Al cation diffuses away from the surface, an oxide film can form which further isolates and intensifies conditions of local acidity, and pitting results. Aluminum pitting occurs only in the pH range of 4.5 to 9.0. Chloride ion is particularly harmful, especially in the presence of traces of copper which readily intensify local cell action. The initiation and growth of pits on Al surfaces has been the subject of a considerable body of research, and the kinetics and mechanism of this type of Al corrosion have been reviewed by Greene and Fontana and Kolytrykin [292, 293], in addition to the work of Godard et al. [289].

Integranular corrosion occurs in Al alloys wherein a local galvanic cell is formed because of some heterogeneity of the alloy structure. These local cells are usually formed by the presence of second-phase precipitates in the grain boundaries. For example, in Mg-Al alloys, Mg_5Al_8 precipitates at the grain boundaries and corrodes since it is less noble than the solid solution and becomes the anode [288]. The degree of susceptibility of an alloy to such intergranular attack can vary appreciably depending on its microstructure (particularly with regard to the amount, size, and distribution of the second phase), which is the result of its metallurgical history and thermal treatment.

Stress corrosion cracking (SCC) of Al is a phenomenon particular to high-strength Al alloys, and it results in the catastrophic failure that often occurs in aluminum structural components at stresses far below the yield strength. It does not occur with pure Al. These stress corrosion cracks propagate perpendicular to the direction of the applied stress and will occur only in the presence of <u>both</u> continuous tensile stress and a corrosive environment. Such stress corrosion cracking is generally assumed to involve an electrochemical and a mechanical process [289]. Local cell action first produces subsurface intergranular fissures. Stress then is concentrated at these fissures and mechanical wearing occurs, exposing fresh metal to further electrochemical action. These electrochemical-mechanical processes occur until cracking of the piece results. The importance of the electrochemical processes has been shown by the effectiveness of cathodic protection in preventing SCC, or in arresting its progress once SCC has begun [294, 295]. While a susceptibility to intergranular corrosion is a prerequisite for SCC, not all Al alloys which are subject to intergranular corrosion are susceptible to SCC. General treatments of SCC of metals are available [296], as are specific reviews of the SCC of Al alloys in the work of Godard et al. [289] and the more recent chapter of Speidel and Hyatt [297].

5.4.2. Effects of Alloying Elements

Pure aluminum is, of course, quite soft and weak and is alloyed to obtain increased strength. Before discussing the effects of specific alloying elements, we will briefly consider the effects of Al purity on its corrosion. In general, the higher the degree of purity of Al, the greater is its corrosion resistance. For example, the corrosion resistance of Al declines appreciably as the purity drops from 99.998 to 99.98%. The decline in corrosion resistance, however, is much less as one proceeds to lower purity Al [289]. One must, of course, restrict these comments to the usual impurities found in commercial Al and the absence of specific media effects. An oft-mentioned exception to this role of Al purity in determining corrosion resistance is the intergranular corrosion of high purity Al which occurs in steam or water above 125°C [298]. In this application (nuclear reactors), the presence of iron in lower purity metal avoids or minimizes this type of attack.

The major and minor alloying elements used in commercial Al alloys are given in Table 5.4.1. Other elements may appear in small amounts in special alloys for particular applications. Rather than discussing the effects of these alloying elements individually, we will briefly comment on their behavior in terms of a particular alloy group, using the alloy designation of the American Aluminum Association for wrought Al alloys. Briefly, this nomenclature identifies wrought alloys by a four-digit number, preceded by the letters AA. These four numbers sequentially identify: 1) the main alloy group, 2) the number of minor elements present, and 3-4) the distinctive alloy number. Unfortunately, cast aluminum alloys do not have a similar standard designation and one must consult the manufacturer's literature.

Alloys in the AA-1000 series consist of 99.0% to 99.99% Al, with the last two digits indicating the purity limits. The second digit indicates any modifications in the impurity limits (0 indicates no special control of individual impurities). For example, AA-1100 is the designation of normal commercially pure Al which may contain up to 1% of impurities, primarily Si and Fe. Corrosion of Al of this purity has been discussed above, although it should perhaps be mentioned here that even 0.005% Cu has a noticeable effect on the corrosion of AA-1099 (99.99% Al).

The Al-Cu group (AA-2XXX) is noted for its relatively poor corrosion resistance, since Cu reduces the corrosion resistance of Al more than any other element. The influence of Cu is greatly dependent on the amount present, its form, and its distribution in the alloy microstructure. For example, in the solution heat-treated condition, immediately after quenching, the copper is present in solid solution. When aged at room temperature, intermediate pre-precipitation phases (θ' and θ'') form. Precipitation heat treatment can transform these phases into the stable θ phase, $CuAl_2$. Errors in heat treatment or subsequent reheating under uncontrolled conditions can lead to more $CuAl_2$, which concentrates in the grain boundaries, and a high susceptibility to intergranular and stress corrosion results [289].

TABLE 5.4.1. Major and Minor Alloying Elements of Aluminum

Major	Minor
Cu	Cr
Mg	Cu
Mn	Fe
Si	Mn
Zn	Si
	Ti

The common Al-Mn alloys (AA-3XXX) contain up to 1.0% Mg in addition to Mn and do have better corrosion characteristics than the AA-1000 series. Manganese can form the intermetallic $MnAl_6$ which has the same corrosion potential as the matrix and prevents local cell formation. Mn can also be taken into solid solution with $FeAl_3$, minimizing the usual detrimental effects of iron [289].

Wrought Al-Si alloys of the AA-4XXX series are not very common and are used mainly as claddings on other alloys. Cast Al-Si alloys have good corrosion resistance if Cu-free [289].

The AA-5XXX series comprises the Al-Mg group and, in general, offers the best combination of strength and corrosion resistance, and is preferred for structural applications. An important consideration in Al-Mg alloys is the tendency for a second Al-Mg phase, Mg_5Al_8, to form at Mg contents greater than 3%. The Mg_5Al_8 phase, called the β-phase, can precipitate at the grain boundaries. When this phase is present in a continuous network form (particularly at high temperatures which accelerate its formation), a susceptibility to intergranular and stress corrosion results since the Mg_5Al_8 phase is strongly anodic of the matrix. Cold work also promotes β-phase precipitation. This susceptibility to β-phase formation is greatly minimized in the alloy AA-5454 which contains only 2.7% Mg in comparison to the more highly alloys AA-5083 (4.5% Mg) [289].

The AA-6XXX group alloys contain both Mg and Si which form a stoichiometric second phase constituent, magnesium silicide (Mg_2Si). Such alloys have good corrosion resistance and are commonly used structural architectural alloys. The corrosion behavior of the two most commonly used alloys of this group has been a controversial subject. In alloy AA-6061 (1.0% Mg-0.6% Si, + Cr and Cu), the magnesium and silicon contents are in correct proportions to form Mg_2Si and are termed "balanced." In alloy AA-6351 (no Cu or Cr), the silicon content (1.0%) exceeds the magnesium content (0.6%) and the alloy is termed "unbalanced." The balanced alloys appear to have slightly better corrosion resistance than the unbalanced alloys, in which the greater excess of Si promotes intergranular corrosion. The presence of Cu tends to increase the incidence of pitting in the AA-6061 alloy. For most practical purposes, both types of Al-Mg-Si alloy appear equal in their corrosion resistance [289].

The Al-Zn-Mg alloy group contains the very high strength aircraft alloys such as AA-7075 and AA-7079. These alloys require protection in severely corrosive environments and are particularly susceptible to stress corrosion cracking. In general, the addition of Zn tends to decrease the corrosion resistance of Al to acids, while increasing the corrosion of Al in alkaline media [289].

5.4.3. Specific Media Effects

As shown earlier in Fig. 5.4.1, the corrosion rate of Al is a minimum at near neutral pH values. In halogen acids (HCl, HBr), severe corrosion occurs. This is contrasted to the

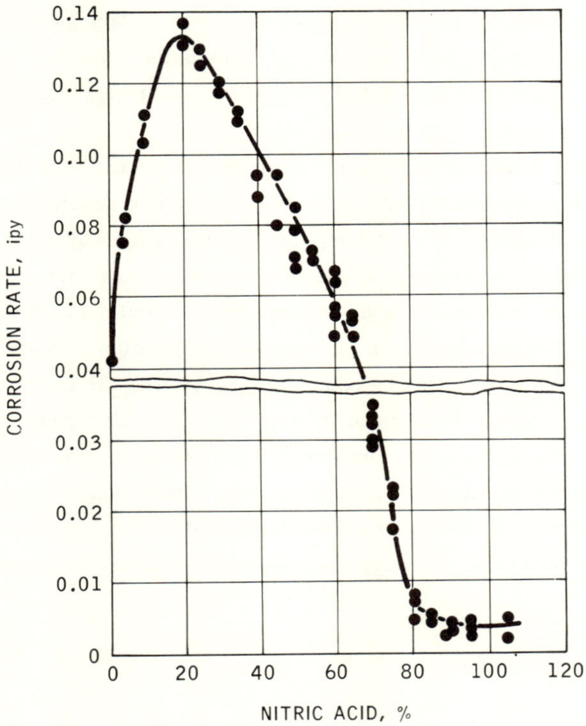

FIG. 5.4.2. Corrosion rate of Al (AA-1100) in nitric acid at room temperature. (Taken from Ref. 299, by permission of NACE.)

good corrosion resistance of Al in strong nitric or sulfuric acids. Al is resistant to corrosion in nitric acid at concentrations $>80\%$ at temperatures up to 50°C as the data of Fig. 5.4.2 illustrate [299]. In general, the presence of oxidizing anions such as chromate or dichromate tends to increase the corrosion resistance of Al in aqueous environments. Severe attack of Al occurs in aqueous solution in the presence of chloride, copper, and mercury ions.

Aluminum is particularly sensitive to corrosion, often violent, with various chlorinated solvents [300]. For example, the reaction of Al with carbon tetrachloride has been shown to involve [301]

$$2Al + 6CCl_4 = 3C_2Cl_6 + 2AlCl_3 \tag{5.4.1}$$

The above reaction is highly exothermic, and if the melting point (660°C) of Al is reached, the reaction may proceed explosively. Traces of water inhibit the reaction, which follows a free radical pathway, as do various ketones, quinones, and amines which serve as radical traps [302]. In addition to this sensitivity to chlorinated solvents, lower anhydrous alcohols (C_2 to C_4) are corrosive to Al, particularly at elevated temperatures [299].

In general, the corrosion resistance of Al-based alloys to various chemical environments is sufficient to allow their use in many applications in the chemical and food industries. For example, Al is used in hydrogen peroxide processing and storage equipment and in the mining of sulfur [290].

Finally, the subject of corrosion inhibitors is pertinent here. Such inhibitors can be classified chemically as 1) inorganic or 2) organic; or by surface reactivity as 1) adsorption type or 2) surface reaction types where a precipitate or film is formed at the metal surface. Among the inorganic type of inhibitors of Al corrosion are such oxidizing anions as chromate, nitrite, and permanganate. Nonoxidizing anions such as molybdate, silicate, and tungstate appear to function primarily by adsorption, although coprecipitation of reaction products between the metal and the inhibitor cannot be completely neglected. The organic-type inhibitors are generally macromolecules, amines, or organic acids. They may function by adsorption on the metal surface, forming a protective layer by chemisorption or mere physical adsorption. An alternative mode of action may involve inhibition of the cathodic reaction of the corrosion cell (usually H_2 evolution). Roebuck and Pritchett have listed a number of inhibitors for corrosion control of Al alloys in aqueous chemical environments [303].

5.5. ELECTROCHEMICAL POWER SOURCES

The high negative potential of aluminum, coupled to its low equivalent weight (9), make it quite attractive for use in electrochemical power sources. Moreover, aluminum is an abundant and relatively inexpensive anode material. In spite of these quite attractive features, commercial batteries utilizing Al anodes do not exist. The practical problems which have inhibited development of Al battery systems are, of course, those of high corrosion rates and appreciable polarization in typical battery electrolytes. While such problems are indigenous to aqueous electrolytes, nonaqueous-electrolyte-based Al batteries (primarily fused salt systems) are still in the developmental stage. Table 5.5.1 lists the theoretical energy densities and other characteristics of possible battery systems using various cathodes coupled to Al. One can discuss Al battery systems on the basis of the nature of the electrolyte: 1) aqueous electrolyte systems and 2) nonaqueous systems, including fused salts.

5.5.1. Aqueous Electrolyte Systems

Glicksman, in 1959, reviewed Al battery systems using aqueous electrolytes [304], which actually date back to the 1855 work of Hulot [305], as referenced in the later but less extensive review of Stokes [306]. The two most commonly investigated aqueous electrolyte aluminum battery systems are Al/O_2 (air) and Al/MnO_2.

TABLE 5.5.1. Theoretical Energy Densities of Various Al Battery Couples

Reaction	Capacity (A-hr/lb)	E° (V)	Energy density (W-hr/lb)
$2Al + \frac{3}{2}O_2 + H_2O + 2OH^- = 2H_2AlO_3^-$	1352	2.75	3718
$Al + \frac{3}{2}Cl_2 = AlCl_3$	273	2.20	600
$2Al + 3S = Al_2S_3$	485	0.86	415
$2Al + 3CuSO_4 = Al_2(SO_4)_3 + 3Cu$	137	1.93	265
$Al + \frac{3}{2}Br_2 = AlBr_3$	137	1.76	241
$2Al + 3CuCl_2 = 2AlCl_3 + 3Cu$	160	1.29	207
$2Al + 3NiSO_4 = Al_2(SO_4)_3 + 3Ni$	141	1.41	200
$2Al + 3Se = Al_2Se_3$	251	0.75	187
$2Al + 3NiCl_2 = 2AlCl_3 + 3Ni$	161	0.79	130
$2Al + 3CuS = Al_2S_3 + 3Cu$	213	0.60	128
$2Al + 3NiS = Al_2S_3 = 3Ni$	225	0.44	99
$Al + CrCl_3 = AlCl_3 + Cr$	196	0.49	96

A primary Al/O_2 or Al/air battery is among the possible high energy density metal/O_2 (air) battery systems, with a theoretical energy density of 3700 W-hr/lb. Zaromb, in particular, has carried out a considerable amount of work on such battery systems which use alkaline electrolytes [307-309].

Zaromb assumed the following basic reaction in an Al/air cell in alkaline electrolytes [307]:

$$4Al + 3O_2 + 4OH^- + H_2O = 4Al(OH)_4^- \quad (5.5.1)$$

Electrolyte regeneration in such cells was also considered by Zaromb via the reaction [308]

$$Al(OH)_4^- = Al(OH)_3 + OH^- \quad (5.5.2)$$

At low (≤3 M) KOH concentrations, Reaction (5.5.2) occurs rapidly enough to permit high current discharge for periods in excess of those predicted without electrolyte regeneration. For small reserve Al-air cells a consumable 10 M KOH electrolyte was recommended to allow high current drains at reasonable voltages and efficiencies.

Two approaches have been taken to the development of usable Al anodes in aqueous electrolytes. The first of these is the use of additives to the electrolyte, while the second is alloying of the aluminum. Often a combination of these approaches has been quite successful. For example, 10 M KOH saturated with ZnO allowed discharge of amalgamated Al anodes at current efficiencies ≥60% at current densities of 3.5 to 140 mA/cm^2 [307]. The zincate apparently increases aluminum oxide solubility, thus preventing passivation and yet minimizing corrosion, perhaps by formation of a zinc or aluminum-zinc layer on the Al surface. Katoh, Shimizu, and Katoh successfully used sodium stannate (0.05 to 0.1 M) in 5 M KOH as an electrolyte for Al-air cells using a pure Al anode [310]. Organic materials such as alkyl dimethylbenzyl ammonium salts (Hyamine 3500) have also been found to be satisfactory corrosion inhibitors with unamalgamated Al anodes [307]. Al-Zn, Al-Mg, and Al-Hg alloys appear to improve Al anode discharge along with alloying aluminum with tin. Of particular note here is the additional role of Al purity in determining improved discharge characteristics.

Turning now to the practical details of Al-air battery systems, the main considerations are: 1) operation at high current densities (≥10 mA/cm^2), 2) effective heat removal, and 3) anode-electrolyte separation during stand-by periods. Operation at high current densities is a necessity for high current efficiencies due to self-discharge of the anode due to corrosion. Effective heat removal is required to avoid electrolyte temperature rises caused by the highly exothermic cell reaction. Due to the coupled parasitic Al corrosion process, this heat product can be critically dependent on current efficiency as discussed by Katoh et al. [310]. High cell temperatures can also lead to anode passivation. The need for anode-electrolyte separation during prolonged stand-by is obviously necessary to avoid excessive self-discharge.

Zaromb has described a 60 to 80 W experimental Al-air alkaline battery which operated at current densities of 40 to 50 mA/cm^2 at a cell voltage of 1.1 to 1.3 V [309]. Based on such performance data, Zaromb presented a design for a 30-kW, 240-kW-hr Al-air power source suitable for electric vehicle applications with an energy density of more than 200 W-hr/lb. Such a battery would be chemically rechargeable by anode and electrolyte replacement. Katoh et al. discussed operation of a five cell aluminum-air battery which had an experimental energy density of 94 W-hr/lb using sodium stannate containing 5 M KOH [310]. Small Al-air cells seem limited in their ampere-hour capacities by the amount of aluminum hydroxide precipitate which must be contained or distributed. In addition to the Al-air system, Zaromb has also described an alkaline electrolyte Al-H_2O_2 power source which could achieve rather high energy densities [311]. Design data indicated that a 30-kW system with an energy density of 280 W-hr/lb was possible, with operation at higher current densities than an Al-air cell.

Leclanché-type cells using aluminum anodes coupled to MnO_2 cathodes were apparently first reported by Brull [312]. Using a 1 M KOH electrolyte, an energy density of 20 W-hr/lb was achieved. Sargent patented a similar cell using a zincate-containing alkaline electrolyte which met with little success due to poor shelf life and power output [313]. Ruben patented an Al/MnO_2 cell with a manganous chloride electrolyte which had an open circuit voltage of 1.53 V but still was not superior to zinc-anode-based cells [314].

Based on earlier work, Stokes reported on aluminum anode Leclanché cells which had the capacity retention and performance of zinc cells [306]. The optimum electrolytes were found to contain chlorides of aluminum or chromium and, usually, chromate corrosion inhibitors. Such electrolytes were acidic (pH 3.0 to 3.5) in nature. Duplex aluminum alloy (Al-Zn and Al-Mn) anode structures were used. Voltage delays were observed, particularly in older cells, although excellent room temperature shelf lives were obtained. The main advantage of Al-MnO_2 cells over Zn-MnO_2 cells appeared to be only in instances of high drain or continuous drain. Belitskus later further characterized and improved the Al/MnO_2 cells described by Stokes [315]. Again such cells showed promise at moderate to high discharge rates, but had lower capacities than zinc cells at lower or more intermittant discharge rates.

In addition to the use of aluminum in primary Al-air and Al-MnO_2 cells, some efforts at reserve battery systems with aluminum anodes in aqueous electrolytes have been reported. An aluminum-anode-based reserve battery using a NaOH electrolyte was reported to have a current efficiency of 99% at a current density of 100 mA/cm^2 [316]. Efficiency dropped to 58% at 10 mA/cm^2. Yokota reported on a number of reserve battery systems with Al anodes with reference to some earlier work [317].

In a more esoteric application, the Al-air cell concept has been applied to an implantable hybrid power source for pacemakers [318-322]. Such cells use an Al anode coupled to an oxygen cathode. Oxygen for cathode operation is supplied either by direct contact with bodily fluids or via an oxygen-permeable membrane. The use of a membrane-coated oxygen cathode system is preferable, since electrode degradation by contact with body fluids is eliminated. Tseung et al. demonstrated that a fully encapsulated (with silicone rubber) Al-O_2 cell, with Pt black cathodes and a 0.9% NaCl electrolyte, was capable of in vitro and in vivo power densities of ~ 20 μW/cm^2 at 0.9 V for periods of over 12 months [319]. Later work by Wan et al. described long-term implantation (to 14 months) studies of such cells [320]. Such cells, when implanted, tend to be encapsulated by a layer of fibrous connective tissue which, however, does not appear to seriously impede O_2 transport.

Despite these various approaches to the use of Al anodes in aqueous battery systems, there seems to be little promise of any commercial application for these systems since the obtainable energy densities are quite low and shelf life is a very marked problem. While the mechanically or chemically rechargeable Al-air cell may be a valid engineering approach, its complexity leaves serious doubts about its practical exploitation.

5.5.2. Nonaqueous and Fused Salt Electrolyte Systems

In an attempt to circumvent the problems associated with the use of aluminum anodes in aqueous systems, nonaqueous-electrolyte-based Al battery systems have been the subject of some recent attention. Despite the technological impetus of successful aluminum electrodeposition from organic-solvent-based electrolyte systems as described in Section 5.2, the use of Al in organic-electrolyte-based battery systems has not met with much success. For example, Farrar et al. found that aluminum anodes showed significantly lower potentials than predicted from thermodynamic data [10]. In addition, such potentials were erratic and not reproducible. A KCNS/DMSO electrolyte gave the most encouraging results, although on discharge Al anode polarization was quite significant and was 0.76 V at a current density of 2 mA/cm^2. Therefore, molten salt electrolytes have seemed more attractive to past investigators, based on the behavior of Al in molten salt electrolytes.

Antipin described a relatively high temperature cell based on an aluminum anode and a O_2/Cu cathode using a 40.5 wt% AlF_3-57 wt% NaF_3-2.5 wt% Al_2O_3 melt which is quite similar to a Hall-Heroult aluminum production cell electrolyte [323]. More recently, there has been significant interest in the use of aluminum anodes in the relatively low melting aluminum chloride-alkali chloride melts. In addition to allowing the effective use of aluminum anodes, such melts would allow low temperature (<200°C) cell operation, and thus minimize the many problems related to materials, construction, and safety of other higher temperature molten salt systems.

The Al/Cl_2 battery system has been of particular interest to a number of investigators. King et al. have shown that little polarization was observed during anodic dissolution of Al in a NaCl-$AlCl_3$ melt at 170°C [68]. Coulometric investigations showed that aluminum dissolution proceeded with 100% current efficiency. Giner and Holleck, working in the ternary $AlCl_3$-NaCl-KCl melt, did find passivation of aluminum on discharge due to formation of a solid salt layer as a result of local melt composition changes. Dendrite formation was observed during aluminum deposition [324]. An experimental test cell, using a noncatalyzed carbon anode, showed a voltage of 1.8 V at a current density of 20 mA/cm^2 in 66 mole % $AlCl_3$-20 mole %-14 mole % KCl at 150°C. Beltzer reported on an Al/Cl_2 battery operating in a 50 mole % $AlCl_3$-35 mole % NaCl-15 mole % KCl electrolyte at 170°C [325]. Operation at a current density of 50 mA/cm^2 and 1.54 V led to a power density of 77 mW/cm^2 of electrode. The main problem with this system would appear to be the problem of Cl_2 storage. To get around this problem, some attention has been given to other cathode materials.

An Al/S battery using a molten $AlCl_3$-$NaCl_3$ electrolyte was described by Greenberg [326]. Such a system showed a voltage of 1.2 V at current densities of 10 to 25 mA/cm^2 at temperatures between 150 and 200°C. Porous carbon with tellurium or tungsten additives was used as the cathode in another patent in an aluminum anode battery system with a molten aluminum halide-alkali halide electrolyte [327]. Cells of the type Al/$AlCl_3$-NaCl/S, Se, or Te have also been investigated [328]. Open circuit voltages of 1.2, 1.0, and 0.92 V were obtained in Al-anode-based cells with sulfur, selenium, and tellurium, respectively. A number of metal halides have been screened as cathode materials for use in $AlCl_3$-$NaCl_3$ melts [329]. However, the high solubility of the metal halides and their high equivalent weights make them unattractive as cathodes for high energy density batteries using aluminum anodes. Fannin et al. have also broached the possibility of an aluminum battery based on an Al concentration cell [25]. Crouch of ESB, Inc. has reported on a rather unusual Al molten salt battery system using a $NaAlCl_4$ electrolyte [330]. The cathode is high surface area porous carbon which has the rather high energy storage capacity of 105 J/g. This secondary Al battery system is being considered for load leveling applications.

Based on these reports, the possibility of successful development of a fused salt Al battery system seems more reasonable than an aqueous electrolyte system, particularly from the standpoint of a high energy density secondary cell.

5.6. ANODIZING

Anodized aluminum surfaces find wide commercial application. In addition to the improved corrosion and abrasion resistance imparted by such anodically formed layers, anodized aluminum can be produced in esthetically attractive colors varying from light champagne through different shades of bronze to black. Alternatively, the adsorptive properties of these anodic layers are readily conducive to coloring by dyes or pigments. A recent estimate projects that the international aluminum anodizing business exceeds $10 (U.S.) million per anum [331]. In Section 4 we have discussed the basic electrochemical processes involved in the anodic oxidation of Al. Here we will discuss the commercial practice of aluminum anodizing, including coloring via reactions inherent in the anodizing bath itself or incorporation of a dye within the porous layer after anodizing.

Aside from the alkaline electrolyte Brytal anodizing process discussed earlier in this section in regard to electropolishing, most industrial Al anodizing processes use acidic electrolytes, typically sulfuric acid or chromic acid [332, 333]. In Japan and Germany, anodizing processes using oxalic acid (or oxalic acid-sulfuric acid mixtures) have been quite popular [133]. Other electrolytes which have attracted some attention are sulfamic acid and phosphoric acid (primarily prior to electroplating) [334, 335]. Opaque coatings can be obtained in oxalic acid baths containing salts of thallium, zirconium, or titanium (the Emetal Process) or other mixed acid electrolytes [336]. In the United States, the Kalcolor process of Kaiser Aluminum [337] and the Duranodic 300 process of Alcoa are the most popular processes used to produce colored surfaces for architectural application [338]. Both these two processes are based on organic-acid-containing electrolytes. Innova Corporation (Seattle, Washington) recently publicized an aluminum anodization process which appears to offer some advantages over the Duranodic and Kalcolor processes and which, again, is an organic-acid-based electrolyte [331].

While most of the processes just mentioned lead to formation of porous anodic oxide films, there are some commercial applications of barrier anodic oxide films on Al. Very thin Al layers are used in reflectors and electrolytic capacitors (see Section 5.7). Ammonium tartrate and boric acid electrolytes are commonly utilized for such applications. Anodization in organic acids (maleic, glycolic, and citric) also can produce barrier Al oxide films.

It is important here to distinguish between conventional and "hard" anodizing of aluminum. While the conventional porous Al anodic films serve mainly as decorative and protective coatings, the hard anodized film on Al is meant primarily for the production of a hard and wear-resistant surface. Such hard anodic films can be obtained in many of the acid electrolytes mentioned by varying the conditions, and their use was first proposed by Tomashov [339].

We will now consider some of these processes in more detail in the following categories: 1) conventional anodizing processes, 2) hard anodizing processes, 3) integral color anodizing, 4) barrier oxide formation processes, and 5) special Al anodizing processes.

5.6.1. Conventional Anodizing Processes

Most conventional aluminum anodizing processes are carried out in sulfuric acid or chromic acid electrolytes, with the sulfuric-acid-based processes being the most widely used for parts fabricated from aluminum alloys. For assembled parts subjected to stress and containing joints or recesses in which the anodizing solution may be retained, the chromic acid process is preferred. Table 5.6.1 gives typical conditions for the conventional sulfuric acid and chromic acid processes.

TABLE 5.6.1. Typical Conditions for Conventional Aluminum Anodizing Processes

Electrolyte	T (°C)	I (mA/cm^2)	t (min)
15% H_2SO_4	20	10	20-30
10% CrO_3	35	2	30
3% $C_2H_2O_4$	30	10	30-40

The usual sulfuric acid electrolyte concentration is 15%, with current densities of ~15 mA/cm^2. Such baths are operating at 15 to 25°C for periods of 15 to 30 min [338, 256]. By varying the anodization time, temperature, and electrolyte, one can form anodic coatings on Al of predetermined characteristics. Generally, the thickness of the anodic layer increases with the time of anodization and decreases with increases in temperature, although temperature effects are not very important for short anodization times. Increasing the temperature at longer anodization times tends to decrease the overall anodic layer formation efficiency due to dissolution of the formed anodic layer. As the electrolyte concentration is decreased, the efficiency of anodic layer formation is increased. The porosity of the layer formed tends to decrease with a decrease in acid concentration, an increase in current density, and shorter anodization times. Higher temperatures and the use of superimposed ac increase porosity. Dilute electrolytes and lower temperatures favor harder coatings.

On pure Al and a number of alloys, the anodic layers formed in sulfuric acid electrolytes are about 2.5 to 40 μm thick and are transparent and colorless. Manganese or silicon alloys lead to greyish or brownish grey coatings, which are also formed on heterogeneous alloys. The films formed in sulfuric acid are quite suitable for dyeing. However, anodized aluminum layers formed in sulfuric acid must be sealed to prevent loss of the absorptive properties of the anodic film.

Such conventional sulfuric acid aluminum anodization procedures are known commercially in the United States and England as the Alumite process and in Germany and most European countries as the Eloxal GS processes (Elektrolytisch Oxidiertes Aluminum, with the GS indicating dc anodizing in sulfuric acid).

Chromic acid processes date back to the early work (1923) of Bengough and Stuart, and are still used primarily due to the advantage of not inducing corrosion [340]. It does, in fact, tend to lead to improved corrosion resistance due to incorporation of chromic acid within the anodic layer. The original chromic acid process of Bengough and Stuart used a 3% chromic acid electrolyte and a rather complex voltage control procedure. Present practice, based on a constant voltage procedure developed at the U.S. National Bureau of Standards, uses a 15 to 10% chromic acid electrolyte at 35°C and 40 V, which corresponds to a final current density of 1 to 5 mA/cm^2 [341].

As determined by Modic, low (<6%) chromic acid concentrations lead to dark films in contrast to the clear, opaque, enamel-like films obtained at 10 to 15 wt% CrO_3 [342]. At temperatures up to 35°C, the films are quite clear and transparent, while at $T > 40°C$, fully opaque films are formed.

Such chromic acid coatings on aluminum are much thinner than sulfuric acid process films and have a different pore structure. While a maximum thickness of ~ 10 μm can be obtained, chromic acid process layers most usually involve thicknesses of 1 to 5 μm. Such chromic acid coatings possess, in general, fewer pores, but their diameter is larger than in sulfuric acid process layers. Since these layers are so thin, they are, of course, less resistant to abrasion. However, they are quite flexible and can better withstand deformation and heating without crazing. The chromic acid process is quite dependent on alloy composition with chromic acid anodizing not recommended for alloys containing greater than 7% Si or 5% Cu, or other Al alloys with a total concentration of alloying elements greater than 7%. With such alloys, anodic dissolution generally occurs unless the process conditions are controlled very carefully.

While there are other types of conventional acidic electrolyte anodizing processes, including those based on oxalic acid, chromic acid-sulfuric acid mixtures, and sulfamic acid, they are not presently employed commercially to any extent. The sulfuric acid process is by far less expensive compared to the chromic acid process. While unsealed chromic acid coatings are more corrosion resistant than unsealed sulfuric acid coatings, they are, as mentioned, relatively thin and soft and cannot be used where wear resistance is necessary. The majority of alloys can be treated by the sulfuric acid processes, which is not true with the chromic acid process. For these and other reasons, the chromic acid anodizing processes for Al and its alloys have been largely supplanted by chemical conversion coatings when chromate-based treatments are desired.

5.6.2. Hard Anodizing Processes

The hard anodizing processes for Al and its alloys rely on finding conditions intermediate between the formation of barrier and duplex anodic films. Such conditions are favored by low temperature, low solubility of the anodic Al layer, good heat removal, and high (100 to 150 V) formation voltages. Such coatings have greater thicknesses and densities than conventional types, typically ranging from 25 to 125 μm in thickness. Such processes are known commercially as Alumite hard coating, Martin hard coatings, Sanford coatings, or Hardas coatings. Typical conditions for such hard Al anodizing processes are given in Table 5.6.2.

TABLE 5.6.2. Typical Conditions for Selected Hard Aluminum Anodizing Processes

Process	Electrolyte	T (°C)	I (mA/cm^2)	t (min)
Martin	15 w/o H_2SO_4	0	20-25	-
Hardas	10 v/o H_2SO_4	0-5	1-4	15
Sanford	7 v/o H_2SO_4 (with various organic additives)	-10	0.1-0.2	-

The Martin hard anodizing process uses a 15% H_2SO_4 electrolyte at 0°C, usually saturated with CO_2, and current densities of 20 to 25 mA/cm^2 are employed [343]. The Martin process can produce relatively thick coatings (>150 μm). The Alumite processes also uses a sulfuric acid electrolyte but with the addition of oxalic acid. The designations 225 or 226 are applied to the Alumite processes for wrought alloys, while 725 or 726 refer to cast alloy processes. The Alumite processes are usually carried out at 8 to 10°C and such coatings are 25 to 50 μm thick. The Hardas process was developed by Campbell and utilizes superimposed ac [344]. While oxalic acid was originally used in the Hardas process, 10% (vol %) sulfuric acid is now used at 0 to 5°C to produce 50 to 75 μm coatings. The ac component serves primarily to raise the allowable current density and the ac/dc ratio varies depending on the alloy used, with current densities of 100 to 350 mA/cm^2. While the Sanford process is used for architectural anodizing along with the Kalcolor and Duranodic processes, it is basically a hard anodizing process. A sulfuric acid based (7%) electrolyte containing various organics is used at -10°C and current densities of 10 to 20 mA/cm^2 [345]. Voltage control (increase) is necessary as the anodization time increases. Lelong, Segonds, and Herenguel have proposed the use of two different electrolytes for the production of hard anodized layers [346]. The first of these is an oxalic acid (80 g/l)-formic acid (55 g/l) mixture, while the second is a sodium bisulfate-citric acid mixture with current densities of 60 mA/cm^2 at 5°C. Both solutions are suitable for wrought Al-Mg alloys. In general, hard anodizing of heterogeneous aluminum alloys gives more abrasion-resistant coatings, probably due to inclusion of intermetallic compounds in the anodic film. Homogeneous alloys tend to give higher porosity coatings of lower wear resistance.

5.6.3. Integral Color Anodizing Processes

While organic dyes have been used to produce colored layers on anodized aluminum, present day aluminum anodizing technology tends toward electrochemical coloring-anodizing processes. Such processes do not require dyeing and produce extremely weather resistant, hard films. While such processes produce hard, thick, anodic coatings, they are not as versatile as the conventional hard-coating process since certain alloys cannot be treated. The first extensive use of these integral color anodizing processes was in Japan in the 1920s and 1930s [133]. The original Japanese process used oxalic acid and superimposed ac to produce yellowish gold coatings primarily for kitchen utensils. Table 5.6.3 lists typical operating conditions for some of the most popular propriety integral color anodizing processes. Basically, these processes use organic-acid-containing electrolytes.

The Kalcolor process developed by Kaiser Aluminum in the late 1950s is perhaps the most popular of these processes. Sulfosalicyclic acid with a small amount of sulfuric acid is used at room temperature to give yellow through brown-to-black films. The color is quite dependent on the alloy composition. The Kalcolor anodic coating must be sealed, usually by a 15 min immersion in boiling water at pH 5.6 to 5.8, which includes some additional compounds such as cobalt acetate, boric acid, and lignosulfonates. The Alcoa Duranodic coating is quite similar but uses sulfophthalic acid instead of sulfosalicylic acid, while the Permalux process also uses sulfosalicyclic acid, but in combination with sulfuric acid and malic acid. Most of the other integral color anodizing processes are based on oxalic acid electrolytes with addition of maleic acid, tartaric acid, and sulfuric acid.

TABLE 5.6.3. Typical Conditions for Integral Color Anodizing Processes for Aluminum (after Taijima [133])

Process	Electrolyte	T (°C)	I (mA/cm^2)	V (volts)
Kalcolor	5-Sulfosalicyclic + sulfuric acid	25	24-30	35-70
Duranodic	4-Sulfophthalic (or 5-sulfolsophthalic) acid + sulfuric acid	25	15-40	35-100
Permalux	5-Sulfosalicyclic acid, sulfuric acid, and maleic acid	25	24-30	35-70
ISML	Oxalic acid, tartaric acid, and sulfuric acid	20-30	15-24	40-80
Veroxal	Oxalic acid, maleic acid, and sulfuric acid	20-30	15-24	40-80
Alcandox	Oxalic acid (satd)	18-22	14-18	40-80

The Innovacolor process is a rather new process which uses a boric acid-hydroxycarboxylic acid-sulfuric acid mixture [331]. Due to savings in chemical cost, power, and time, this process is touted as a contender to the Kalcolor and Duranodic processes.

5.6.4. Barrier Oxide Formation Processes

Nonporous, barrier-type anodic layers on Al are important in electrolytic capacitors and are sometimes used to protect very thin Al layers such as those produced by vacuum deposition. Such layers are normally formed in boric acid, ammonium tartrate (or citrate), or very dilute oxy-acids with low dissolution properties [338]. Such deposits are less than 1 μm in. thickness and can be applied by a continuous strip method to Al foil for use in capacitors.

Such processes use voltages which can exceed 200 V, and thus the purity of the electrolyte is important since trace impurities can cause burning and perforation of the thin Al foil.

5.6.5. Coloring and Sealing

While integral color anodizing processes are of considerable importance, the use of alternative procedures to color anodized aluminum is quite commonly practiced. For the most part such coloring is achieved today by the use of special organic dyes rather than the older use of various insoluble metal compounds. These older treatments consisted of alternatively immersing the anodized surface in concentrated metal salt solutions until a sufficient amount of the pigment was precipitated to produce the desired color. The present practice of dyeing an aluminum oxide layer is quite simple and involves immersion of the anodized specimen into a dilute ($\leq 1\%$) solution of the dye at somewhat elevated temperatures ($\sim 65°C$) for a brief period of time (~ 10 min). A wide number of factors influence the shade and fastness of such dyed anodic Al layers, including the variables of anodic layer formation, the properties of the dye itself, and the sealing of the dyed anodic coating.

Sulfuric-acid-produced coatings are preferred for dyeing of anodized aluminum. While a more porous coating, such as that achieved at higher temperatures, improves dye absorption, the coating does lose its hardness and luster. If the pore size is too large, sealing becomes a problem and significant amounts of dye can be bled out in the sealing process. The effects of the type of Al alloy on the properties of the anodized layer have been considered earlier in this section.

The usual organic dyes for aluminum are acidic in nature, and the dye is usually present as an anion. Sodium salts of sulfonated compounds are commonly used. The interaction between the dye and the anodic film has been postulated by Giles to be similar to that between activated alumina and an organic absorbent [347]. The mechanism by which the dye becomes incorporated into the anodic film can be by covalent hydrogen bond formation, chelate bonding, or ion exchange, depending on the nature of the dye. While a larger number of dyes can be used to color anodized aluminum, very few possess sufficient resistance to direct sunlight to prevent eventual fading. The rate at which this color fading occurs can be influenced by such additional factors as the type and degree of sealing, the depth of shade (degree of color), and the depth of penetration of the dye.

As a final step in the finishing of an anodized Al piece, it is sealed. (Barrier-film-type coatings do not require sealing since they are nonporous.) The purpose of sealing is to close the pores of the anodic film and thus increase resistance to corrosion, staining, and damage in handling. During sealing, the oxide film is converted from a relatively amorphorous form to AlOOH or $Al_2O_3 . H_2O$ (boehmite), and an increase in volume occurs which plugs the pores. Sealing can be accomplished by immersion in boiling water (or steam treatment), nickel (or cobalt) acetates, or dichromate solutions. The metal salt sealing solutions are most often used with dyed anodic films on Al and appear to increase dye fastness. It should be noted here that sealing does tend to reduce the abrasion resistance of most anodic coatings.

As the reader may infer, the scope of commercial Al anodizing practice is much too broad to be dealt with in complete detail in this subsection, and the interested are advised to consult the rather extensive literature as referenced herein, with the review of Taijima and the work of Wernick and Pinner of particular interest [133, 256].

5.7. CAPACITORS

Aluminum-based electrolytic capacitors have been the most widely used device of this type, due to both low cost and weight of aluminum compared, for example, to Ta. Such capacitors usually employ etched Al foil for both anode and cathode. Porous paper with an ethylene glycol-borate electrolyte usually serves as the separator/electrolyte system. It is the capability of Al to form an oxide and the properties of such oxides which characterize the capabilities of Al-based capacitors. The essential electrochemistry of anodic aluminum oxide films has been discussed in Section 4.

In a typical Al electrolytic capacitor, the anode carries the dielectric film (200 to 5000 μm). The oxide here is of the barrier type and is usually formed on pre-etched Al foil. It should be mentioned here that while hard-rolled Al foil could be used to provide capacitor anodes of improved capacitance, no use of this property was made for almost 25 years [348]. In 1962, this superior behavior of work-hardened foil was traced to its larger number of dislocations which act as preferred sites for etch attack. The greater area formed by etching leads to greater capacitance [349]. The cathode is normally oxide-free except for the normal thin film present on Al exposed to air. In such capacitors (polar), anode and cathode are not interchangeable and their polarity must be of the proper sign in the electronic circuit. In contrast, in bipolar capacitors, both anode and cathode have anodically-grown oxide films and the poles of such capacitors are interchangeable in any circuit.

Vijh has summarized the relationships between the solid-state properties of oxides and their dielectric behavior [350]. Low dissipation factors (low losses) and high electric strengths require a high band gap in the oxide, usually found in materials with a low dielectric constant. High dielectric strengths allow high fields to be tolerated without dielectric breakdown. According to Vijh, high band gaps empirically result in oxide dielectrics with low losses, high dielectric strengths, and low dielectric constants.

In a practical capacitor, the anode and cathode are usually separated by a porous separator which contains the capacitor electrolyte. One of the roles of this electrolyte is to enable healing of the oxide film to occur when damaged. The most widely used electrolytes for Al capacitors are of the ethylene glycol/borate type. While such solutions are well-suited for maintenance of the oxide film, they do possess certain temperature limitations. At low temperatures the solution conductivity decreases to the point where the dissipation factor increases significantly and a reduction in capacitance occurs. At elevated temperatures loss of water vapor may occur (since such capacitors are not hermetically sealed) and changes in electrolyte composition can result.

Barnard has discussed some mechanistic aspects of the charge and discharge processes in an Al electrolytic capacitor [351]. During charging of an Al capacitor, a charge, Q_c, which equals C_AV, where C_A and V are the anode capacitance and voltage, respectively, builds up on the anode. Of course, at the cathode an electrode reaction equivalent to C_AV/F units of charge must occur. Since the expected cathode reaction is evolution of hydrogen, gas evolution and electrolyte concentration changes would be expected. Since 10^6 cycles of service are normally attained by such devices without signs of failure, such cathode processes are apparently not in operation. As demonstrated by Bernard, significant hydrogen evolution does not appear to occur at the cathode due to deposition of hydrogen atoms on the Al surface, and little or no recombination of hydrogen atoms occurs. On discharge, the oxidation of these H-atoms occurs. Etching of the Al cathode foil supplies a much larger real surface for deposition of absorbed hydrogen.

The main advantage that Al capacitors have in comparison to tantalum-based devices is, of course, the lower cost of Al. This is an important consideration in large capacitors designed for high voltage ($\sim$ 500 V) applications. Due to the enormous variations in Al capacitor design, any further discussion of the complexities of Al capacitors seems best reserved for manufacturers' bulletins.

ACKNOWLEDGMENTS

The author would like to express his appreciation to Professor Bard for his patience and understanding during the preparation of this chapter, and to Exxon R&E for their assistance in the final preparation of this manuscript. Finally, special recognition must be given to my wife, Ginny, for her masterful typing of this manuscript.

REFERENCES

1. A. J. de Bethune and N. A. S. Loud, Encyclopedia of Electrochemistry (C. A. Hampel, ed.), Reinhold, New York, 1964.
2. W. M. Latimer, Oxidation Potentials, Prentice-Hall, New York, 1952.
3. J. Heyrovsky, J. Chem. Soc., 117, 27 (1920).
4. R. C. Plumb, J. Phys. Chem., 66, 866 (1962).
5. R. C. Plumb and J. W. Swaine, Jr., ibid., 68, 2057 (1964).
6. T. Hagyard, W. B. Earl, K. J. Kirkpatrick, and I. G. Watson, J. Electrochem. Soc., 113, 962 (1966).
7. T. Hagyard and W. B. Earl, ibid., 114, 694 (1967).
8. I. M. Kolthoff and C. J. Sambucetti, Anal. Chim. Acta, 21, 17 (1959).
9. S. I. Yakubson and M. A. Abramova, Ukr. Khm. Zh., 15, 362 (1949); Chem. Abstr., 47, 73481 (1953).
10. J. Farrar, R. Keller, and C. J. Mazac, Proc. Power Sources Conf., 18, 92 (1964).
11. R. A. Wallace and P. F. Bruins, J. Electrochem. Soc., 114, 209 (1967).

12. Y. Delimarski and B. Markov, Electrochemistry of Fused Salts, Sigma, New York, 1961.

13. J. Thonstad, Electrochim. Acta, 13, 449 (1968).

14. H. C. Gaur and H. L. Jindal, ibid., 13, 835 (1968).

15. H. A. Laitinen and C. H. Liu, J. Amer. Chem. Soc., 80, 1015 (1958).

16. K. Grjotheim, Z. Phys. Chem. (Frankfurt), 11, 150 (1957).

17. A. J. Arvia and H. A. Videla, Electrochim. Acta, 11, 537 (1966).

18. E. M. Skobets and N. S. Kavetskii, J. Gen. Chem. (USSR), 10, 1858 (1940).

19. D. R. Stull, H. Prophet, et al., Editors, JANAF Thermochemical Tables, 2nd ed., NSRDS-NBS37, June 1971.

20. E. Deltome, C. Vanleugenhaghe, and M. Pourbaix, Atlas of Electrochemical Equilibrium in Aqueous Solution (M. Pourbaix, ed.), Pergamon, New York, p. 168.

21. D. D. MacDonald and P. Butler, Corros. Sci., 13, 259 (1973).

22. P. A. Brook, ibid., 12, 297 (1972).

23. R. E. Mesmer and C. F. Baes, Inorg. Chem., 10, 2290 (1971).

24. L. G. Boxall, H. L. Jones, and R. A. Osteryoung, J. Electrochem. Soc., 120, 223 (1973).

25. A. A. Fannin, Jr., L. A. King, and D. W. Seegmiller, ibid., 119, 801 (1972).

26. B. Tremillon and G. Letisse, J. Electroanal. Chem., 17, 371 (1968).

27. G. Torsi and G. Mamantov, Inorg. Chem., 10, 1900 (1971).

28. M. Rey, Electrochim. Acta, 14, 991 (1969).

29. G. Torsi, G. Mamantov, and G. Begun, Inorg. Nucl. Chem. Lett., 6, 553 (1970).

30. I. M. Kolthoff and J. J. Lingane, Polarography, 2nd ed., Wiley-Interscience, New York, 1962, p. 513.

31. J. Prajzler, Collect. Czech. Chem. Commun., 3, 407 (1931).

32. H. C. Gull, J. Soc. Chem. Ind., 56, 177 (1937).

33. T. D. Parks and L. Lykken, Anal. Chem., 20, 1102 (1948).

34. H. H. Willard and J. A. Dean, ibid., 22, 1264 (1950).

35. M. Heyrovsky, Collect. Czech. Chem. Commun., 25, 3120 (1960).

36. M. Heyrovsky, Advan. Polarog., 3, 845 (1960).

37. J. P. Stradins and L. K. Liepina, Zh. Fiz. Khim., 32, 196 (1958).

38. M. Galova and V. Szmereková, Chem. Zvesti, 27, 437 (1973).

39. G. W. Latimer, Talanta, 15, 1 (1968).

40. T. M. Florence, Anal. Chem., 34, 496 (1962).

41. T. M. Florence, F. J. Miller, and H. E. Zittel, ibid., 38, 1065 (1966).

42. B. Lendermann, H. Monien, and H. Specker, Anal. Lett., 5, 837 (1972).

43. L. Holleck, J. M. Abd El Kader, and A. M. Shams El Din, J. Electroanal. Chem., 20, 287 (1969).

44. J. M. Abd El Kadar, A. M. Shams El Din, B. Kastening, and L. Holleck, ibid., 26, 41 (1970).

45. M. E. Hall and C. T. Skoumbourdis, Anal. Chem., 38, 64 (1966).

46. A. Cizak and P. J. Elving, Electrochim. Acta, 10, 935 (1965).

47. H. Dehn, G. Gritzner, and V. Gutmann, Mikrochim. Acta, 3, 422 (1967).

48. H. Dehn, V. Gutmann, and G. Schober, Monatsh. Chem., 93, 453 (1962).

49. A. D. McElroy and H. A. Laitinen, J. Phys. Chem., 57, 564 (1953).

50. G. P. Kumar and D. A. Pantony, J. Polarog. Soc., 14, 84 (1968).

51. I. M. Kolthoff and J. F. Coetzee, J. Amer. Chem. Soc., 79, 1852 (1957).

52. A. I. Popov and D. H. Geske, ibid., 79, 2074 (1957).

53. R. C. Larson and R. T. Iwamoto, ibid., 82, 3239 (1960).

54. V. Gutmann, M. Kogelnig, and M. Michlmayr, Monatsh. Chem., 99, 699 (1968).

55. J. B. Headridge, D. Pletcher, and M. Callingham, J. Chem. Soc., A, 1967, 684.

56. V. Gutmann and G. Schober, Monatsh. Chem., 93, 1353 (1962).

57. G. F. Reynolds and T. J. Webber, Anal. Chim. Acta, 19, 293 (1958).

58. W. Kemula and J. Zawadowska, Rocz. Chem., 47, 173 (1973).

59. I. M. Kolthoff and C. J. Sambucetti, Anal. Chim. Acta, 21, 17 (1959); 233 (1959).

60. M. Galova and D. A. Pantony, Talanta, 18, 1209 (1971).

61. W. S. Harris, UCRL Report 8381, July 1958.

62. V. F. Toropova, G. K. Budnikov, and V. N. Maistrenko, Zh. Obshch. Khim., 42, 1207 (1972).

63. M. Eisenberg, K. Wong, and K. C. Tsai, J. Electrochem. Soc., 120, 1686 (1973).

64. M. Fleischmann, D. Pletcher, and C. J. Vance, J. Organometal. Chem., 40, 1 (1972).

65. H. A. Laitinen, C. H. Liu, and W. S. Ferguson, Anal. Chem., 30, 1266 (1958).

66. I. I. Naryshkin and N. A. Minin, Zh. Priklad. Khim., 34, 2353 (1961).

67. R. Piontelli, G. Sternheim, and M. Francini, J. Phys. Chem., 24, 1113 (1956).

68. L. A. King, A. D. Brown, and F. H. Frayer, Proceedings of the 1968 OAR Research Applications Conference, Vol. 1, Rep. OAR-68-001, Institute for Defense Analyses, 1968.

69. B. S. Del Duca, J. Electrochem. Soc., 118, 405 (1971).

70. R. Piontelli and U. Ducati, ibid., 118, 1966 (1971).

71. G. L. Holleck and J. Giner, ibid., 119, 1161 (1972).

72. K. Schulze and H. Hoff, Ber. Bunsenges. Phys. Chem., 74, 687 (1970).

73. K. Schulze and H. Hoff, Electrochim. Acta, 17, 119 (1972).

74. K. Schulze and H. Hoff, ibid., 1783 (1972).

75. M. M. Vetyukov, B. S. Dyblin, and Yu. V. Boriosglebskii, Elektrokhimiya, 8, 343 (1972).

76. A. K. Vijh, J. Phys. Chem., 72, 1148 (1968).

77. B. Jakuszewski and Z. Kozlowski, Rocz. Chem., 36, 1873 (1962).

78. T. N. Andersen, J. L. Anderson, and H. Eyring, J. Phys. Chem., 73, 3562 (1969).

79. B. Jakuszewski and Z. Kozlowski, Rocz, Chem., 38, 93 (1964).

80. B. Jakuszewski and Z. Kozlowski, Soc. Sci. Lodz. Acta Chim., 10, 5 (1965).

81. E. A. Ukshe, N. G. Bukun, and D. I. Leikis, Zh. Fiz. Khim., 36, 2322 (1962).

82. S. Karpachev and A. Stromberg, ibid., 18, 47 (1944).

83. A. D. Graves, G. J. Hills, and D. Inman, Advan. Electrochem. Electrochem. Eng., 4, 153 (1966).

84. B. N. Bhaduri and A. Fowler, Proc. Roy. Soc. (London), A145, 321 (1934).

85. H. H. Emons and H. Hasselbarth, Chem. Tech. (Berlin), 14, 621 (1962).

86. B. J. Chai, H. C. Ko, M. A. Greenbaum, and M. Farber, J. Phys. Chem., 71, 3331 (1967).

87. W. C. Schumb and H. H. Rogers, J. Amer. Chem. Soc., 73, 5860 (1951).

88. F. A. Cotton and G. Wilkinson, Advanced Inorganic Chemistry, Wiley-Interscience, New York, 1962, p. 345.

89. F. Wohler and H. Buff, Justus Liebigs Ann. Chem., 103, 218 (1857).

90. J. W. Turrentine, J. Phys. Chem., 12, 448 (1908).

91. A. W. Davidson and F. Jirik, J. Amer. Chem. Soc., 72, 1700 (1950).

92. P. Brouillet, I. Epelboin, and F. Froment, C.R. Acad. Sci., Paris, 239, 1795 (1954).

93. M. C. del Boca, Helv. Chim. Acta, 16, 565 (1933).

94. W. E. Bennett, A. W. Davidson, and J. Kleinberg, J. Amer. Chem. Soc., 74, 732 (1952).

95. U. Sborgi and P. Marchetti, Nuovo Cimento, 22, 151 (1921).

96. T. T. Tsai, W. E. McEwen, and J. Kleinberg, J. Amer. Chem. Soc., 82, 3326 (1960).

97. E. Raijola and A. W. Davidson, ibid., 78, 556 (1956).

98. M. E. Straumanis, J. Electrochem. Soc., 108, 1087 (1961).

99. M. E. Straumanis and K. Poush, ibid., 112, 1185 (1965).

100. L. Kiss, M. L. Varsanyi, and E. Dudas, Acta Chem. (Budapest), 79, 73 (1973).

101. Yu. K. Delimarskii, V. F. Makogon, and A. Ya. Zhigailo, Elektrokhimiya, 5, 108 (1969).

102. V. N. Storozhenko, Vop. Khim. Khim. Tekhnol., 1972, 67; Chem. Abstr., 78, 105368g (1973).

103. R. Midorikawa, Denki Kagaku, 25, 13 (1957).

104. T. Notoya, Electrochim. Acta, 13, 2194 (1968).

105. R. J. Gale and R. A. Osteryoung, J. Electrochem. Soc., 121, 983 (1974).

106. A. I. Belyaev, Sb. Nuach. Tr. Mosk. Inst. Tswet. Metal. Zolota, 31, 46 (1958).

107. A. Brenner, Advan. Electrochem. Electrochem. Eng., 5, 205 (1967).

108. D. E. Couch and A. Brenner, J. Electrochem. Soc., 99, 234 (1952).

109. A. Brenner and D. Couch, U.S. Patent 2,651,608 (September 1953).

110. F. A. Clay, W. B. Harding, and C. J. Stimetz, Plating, 56, 1027 (1969).

111. E. C. Ashby and J. Prather, J. Amer. Chem. Soc., 88, 729 (1966).

112. N. Ishibashi and M. Yoshio, Electrochim. Acta, 17, 1343 (1972).

113. M. Yoshio and N. Ishibashi, J. Appl. Electrochem., 3, 321 (1973).

114. M. Yoshio, N. Ishibashi, H. Waki, and T. Seiyama, J. Inorg. Nucl. Chem., 34, 2439 (1972).

115. G. A. Capuano and W. G. Davenport, J. Electrochem. Soc., 118, 1638 (1971).

116. G. G. Gustavson, J. Prakt. Chem., 42, 250 (1890).

117. H. C. Brown and H. W. Pearsall, J. Amer. Chem. Soc., 74, 191 (1952).

118. H. C. Brown and W. J. Wallace, ibid., 75, 6268 (1953).

119. H. C. Brown and W. J. Wallace, ibid., 75, 6265 (1953).

120. L. Simanavicius and A. Levinskiene, Trans. Acad. Sci. Lithuanian SSR, B4(47), 39 (1966).

121. T. Hisano, T. Terazawa, I. Takeuchi, S. Inohara, and H. Ikeda, Bull. Chem. Soc. Japan, 44, 599 (1971).

122. W. J. Aalbersbery, G. J. Hoijtink, E. L. Mackor, and W. P. Weijland, J. Chem. Soc., 1959, 3049.

123. D. B. Keyes, S. Swann, Jr., W. Klabunde, and S. T. Shicktanz, Ind. Eng. Chem., 20, 1068 (1928).

124. F. H. Hurley and T. P. Weir, Jr., J. Electrochem. Soc., 98, 203 (1951).

125. K. Ziegler and H. Lehmkuhl, Angew. Chem., 67, 424 (1955).

126. K. Ziegler and H. Lehmkuhl, Z. Anorg. Allg. Chem., 283, 414 (1956).

127. S. Senderoff, U.S. Patent 3,751,298 (August 1973).

128. R. Midorikawa, Denki Kagaku, 24, 366, 419, 457, 511, 562 (1956).

129. R. Midorikawa, ibid., 25, 13 (1957).

130. R. C. Howie and D. W. MacMillan, J. Appl. Electrochem., 2, 217 (1972).

131. T. Sato, T. Ishikawa, and R. Midorikawa, Denki Kagaku, 41, 446 (1973).

132. F. H. Hurley and T. P. Wier, Jr., J. Electrochem. Soc., 98, 207 (1951).

133. S. Tajima, Advan. Corros. Sci. Technol., 1, 229 (1970).

134. G. C. Wood, in Oxides and Oxide Films (J. W. Diggle, ed.), Dekker, New York, 1973.

135. J. W. Diggle, T. C. Downie, and C. W. Goulding, Chem. Rev., 69, 365 (1969).

136. A. K. Vijh, Electrochemistry of Metals and Semiconductors, Dekker, New York, 1973.

137. E. J. Verwey, J. Chem. Phys., 3, 592 (1935).

138. H. Ginsberg and K. Wefers, Metalloberflaeche, 17, 202 (1963).

139. G. A. Dorsey, J. Electrochem. Soc., 113, 169, 172, 284 (1966).

140. I. T. Kormany, Chem. Abstr., 64, 1619f (1966).

141. W. G. Burgers, A. Classen, and J. Zernicke, Z. Phys., 74, 593 (1932).

142. E. Lichtenberger, Metalloberflaeche, 15, 38 (1961).

143. K. Norden, Z. Elektrochem., 6, 159 (1899).

144. F. Liechti and W. D. Treadwell, Helv. Chim. Acta, 30, 1204 (1947).

145. W. J. Bernard and J. J. Randoll, Jr., J. Electrochem. Soc., 108, 822 (1961).

146. R. C. Plumb, ibid., 105, 498 (1958).

147. S. Taijama, N. Baba, and M. Shimura, Electrochim. Acta, 12, 955 (1967).

148. D. A. Vermilyea, Acta Met., 2, 482 (1954).

149. J. E. Lewis and R. C. Plumb, J. Electrochem. Soc., 105, 498 (1958).

150. G. Amsel and D. Samuel, J. Phys. Chem. Solids, 23, 1707 (1965).

151. A. Guntherschultze and H. Betz, Z. Phys., 68, 145 (1931).

152. N. Cabrera and N. F. Mott, Rept. Progr. Phys., 12, 163 (1948-1949).

153. D. A. Vermilyea, Acta Met., 1, 282 (1953).

154. L. Young, Trans. Faraday Soc., 50, 159 (1954).

155. L. Young, Proc. Royal Soc., A263, 395 (1961).

156. J. F. Dewald, Acta Met., 2, 340 (1954).

157. J. F. Dewald, J. Electrochem. Soc., 102, 1 (1955).

158. B. Domeij, R. Brown, J. A. Davies, and M. McCargo, Can. J. Phys., 42, 1624 (1964).

159. D. M. Cheseldine, J. Electrochem. Soc., 111, 128 (1964).

160. J. J. Randall, Jr., W. J. Bernard, and R. R. Wilkinson, Electrochim. Acta, 10, 183 (1965).

161. J. Klerer, J. Electrochem. Soc., 112, 896 (1965).

162. F. Brown and W. D. Mackintosh, ibid., 120, 1096 (1973).

163. J. Frenkel, Kinetic Theory of Liquids, Dover, New York, 1955, p. 40.

164. L. Young and F. G. R. Zobel, J. Electrochem. Soc, 113, 277 (1966).

165. M. J. Dignam, Can. J. Chem., 42, 1155 (1964).

166. M. J. Dignam, D. Goad, and M. Sole, ibid., 43, 800 (1965).

167. M. J. Dignam and D. Goad, J. Electrochem. Soc., 113, 381 (1966).

168. M. J. Dignam, ibid., 112, 722 (1965).

169. M. J. Dignam, ibid., 112, 729 (1965).

170. M. J. Dignam and P. J. Ryan, Can. J. Chem., 46, 535 (1968).

171. M. J. Dignam and P. J. Ryan, ibid., 46, 549 (1968).

172. M. J. Dignam, in Oxides and Oxide Films, Vol. 1 (J. W. Diggle, ed.), Dekker, New York, 1973.

173. S. Setoh and A. Miyata, Sci. Pap. Inst. Phys. Chem. Res. (Tokyo), 19, 237 (1932).

174. F. Keller, M. S. Hunter, and D. L. Robinson, J. Electrochem. Soc., 100, 411 (1953).

175. C. J. L. Booker, J. L. Wood, and A. Walsh, Nature, 176, 222 (1955).

176. H. Grubitsch, W. Geymeyer, and E. Buvik, Aluminum, 37, 569 (1961).

177. M. Paganelli, Alumino, 27, 3 (1958).

178. R. L. Burwell, P. A. Smudski, and T. P. May, J. Amer. Chem. Soc., 69, 1525 (1947).

179. G. Paolini, M. Masaero, R. Sacchi, and M. Paganelli, J. Electrochem. Soc., 112, 32 (1965).

180. G. C. Wood, J. P. O'Sullivan, and B. Vaszko, ibid., 115, 618 (1968).

181. R. W. Franklin and D. J. Stirland, ibid., 110, 262 (1963).

182. P. Csokan and M. Gy. Hollo, Werkst. Korros., 12, 288 (1961).

183. P. Csokan, Electroplating Metal Finishing, 15, 75 (1962).

184. P. Csokan, Trans. Inst. Metal Finishing, 41, 51 (1964).

185. J. F. Murphy and C. E. Michelson, Proceedings of the Conference on Anodizing Aluminum (Nottingham, 1961), Aluminum Federation, London, 1962, p. 83.

186. J. F. Murphy, Proceedings of the Symposium on Anodizing Aluminum (Birmingham, 1967), Aluminum Federation, London, 1967, p. 3.

187. J. F. Murphy, Plating, 54, 1241 (1967).

188. A. F. Bogoyavlenskii, paper presented at the First Interuniversity Conference, Moscow, 1964 (cited in Ref. 135).

189. A. V. Shreider, J. Appl. Chem. USSR, 39, 2533 (1966).

190. M. S. Hunter and P. E. Fowle, J. Electrochem. Soc., 101, 481 (1954).

191. G. H. Kissin, Anodized Aluminum, ASTM Special Technical Publication No. 388, Philadelphia, 1965.

192. M. Nagayama and K. Tamura, Electrochim. Acta, 12, 1097 (1967).

193. T. P. Hoar and N. F. Mott, J. Phys. Chem. Solids, 9, 97 (1959).

194. T. P. Hoar and J. Yahalom, J. Electrochem. Soc., 110, 614 (1963).

195. M. Fleischman and H. R. Thirsk, Advan. Electrochem. Electrochem. Eng., 3, 123 (1963).

196. S. Ya. Grilikhes and N. P. Fedot'ev, Zh. Prikl. Khim., 40, 841 (1967).

197. P. Neufield and H. O. Ali, J. Electrochem. Soc., 120, 479 (1973).

198. W. Vedder and D. A. Vermilyea, Trans. Faraday Soc., 65, 561 (1969).

199. R. S. Alwitt, J. Electrochem. Soc., 114, 843 (1967).

200. R. S. Alwitt, A. J. Breen, and J. S. L. Leach, Proceedings of the Symposium on Oxide-Electrolyte Interfaces (R. S. Alwitt, ed.), The Electrochemical Society, Princeton, New Jersey, 1973, p. 265.

201. A. Conte and L. Campanella, Electrochim. Metall., 3, 183 (1968).

202. L. Campanella and A. Conte, J. Electrochim. Soc., 116, 144 (1969).

203. S. Tajima, M. Soda, and T. Mori, Electrochim. Acta, 1, 205 (1959).

204. Chemical and Engineering News, April 30, 1973, p. 12.

205. J. H. Dunn and L. S. Sewell, Jr., in Aluminum, Vol. 2, American Society for Metals, Metals Park, Ohio, 1966, p. 1.

206. Chemical and Engineering News, January 15, 1973, p. 12.

207. Chemical and Engineering News, February 26, 1973, p. 11.

208. C. L. Mantell, Electrochemical Engineering, 4th ed., McGraw-Hill, New York, 1960, p. 371.

209. T. G. Pearson and J. Waddington, Discussions Faraday Soc., 1, 307 (1947).

210. J. Thonstad, J. Electrochem. Soc., 111, 959 (1964).

211. R. Piontelli, Proceedings of the 1st Australian Conference on Electrochemistry (J. A. Friend and F. Gutmann, eds.), Pergamon, London, 1965, p. 932.

212. T. G. Pearson, Roy. Inst. Chem. (London), Lectures, Monographs Rep., 1955, 3.

213. S. Gjerstad and B. J. Welch, J. Electrochem. Soc., 111, 976 (1964).

214. O. Bersimenko and M. M. Vetyukov, Zh. Prikl. Khim., 39, 1696 (1966).

215. W. E. Haupin, J. Electrochem. Soc., 120, 83 (1973).

216. N. W. F. Phillips, R. H. Singleton, and E. A. Hollingshead, ibid., 102, 648 (1955).

217. K. Grjotheim, K. Matiasovsky, and M. Malinovsky, Electrochim. Acta, 15, 259 (1970).

218. W. B. Frank and L. M. Foster, J. Phys. Chem., 61, 1531 (1957).

219. J. Thonstad, Electrochim. Acta, 14, 127 (1969).

220. P. A. Foster, Jr., and W. B. Frank, J. Electrochem. Soc., 107, 997 (1960).

221. J. Brynestad, K. Grjotheim, F. Gronwold, J. L. Holm, and S. Urnes, Discussions Faraday Soc., 32, 90 (1961).

222. N. E. Richards and B. J. Welch, J. Electrochem. Soc., 109, 860 (1962).

223. J. Thonstad, Electrochim. Acta, 12, 1219 (1967).

224. J. Thonstad, ibid., 15, 1569 (1970).

225. J. Thonstad, F. Nordmo, and K. Vee, ibid., 18, 27 (1973).

226. T. R. Beck, J. Electrochem. Soc., 106, 710 (1959).

227. T. R. Beck, ibid., 107, 577 (1960).

228. V. A. Cammarota, Jr., and D. Schlain, ibid., 117, 282 (1970).

229. W. E. Haupin, ibid., 120, 83 (1973).

230. A. T. Kuhn, Industrial Electrochemical Processes (A. T. Kuhn, ed.), Elsevier, New York, 1971.

231. N. E. Richards, in Proceedings of the 1st Australian Conference on Electrochemistry (J. A. Friend and F. Gutmann, eds.), Pergamon, London, 1965, p. 867.

232. Gy. Szeker, Acta Tech. (USSR), 10, 91 (1954).

233. H. Schmitt, Proceedings of the International Symposium on the Extraction of Aluminum, AIME, New York, 1962.

234. W. Schmitt-Hatting, Chem.-Ing.-Tech., 41, 163 (1969).

235. A. Russell, L. L. Knapp, and W. E. Haupin, U.S. Patent 3,725,222.

236. U.S. Patents 1,534,317; 1,534,318; 1,534,322 (1925).

237. French Patent 759,588 (1934).

238. British Patent 405,596 (1934).

239. British Patent 469,361 (1937).

240. K. Akerman, M. Brafman, D. Kozminska, N. Biala, J. Kierzek, W. Tabor, T. Piechoia, Chem. Abstr., 68, 80648 (1968).

241. R. Gaugin and G. Nury, Bull. Soc. Franc. Electriciens, 8, 733 (1956).

242. V. A. Plotnikov, J. Russ. Phys. Chem. Soc., 34, 466 (1902).

243. R. D. Blue and F. C. Mathers, Trans. Electrochem. Soc., 65, 339 (1934).

244. R. D. Blue and F. C. Mathers, ibid., 69, 519 (1936).

245. F. H. Hurley and T. P. Weir, Jr., J. Electrochem. Soc., 98, 203 (1951).

246. W. H. Safranek, W. C. Shickner, and C. L. Faust, U.S. Patent 2,692,850 (1954).

247. J. H. Connor and A. Brenner, J. Electrochem. Soc., 103, 657 (1956).

248. J. B. Berkowitz and N. H. Emerson, Plating Methods, NASA Special Publication NASA SP-5114, U.S. Government Printing Office, Washington, D.C., 1972, p. 31.

249. R. J. Rothlein, J. Electrochem. Soc., 117, 714 (1970).

250. B. O. Holland, J. Aust. Inst. Metals, 6, 212 (1961).

251. J. J. Utz and S. Kritzer, Mater. Design Eng., 49, 88 (1959).

252. J. G. Beach, L. D. McGrae, and C. L. Faust, Plating, 55, 936 (1968).

253. F. R. Collins, Iron Age, 169, 100 (1952).

254. K. Matiasovsky and M. Paucirova, Povrchove Upravy, 13, 10 (1973); Chem. Abstr., 80, 22000d (1974).

255. R. Mason, U.S. Patent 2,108,603 (1938).

256. S. Wernick and R. Pinner, The Surface Treatment of Aluminum and Its Alloys, Draper, London, 1956.

257. N. D. Pullen, J. Inst. Metals, 59, 151 (1936).

258. D. J. George and R. V. Vandenberg, in Aluminum, Vol. 3, American Society for Metals, Metals Park, Ohio, 1967, p. 623.

259. J. M. Andrus and R. Pettit, ASTM Special Technical Publication 388, Philadelphia, 1965, p. 1.

260. A. Jackson and A. T. Kuhn, in Industrial Electrochemical Processes (A. T. Kuhn, ed.), Elsevier, Amsterdam, 1971, p. 507.

261. M. Finkelstein, R. C. Petersen, and S. D. Ross, Electrochim. Acta, 10, 465 (1965).

262. S. D. Ross, M. Finkelstein, and R. C. Petersen, J. Amer. Chem. Soc., 82, 1582 (1960).

263. S. D. Ross, M. Finkelstein, and R. C. Petersen, ibid., 92, 6003 (1970).

264. O. R. Brown and E. R. Gonzalez, J. Electroanal. Chem., 35, 13 (1972).

265. O. R. Brown and E. R. Gonzalez, ibid., 43, 215 (1973).

266. N. E. Khomutov and V. V. Tsodikov, Elektrokhimiya, 1, 482 (1965).

267. H. French and M. Drane, J. Amer. Chem. Soc., 52, 4904 (1930).

268. W. Evans and F. Lee, ibid., 56, 654 (1934).

269. E. M. Marlett, Ann. N.Y. Acad. Sci., 125, 12 (1965).

270. A. P. Tomilov and I. N. Brago, in Progress in the Electrochemistry of Organic Compounds (A. N. Frumkin and A. B. Ershler, eds.), Plenum, New York, 1971.

271. E. H. Mottus and M. R. Ort, J. Electrochem. Soc., 117, 885 (1970).

272. E. H. Mottus and M. R. Ort, Extended Abstracts, ECS Spring Meeting, Chicago, May 1973.

273. R. Ercoli, M. Guainazzi, and G. Silvestri, Ger. Offen. 2,223,335 (1972).

274. W. Evans and R. Pearson, J. Amer. Chem. Soc., 64, 2865 (1942).

275. K. Ziegler and H. Lehmkuhl, German Patent 1,150,078 (1963).

276. Ethyl Corporation, British Patent 842,090 (1960).

277. K. Ziegler and H. Lehmkuhl, Chem.-Ing.-Tech., 35, 325 (1963).

278. K. Ziegler, Belgian Patent 617,628 (1962).

279. U.S. Patent 3,254,008 (1966).

280. U.S. Patent 3,254,009 (1966).

281. U.S. Patent 2,960,450 (1960).

282. A. P. Giraitis, U.S. Patent 2,944,948 (1960).

283. K. Ziegler and H. Lehmkuhl, U.S. Patent 2,985,568 (1961).

284. U.S. Patent 3,028,318 (1962).

285. K. Ziegler and H. Lehmkuhl, German Patent 1,153,754 (1963).

286. K. Ziegler, British Patent 814,609 (1959).

287. J. R. Chadwick and E. Kinsella, J. Organometal. Chem., 4, 334 (1965).

288. N. D. Tomashov, Theory of Corrosion and Protection of Metals (B. H. Tytell, I. Geld, and H. S. Preiser, eds.), Macmillan, New York, 1966, p. 614.

289. H. P. Godard, W. B. Jepson, M. R. Bothwell, and R. L. Kane, The Corrosion of Light Metals, Wiley, New York, 1967.

290. H. H. Uhlig, Corrosion and Corrosion Control, Wiley, New York, 1963.

291. J. C. Bailey and F. C. Porter, in Corrosion, Vol. I, (L. L. Shreir, ed.), Wiley, New York, 1963, p. 413.

292. N. D. Greene and M. Fontana, Corrosion, 15, 25 (1959).

293. J. M. Kolytyrkin, ibid., 20, 261 (1963).

294. R. B. Mears, R. H. Brown, and E. H. Dix, Jr., Symposium on Stress Corrosion Cracking of Metals, ASTM-AIMME, 1945, p. 323.

295. H. K. Farmery and U. R. Evans, J. Inst. Metals, 84, 413 (1955-56).

296. H. L. Logan, The Stress Corrosion of Metals, Wiley, New York, 1966.

297. M. O. Speidel and M. V. Hyatt, Advan. Corros. Sci. Technol., 2, 115 (1972).

298. J. Draley and W. Ruther, Corrosion, 12, 480t (1956).

299. E. Cook, R. Horst, and W. Binger, ibid., 17, 25t (1961).

300. A. Hamstead, G. Elder, and J. Canterbury, ibid., 14, 189t (1958).

301. M. Stern and H. H. Uhlig, J. Electrochem. Soc., 99, 381, 389 (1952).

302. M. Stern and H. H. Uhlig, ibid., 100, 543 (1953).

303. A. H. Roebuck and T. R. Pritchett, Mat. Protection, 1966, 16.

304. R. Glicksman, J. Electrochem. Soc., 106, 457 (1959).

305. M. Hulot, C.R. Acad. Sci., Paris, 40, 1148 (1855).

306. J. J. Stokes, Jr., Electrochem. Technol., 6, 36 (1968).

307. S. Zaromb, J. Electrochem. Soc., 109, 1125 (1962).

308. S. Zaromb, ibid., 109, 1191 (1962).

309. S. Zaromb, Power Systems for Electric Vehicles (H. B. Linford, H. P. Gregor, and B. J. Steigerwald, eds.), U.S. Dept. of Health, Education, and Welfare, Cincinnati, Ohio, 1967, p. 255.

310. M. Katoh, K. Shimizu, and S. Katoh, Denki Kagaku, 38, 753 (1970).

311. S. Zaromb, Proceedings of the 4th Intersociety Energy Conversion Engineering Conference, Washington, D.C., American Institute of Chemical Engineers, New York, 1969, p. 904.

312. L. Brull, Ric. Sci., 1, 525 (1937).

313. D. E. Sargent, U.S. Patent 2,554,447 (1951).

314. S. Ruben, U.S. Patent 2,638,489 (1953).

315. D. Belitskus, J. Electrochem. Soc., 119, 295 (1972).

316. A. Kuzmina and L. Kuzmin, Zh. Prikl. Khim., 36, 335 (1963).

317. P. L. Joseph, V. Balasubramanian, and B. A. Shenoi, J. Electrochem. Soc. Japan, 35, 8 (1967).

318. M. Schaldach, Ann. N.Y. Acad. Sci., 67, 1016 (1969).

319. A. C. C. Tseung, W. J. King, and B. Y. C. Wan, Med. Biol. Eng., 9, 175 (1971).

320. B. Y. C. Wan, A. C. C. Tseung, J. Kenny, and A. Wilds, Proceedings of the 7th Intersociety Energy Conversion Engineering Conference, San Diego, California, American Chemical Society, Washington, D.C., 1972, p. 745.

321. J. B. Cooper and A. W. Hahn, IEEE Trans. Bio-Med. Eng., BME-20, 336 (1973).

322. J. F. DeRosa, R. B. Beard, H. M. Carim, and S. E. Dubin, ibid., BME-20, 345 (1973).

323. L. Antipin, Zh. Fiz. Khim., 30, 1425 (1956).

324. J. Giner and G. Holleck, Final Report by Tyco Laboratories, Inc., on Contract No. NAS 12-688, March 1970 (NASA-CR-1541).

325. M. Beltzer, U.S. Patent 3,632,448 (1972).

326. J. Greenberg, U.S. Patent 3,635,765 (1972).

327. E. Buzzelli, U.S. Patent 3,650,834 (1972).

328. W. K. Behl, Interagency Advanced Power Group Project Brief, PIC No. 2393 (1972).

329. L. E. Topol, W. A. McCollum, Jr., and S. J. Yosim, Final Report by North American Rockwell Corp. on Contract No. NAS 3-12970, December, 1970.

330. D. A. Crouch, ECS Extended Abstracts, Fall Meeting, Miami Beach, October 1972, p. 89.

331. Chemical and Engineering News, July 23, 1973, p. 12.

332. G. E. Best, in Electroplating Engineering Handbook (A. K. Graham, ed.), Van Nostrand Reinhold, New York, 1973, p. 447.

333. R. A. Wodehouse, ibid., p. 456.

334. W. E. Gorden and M. E. Cupery, Ind. Eng. Chem., 31, 1237 (1934).

335. J. Fisher, U.S. Patent 2,095,519 (1937).

336. M. Schenk, Swiss Patents 182,415 (1936); 221,938 (1942); 232,613 (1944).

337. B. E. Diel and L. Swanson, U.S. Patent 3,031,387 (1962).

338. W. C. Cochran, in Aluminum, Vol. 3, American Society for Metals, Metals Park, Ohio, 1966, p. 641.

339. N. D. Tomashov, Light Metals, 9, 429 (1948).

340. G. Bengough and J. M. Stuart, British Patent 223,994 (1924).

341. R. W. Buzzard, J. Res. Nat. Bur. Stand., A, 18, 25 (1937).

342. F. Modic, Aluminum, 39, 169 (1963).

343. G. L. Martin, U.S. Patents 2,692,851; 2,692,852 (1950).

344. W. J. Campbell, Light Metals, 15, 46 (1952).

345. Sanford Process Co., U.S. Patents 2,897,125; 2,692,852.

346. A. Lelong, P. Segonds, and J. Herenguel, Rev. Aluminium, 37, 67 (1960).

347. G. H. Giles, Proceedings of the Conference on Anodizing Aluminum, Aluminum Federation, London, 1962, p. 174.

348. F. Pavelka, U.S. Patent 2,168,135 (1939).

349. M. S. Hunter, Electrochem. Technol., 1, 151 (1963).

350. Ref. 136, p. 261.

351. W. J. Bernard, J. Electrochem. Soc., 113, 749 (1966).

352. Application Brief A-2, Princeton Applied Research Corporation, Princeton, New Jersey, 1974.

353. R. E. Iversen, J. Metals, 1973, 19.

Chapters VI-4–VI-9

PLATINUM METALS

J. F. LLOPIS,* I. M. TORDESILLAS, and F. COLOM

Department of Surface Chemistry and Electrochemistry
Institute of Physical Chemistry, C.S.I.C.
Madrid, Spain

---

*Deceased.

Chapter VI-4

PLATINUM

J. F. LLOPIS* and F. COLOM

Department of Surface Chemistry and Electrochemistry
Institute of Physical Chemistry, C.S.I.C.
Madrid, Spain

---

*Deceased.

The References for Chapters VI-4 through VI-7 appear on pages 267-275. A General Bibliography appears on pages 328 and 329.

Platinum is part of a great number of compounds in which it is mainly present in the II and IV oxidation state. Although some III and I compounds have been reported, they are poorly characterized. As is well known, transition elements have a great capacity to form complex compounds, and the aqueous chemistry of platinum deals almost exclusively with these complexes.

1. STANDARD AND FORMAL POTENTIALS

1.1. AQUEOUS SOLUTIONS

The theoretical potential-pH diagram of platinum at 25°C, Fig. 1.1.1 [1], shows that the metal is stable in aqueous solutions at almost all pH levels in the absence of complexing agents or chemicals producing insoluble salts. Only in very acidic solutions is there formation of the hydrated oxide $PtO_2.nH_2O$ or unstable $Pt(OH)_2$ at high potentials. The main equilibrium reactions of this system and their corresponding potentials are [1]:

1. $PtO + 2H^+ + 2e = Pt + H_2O$ $\quad E_0 = 0.980 - 0.0591\ pH$
2. $PtO_2 + 2H^+ + 2e = PtO + H_2O$ $\quad E_0 = 1.045 - 0.0591\ pH$
3. $PtO_3 + 2H^+ + 2e = PtO_2 + H_2O$ $\quad E_0 = 2.000 - 0.0591\ pH$
4. $PtO + 2H^+ = Pt^{2+} + H_2O$ $\quad \log(Pt^{2+}) = -7.06 - 2\ pH$
5. $Pt^{2+} + 2e = Pt$ $\quad E_0 = 1.188 + 0.0295 \log(Pt^{2+})$
6. $PtO_2 + 4H^+ + 2e = Pt^{2+} + 2H_2O$ $\quad E_0 = 0.837 - 0.1182\ pH - 0.0295 \log(Pt^{2+})$

The most extensive and typical series of Pt(IV) and Pt(II) complexes, which possess the coordination numbers 6 and 4, respectively, are the halo compounds. They encompass the entire range from the tetra-, $Pt(Am_4)X_2$, and hexamines, $Pt(Am)_6X_4$, through all intermediates such as $(PtAm_3X)X$ and $(PtAmX_5)M^I$, to $PtX_4{}^{2-}$ and $PtX_6{}^{2-}$, where X stands for Cl^-, Br^-, or I^- and Am for ammonia, hydrazine, hydroxylamine, ethylenediamine, etc. The Pt(II) complexes are square in shape and can be converted to Pt(IV) complexes by oxidation with a variety of agents. The Pt(IV) complexes tend to be octahedral and kinetically inert.

The standard potentials of the redox systems of platinum complexes have usually been measured by potentiometric methods against SCE at 25°C, usually at 10^{-3} M. Those complexes sensitive to light have been studied in the dark [2-6].

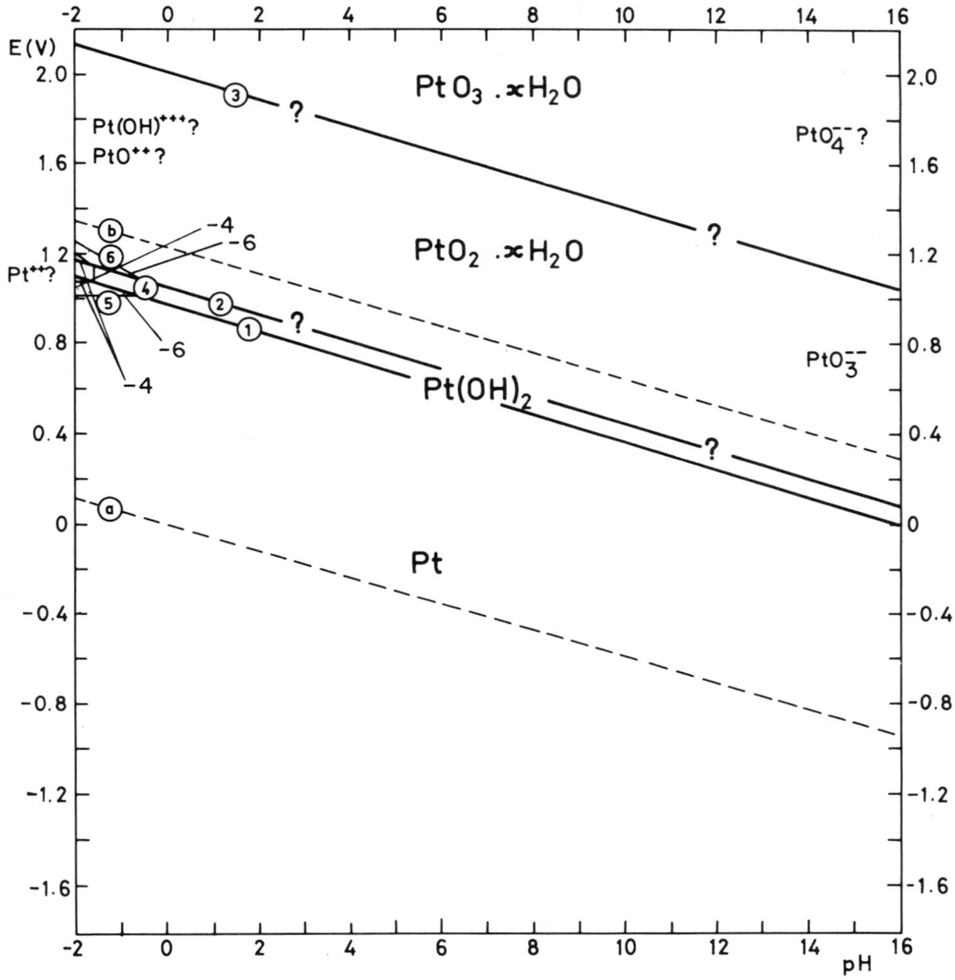

FIG. 1.1.1. Potential-pH diagram of the platinum-water system at 25°C. (Reprinted from Ref. 1, p. 380, by courtesy of the Centre Belge de la Corrosion.)

The standard and formal potentials in aqueous solutions of the main Pt-complex systems known are listed in Table 1.1.1, where potentials are given vs NHE at 25°C and measured with platinum electrodes, unless otherwise stated.

The reversible potential of the Pt halo-complex systems appears to decrease with the increasing atomic number of the halogen atom. The high oxidation potential of the hexachloro complex points to the greater instability of the Pt(IV) complexes in relation to the Pt(II) compounds. The successive substitution of NH_3 for chloro ligands raises the stability of the complex as the redox potential declines.

TABLE 1.1.1. Standard and Formal Potentials in Aqueous Solutions

Half-reaction	Standard or formal potential (V)	Conditions	Refs.
1. $PtCl_4^{2-} + 2e = Pt + 4Cl^-$	+0.579	vs Ag/AgCl; 60°C; calcd 3 M HCl; 3 M $HClO_4$	[7]
	+0.574	calcd 3 M NaCl; $NaClO_4$	[7]
	+0.749	SCE; calcd 1 N HCl; 35°C	[8]
2. $PtI_4^{2-} + 2e = Pt + 4I^-$	-0.16	SCE; 0.5 M NaI	[9]
3. $PtCl_6^{2-} + 4e = Pt + 6Cl^-$	+0.557	vs Ag/AgCl; 60°C; calcd 3 M HCl; 3 M $HClO_4$ 3 M NaCl; 3 M $NaClO_4$	[7]
4. $PtCl_6^{2-} + 2e = PtCl_4^{2-} + 2Cl^-$	+0.708 ± 0.003	1 M NaCl + 3 M H_2SO_4	[10]
	+0.757	1 M HCl; 90°C	[8]
	+0.747	1 M HCl; 70°C	[8]
	+0.730	1 M HCl; 35°C	[8]
5. $PtBr_6^{2-} + 2e = PtBr_4^{2-} + 2Br^-$	+0.581 ± 0.003	1 M NaBr + 3 M H_2SO_4	[11]
6. $PtI_6^{2-} + 2e = PtI_4^{2-} + 2I^-$	+0.329 ± 0.003	1 M NaI + 3 M H_2SO_4	[10]
7. $Pt(SCN)_6^{2-} + 2e = Pt(SCN)_4^{2-} + 2SCN^-$	+0.468	1 M NaSCN	[12]
8. $PtNH_3Cl_5^- + 2e = PtNH_3Cl_3^- + 2Cl^-$	+0.731	1 M NaCl	[3]

9.	$Pt(NH_3)_2Cl_4 + 2e = Pt(NH_3)_2Cl_2 + 2Cl^-$	+0.669	1 <u>M</u> NaCl	[2]
10.	$Pt(NH_3)_3Cl_3^+ + 2e = Pt(NH_3)_3Cl^+ + 2Cl^-$	+0.651	1 <u>M</u> NaCl	[3]
11.	$Pt(NH_3)_4Cl_2^{2+} + 2e = Pt(NH_3)_4^{2+} + 2Cl^-$	+0.619	1 <u>M</u> NaCl	[4]
12.	$Pt(NH_3)_6^{4+} + 2e = Pt(NH_3)_4^{2+} + 2(NH_3)$	+0.401	1 <u>M</u> NH_4Cl	[12]
13.	$Pt(NH_3)_2Br_2Cl_2 + 2e = Pt(NH_3)_2Br_2 + 2Cl^-$	+0.650	1 <u>M</u> NaCl	[2]
14.	$Pt(NH_3)_2(NO_2)_2Cl_2 + 2e = Pt(NH_3)_2(NO_2)_2 + 2Cl^-$	+0.707	1 <u>M</u> NaCl	[2]
15.	$Pt(NH_3)_2(SCN)_2Cl_2 + 2e = Pt(NH_3)_2(SCN)_2 + 2Cl^-$	+0.593	1 <u>M</u> NaCl	[2]
16.	cis-$Pt(NH_3NO_2)_2Br_2 + 2e =$ cis-$Pt(NH_3NO_2)_2 + 2Br^-$	+0.782	aq soln	[13]
17.	trans-$Pt(NH_3NO_2)_2Br_2 + 2e =$ trans-$Pt(NH_3NO_2)_2 + 2Br^-$	+0.786	aq soln	[13]
18.	$PtNH_3NO_2NH_3BrNO_2Br + 2e = PtNH_3NO_2NH_3Br + NO_2^- + Br^-$ (cis-assym)	+0.807	aq soln	[13]
19.	$Pt(CH_3NH_2)_4Cl_2^{2+} + 2e = Pt(CH_3NH_2)_4^{2+} + 2Cl^-$	+0.617	1 <u>M</u> NaCl	[6]
20.	$Pt(Et{-}NH_2)_4Cl_2^{2+} + 2e = Pt(Et{-}NH_2)_4^{2+} + 2Cl^-$	+0.681	1 <u>M</u> NaCl	[6]
		+0.678	1 <u>M</u> Na_2SO_4	
21.	$Pt(Pr{-}NH_2)_4Cl_2^{2+} + 2e = Pt(Pr{-}NH_2)_4^{2+} + 2Cl^-$	+0.712	1 <u>M</u> Na_2SO_4	[6]
22.	$Pb(Bu{-}NH_2)_4Cl_2^{2+} + 2e = Pt(Bu{-}NH_2)_4^{2+} + 2Cl^-$	+0.741	1 <u>M</u> Na_2SO_4	[6]

(continued)

TABLE 1.1.1 (continued)

Half-reaction	Standard or formal potential (V)	Conditions	Refs.
23. $Pt(OH{-}(CH_2)_2{-}NH_2)_4Cl_2^{2+} + 2e = Pt(OH{-}(CH_2)_2{-}NH_2)_4^{2+} + 2Cl^-$	+0.701	1 M NaCl	[12]
24. $Pt(OH{-}(CH_2)_3{-}NH_2)_4Cl_2^{2+} + 2e = Pt(OH{-}(CH_2)_3{-}NH_2)_4^{2+} + 2Cl^-$	+0.647	1 M NaCl	[12]
	+0.637	1 M HCl	
25. $Pt(py)_4Cl_2^{2+} + 2e = Pt(py)_4^{2+} + 2Cl^-$	+0.932	1 M NaCl	[5]
	+0.982	1 M H_2SO_4	
26. cis-$Pt(Py)_2(NH_3)_2Cl_2^{2+} + 2e =$ cis-$Pt(Py)_2(NH_3)_2^{2+} + 2Cl^-$	+0.776	1 M NaCl	[4]
27. trans-$Pt(Py)_2(NH_3)_2Cl_2^{2+} + 2e =$ trans-$Pt(Py)_2(NH_3)_2^{2+} + 2Cl^-$	+0.795	1 M NaCl	[4]
	+0.731	1 M HCl	
28. $Pt(\alpha\text{-Pic})_2(NH_3)_2Cl_2^{2+} + 2e = Pt(\alpha\text{-Pic})_2(NH_3)_2^{2+} + 2Cl^-$ [a]	+0.801	1 M NaCl	[12]
	+0.774	1 M HCl	
29. $Pt(\beta\text{-Pic})_4Cl_2^{2+} + 2e = Pt(\beta\text{-Pic})_4^{2+} + 2Cl^-$	+0.930	1 M NaCl	[5]
	+0.877	1 M HCl	
	+1.042	1 M H_2SO_4	

30.	$Pt(\beta\text{-}Pic)_2(NH_3)_2Cl_2^{2+} + 2e = Pt(\beta\text{-}Pic)_2(NH_3)_2^{2+} + 2Cl^-$	+0.726	1 M NaCl	[12]
31.	$Pt(\gamma\text{-}Pic)_4Cl_2^{2+} + 2e = Pt(\gamma\text{-}Pic)_4^{2+} + 2Cl^-$	+0.896	1 M NaCl	[5]
		+0.823	1 M HCl	
		+1.021	1 M H_2SO_4	
32.	$Pt(\gamma\text{-}Pic)_2(NH_3)_2Cl_2^{2+} + 2e = Pt(\gamma\text{-}Pic)_2(NH_3)_2^{2+} + 2Cl^-$	+0.713	1 M NaCl	[12]
33.	$Pt(quin)_2(NH_3)_2Cl_2^{2+} + 2e = Pt(quin)_2(NH_3)_2^{2+} + 2Cl^-$ [b]	+0.746	1 M NaCl	[12]
34.	$Pt(Phen)_2(NH_3)_2Cl_2^{2+} + 2e = Pt(Phen)_2(NH_3)_2^{2+} + 2Cl^-$ [c]	+0.718	1 M NaCl	[12]
35.	$Pt(An)_2(NH_3)_2Cl_2^{2+} + 2e = Pt(An)_2(NH_3)_2^{2+} + 2Cl^-$ [d]	+0.569	1 M NaCl	[12]
36.	$Pt(Tol\text{-}am)_2(NH_3)_2Cl_2^{2+} + 2e = Pt(Tol\text{-}am)_2(NH_3)_2^{2+} + 2Cl^-$ [e]	+0.648	1 M NaCl	[12]
37.	$Pt(en)Cl_4 + 2e = Pt(en)Cl_2 + 2Cl^-$	+0.654	1 M NaCl	[2]
38.	$Pt(en)_2Cl_2^{2+} + 2e = Pt(en)_2^{2+} + 2Cl^-$	+0.579	1 M NaCl	[6]

[a] Pic = picoline.

[b] quin = quinoline.

[c] Phen = phenanthroline.

[d] An = Aniline.

[e] Tol-am = benzylamine.

The introduction of different acid groups as ligands in the amine complexes increases the standard potential according to the order I < CN < SCN < Br < Cl < NO_2 in solutions of 1 M NaCl (systems 8 to 15).

The halonitrodiamine isomers of Pt complexes seem to be very sensitive to electromagnetic radiation; therefore, the potentials of these systems (16 to 18) were measured in the dark with solutions of the salts at 0.25×10^{-3} M concentrations, the cis-asymmetrical isomers being the more unstable [13]. The similarity of the standard potentials of these systems implies a very similar stability for the isomeric halonitroamine platinum complexes.

A gradual substitution of ammonia by alkylamine ligands leads to a rise in the oxidizing power of the compound as the length of the amine chain increases (systems 19 to 22). In contrast, long-chain amine alcohols stabilize Pt(IV) complexes (systems 23 and 24), although a higher number of alcohol ligands gives rise to higher potentials. Dkhara et al. [6, 12] argue through these results that the electron donor capacity of the amines is enhanced by the longer chain of the substituted amine, increasing in this way the electronic charge of the bonding N and hence favoring the transition of Pt(IV) to Pt(II). Amine alcohols can form rings with octahedral coordination for Pt(II) so that the complex will become relatively more stable in accordance with the type of ring linked. Because the five-membered ring is more stable than the six-membered cyclic structure, it would be expected that Pt(II) ethanol amine complexes would be more stable than the propanol amine compounds, and indeed, the redox potentials of the ethanol amines are higher than the propanol systems.

The redox potential series of Pt complexes with aryl ligands show higher values than those with alkyl ligands (systems 25 to 36). The position of substituents in the aryl ligand exerts an inductive effect on the donor capacity of this ligand and subsequently affects the stability of the Pt(IV) complexes, though no clear rule can be established [12]. The Pt(II) ions have more d electrons able to form π-metal-ligand bondings than the Pt(IV) ions; this capacity accounts for the higher stability of the Pt(II) complexes [12].

The molecular orbital treatment of square planar Pt(II) complexes reveals that the square planar configuration is stabilized to a greater extent by ligands with a π-orbital system (CN^-, SCN^-) than by those with no π-system (Cl^-, Br^-, H_2O, NH_3), due to influence of the low-lying π-levels of the unsaturated ligand [272-274]. Also, molecular orbital calculations on the kinetic trans-effect of σ-bonding ligands ($H_2O < NH_3 < Cl^- < H_2S < CH_3^- < PH_3 < H^-$) in square planar Pt(II) complexes [275] disclose an increase in the bonding strength of these ligands with the increasing π-character of the bonding. This is accompanied by a weakening of all the other bondings in the complex with a major effect on the trans-position to the director ligand.

These results allowed Hubbard and Lai [276] to explain in a qualitative way the decelerating influence of unsaturated ligands on the electrode reactions of Pt complexes as due to stabilization by metal-ligand π-back-bonding of the molecular orbitals directly involved in the electron transfer process.

Insertion of bidentate instead of monodentate ligands into platinum complexes lowers the oxidizing potential of Pt(IV) halocompounds systems (37 and 38). The increase in bidentate substituents further stabilizes the complex.

From this summary, we may state that in spite of tentative applications of the ligand field theory to relate at least in a qualitative way the standard potentials with the structure of the complexes, a comprehensive theory on this subject is still completely lacking.

1.2. NONAQUEOUS SOLVENTS

No data have as yet been found for electrode potentials in nonaqueous solvents.

1.3. FUSED SALTS

Research on electrode potentials of platinum or Pt redox systems in molten compounds is still very scant. The only system that has been widely investigated in different melts is the $Pt/PtCl_2$ electrode. This system is shown to be very reversible and to attain equilibrium very rapidly in molten chlorides. These qualities, along with its easy preparation, make this electrode highly commendable as a reference electrode [14]. Laitinen and coworkers have used this system extensively in their work in molten salts, although some difficulties encountered at temperatures above 500°C may be due to volatilization of the $PtCl_2$ [15]. The $Pt/PtCl_2$ system is easily prepared by anodic oxidation of a platinum wire in fused chlorides at a low current density (about 1 to 10 $\times$ 10^{-3} A/cm^2), the concentration of Pt^{2+} being carefully controlled and accurately known. Concentrations usually range from mole fractions of 10^{-2} to 10^{-5}. These mole fraction values are extrapolated to 1 M $PtCl_2$.

In Table 1.3.1, potentials of the $Pt/PtCl_2$ system are given for different fused salts, the compositions of which are expressed in M%. Standard potentials are referred mainly to Cl_2(graphite)/Cl^- or Ag/AgCl electrodes, except for systems 5 and 6 where other types of reference electrodes are indicated. Concentrations of Pt^{2+} are expressed in mole fraction. These potentials include the activity coefficients which are shown to be constant or equal to unity in the concentration range studied. Among the systems measured, the $Pt/PtCl_2$ potential in NaCl-KCl melts [16] assumes a value very near that calculated from thermodynamic data for the formation of $PtCl_2$, -0.170 V, at the same temperature (system 1); this points to an activity coefficient of 1 and the absence of complex formation. Potential values of other alkali chloride melts (systems 2 to 4, 7) show an appreciable effect of the activity coefficients [17] which is not explained by the influence of the solvent as given by Senderoff. The use of fused diamine chloride as solvent (system 5) allows the measurement of the Pt/$PtCl_2$ potential against the hydrogen reference electrode. The difference of normal potential values from those in the molten alkali chloride bath is ascribed by the author to the

TABLE 1.3.1. Electrode Potentials in Fused Salts

Fused salts	Electrode	Potential (V)	Reference electrode	Conditions	Refs.
1. NaCl-KCl (50:50 M%)	Pt/$PtCl_2$ (10^{-3} to 10^{-5} X_{PtCl_2})	-0.167	Cl_2(graphite)/Cl^-	727°C	[16]
		±0.008			
		-0.170		calcd	
2. $MgCl_2$-KCl (32.5:67.5 M%)	Pt/$PtCl_2$ (0.2 to 4×10^{-3} X_{PtCl_2})	+0.580	Ag/AgCl (1.24 to 5×10^{-3} X_{AgCl})	475°C	[17]
		-0.228	Cl_2(graphite)/Cl^-		
3. $MgCl_2$-NaCl-KCl (50:30:20 M%)	Pt/$PtCl_2$ (0.67 to 3.8×10^{-3} X_{PtCl_2})	-0.076	Cl_2(graphite)/Cl^-	475°C	[18]
4. KCl-LiCl (41:59 M%)	Pt/$PtCl_2$ (33.7×10^{-3} X_{PtCl_2})	-0.216	Cl_2(graphite)/Cl^-	450°C	[14]
		+0.637	Ag/AgCl (2.8 to 0.5×10^{-2} X_{AgCl})		
5. $(CH_3)_2NH_2Cl$	Pt/$PtCl_2$ (2 to 0.5×10^{-2} X_{PtCl_2})	+0.223	H_2(Pt)/H^+	180°C $E° = 0.597 - 8.25 \times 10^{-4}$ T	[19]
		+0.340	Ag/AgCl (2.8 to 0.5×10^{-2} X_{AgCl})		
6. NaOH + KOH (61.3:38.7 M%)	Pt/PtO	+1.8 ± 0.1	Na(1 M)/Na^+	227°C; E referred to M	[20]
7. $PtCl_2$	Pt/$PtCl_2$	-0.284	Cl_2(graphite)/Cl^-	475°C; calcd from thermodynamic data	[21]

different solvation of the Pt^{2+} ions. Finally, the Pt electrode is oxidized in molten alkali hydroxides (system 6), forming PtO and PtO_2 which are sparingly soluble in the melt. The standard potential of the Pt/PtO system was measured vs the Na (1 M)/Na^+ electrode.

2. VOLTAMMETRIC CHARACTERISTICS

2.1. POLAROGRAPHIC CHARACTERISTICS

2.1.1. Aqueous Solutions

The polarographic reduction of inorganic Pt(II) complexes in aqueous solutions revealed an anomalous behavior very similar to that observed in the reduction of other anions; i.e., they show a decrease in the diffusion current before the zero charge potential of mercury (-0.5 V_{SCE}), e.g., Fig. 2.1.1. This anomaly induced Frumkin to develop his theory to explain the polarographic reduction of some anions (persulfate and ferricyanides), but the polarographic behavior of platinum complexes is rather in contradiction with the Frumkin theory (the platinum complexes show minima at high concentrations of supporting electrolyte and so do even some uncharged complexes). Alternate explanations have been proposed for interpretation of these facts by taking the structure of the reacting complex into account [22].

The polarographic characteristics of some of the Pt compounds studied are listed in Table 2.1.1. The reduction of Pt(II) amine complexes [23] shows definite diffusion waves

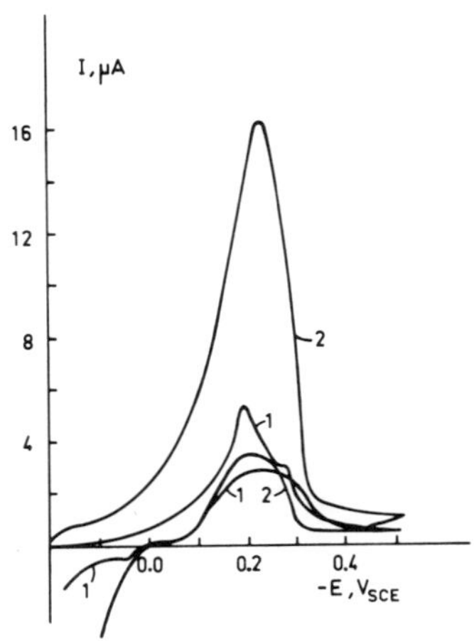

FIG. 2.1.1. Stationary electrode polarograms of a solution of 10^{-3} M cis-$Pt(NH_3)_2Cl_2$ in 1 M $NaClO_4$. Starting potential +0.2 V (SCE). Sweep rate: (1) 0.01 and (2) 0.1 V/sec. (From Ref. 32 by courtesy of the Finnish Chemical Society.)

TABLE 2.1.1. Polarographic Characteristics

Substance	Product	Conditions	$E_{1/2}$ (V vs SCE)	I	$E_{3/4} - E_{1/4}$	Remarks	Refs.
1. trans-$Pt(NH_3)_2(C_6H_5{-}NH_2)_2^{2+}$	Pt	0.1 M KCl	-0.670w	-	$D° = 7.71 \times 10^{-6}$ irr	Diffusion; 0.01% gelatin	[23]
2. cis-$Pt(NH_3)_2(C_6H_5{-}NH_2)_2^{2+}$	Pt	0.1 M KCl	-0.930w	-	irr	Diffusion; 0.01% gelatin	[23]
3. $Pt(py)_4^{2+}$	Pt	0.1 M KCl	-1.130w	-	irr	0.01% gelatin; diffusion	[23]
4. trans-$Pt(NH_3)_2Cl_2$	Pt	0.1 M KCl	-0.165w	-	irr	35.5°C; 0.005% gelatin	[24]
5. cis-$Pt(NH_3)_2Cl_2$	Pt	0.1 M KCl	-0.100w	-	irr	35.5°C; 0.005% gelatin	[24]
6. $Pt(en)Cl_2$	Pt	0.1 M KCl	-0.110w	-	irr	35.5°C; 0.005% gelatin	[24]
7. trans-$Pt(NH_3)_2(H_2O)_2^{2+}$	Pt	0.1 M $NaClO_4$ + 0.01 M $HClO_4$	-0.05i	-	irr	35.5°C; 0.005% gelatin	[24]
8. $Pt(o\text{-}phen)_2^{2+}$	A Hg surface compound	1 M NaCl	(-0.1)w	-	irr	Independent of pH and concn	[26]
$Pt(o\text{-}phen)^{2+}$	$Pt(o\text{-}phen)^{+}$	1 M NaCl	-0.43w	-	irr	Diffusion; independent of pH	
$Pt(o\text{-}phen)^{2+}$	Pt		-0.83i	-	irr		

9.	$Pt(en)_3^{4+}$	$Pt(en)_2^{2+}$	0.1 M KNO_3 or 0.1 M KCl	-0.2w -0.3w	3.07	irr $D° = 6.71 \times 10^{-6}$	Diffusion; two waves	[27]
10.	$Pt(NH_3)_6^{4+}$	$Pt(NH_3)_4^{2+}$	0.1 M KNO_3 or 0.1 M KCl	-0.18w -0.28w	3.42	irr $D° = 7.16 \times 10^{-6}$	Diffusion; two waves	[27]
11.	$Pt(en)_3^{4+}$	$Pt(en)_2^{2+}$	0.1 M KNO_3 + 1 M NH_4OH	-0.59w	3.06	irr	Diffusion	[27]
12.	$Pt(NH_3)_6^{4+}$	$Pt(NH_3)_4^{2+}$	0.1 M KNO_3 + 1 M NH_4OH	-0.54w	3.4	irr	Diffusion	[27]
13.	$Pt(NH_3)_5OH^{3+}$	$Pt(NH_3)_4^{2+}$	0.1 M KNO_3	-0.438w	3.56	irr $D° = 8.22 \times 10^{-6}$	Diffusion	[28]
			0.1 M KCl	-0.475w	3.58	irr	Diffusion	
			0.1 M KNO_3 + 1 M NH_4OH	-0.560w	3.6	irr	Diffusion	
14.	$Pt(NH_3)_5Cl^{3+}$	$Pt(NH_3)_4^{2+}$	0.1 M KNO_3	+0.055w	3.4	irr	Diffusion	[28]
15.	$(Pt(NH_3)_4NOCl)^{2+}$	$Pt(NH_3)_4^{2+}$	0.75 M $NaClO_4$	-0.97w	-	irr	Diffusion	[29]
	$(Pt(CH_3\text{-}NH_2)_4NOCl)^{2+}$	$Pt(CH_3-NH_2)_4^{2+}$	0.75 M $NaClO_4$	-0.97w	-	irr	Diffusion	[29]
	$(Pt(en)_2NOCl)^{2+}$	$Pt(en)_2^{2+}$	0.75 M $NaClO_4$	-0.97w	-	irr	Diffusion	[29]
16.	$Pt(en)_3^{4+}$	$Pt(en)_2^{2+}$	0.1 M ClK + 0.25 M borax buffer	-0.04w	-	irr	$(0 \text{ to } 2 \times 10^{-2})$ M 1,2-Cyclohex-andione-dioxime	[33]
	$Pt(en)_3^{4+}$	Pt	0.1 M ClK + 0.25 M borax buffer	-0.250w	-	irr	$(0 \text{ to } 2 \times 10^{-2})$ M 1,2-Cyclohex-andione-dioxime	[33]

after addition of gelatin only for cis- and trans-$Pt(NH_3)_2(C_6H_5—NH_2)_2^{2+}$ and pyridine complexes (systems 1 to 3). The evolution of H_2 from the electrolysis of water molecules interferes with the reduction of the other amine complexes investigated. As a rule, the trans-isomers are reduced more easily than the cis-isomers, pointing to a higher stability of the cis-compounds.

Polarograms of Pt(II) complexes with the cis- and trans-Cl^-, H_2O, OH^-, NH_3, and ethylenediamine ligands [24] are obtained only for a few compounds (systems 4 to 7) where the half-wave potentials for trans-isomers with two acid group ligands (Cl^-, NO_2^-, OH^-) are more negative than those for the cis-configuration.

Attempts have been made to detect the existence of unusual oxidation states of the platinum ion. Early reports [25] on the polarographic reduction of several amine and 1,10-phenanthroline Pt(II) complexes revealed that polarographic wave shapes were dependent on the type of ligand and on complex ion concentration. Only the $Pt(o\text{-}phen)_2^{2+}$ complex exhibited a two-step wave which was discussed as evidence of formation of an intermediate Pt(I) compound, but further investigations by EPR of the suggested Pt^+ intermediate were unsuccessful [26].

Research on the reduction of Pt(II) halides with a DME is rather scanty; Brand et al. [30] recorded two-step waves, very similar to those observed for the Pt(IV) halides, in the polarographic reduction of these halo complexes in pH 7 solutions. The authors, therefore, assumed that a disproportionation of the type

$$2PtCl_4^{2-} \rightleftharpoons Pt(0) + PtCl_6^{2-} + 2Cl^-$$

occurs which is catalyzed by the mercury electrode, the hexahalide being the species reduced in every case.

The extensive work of Sundholm [22] on the electrochemical reduction of some inorganic Pt(II) complexes, and particularly those of formula $Pt(NH_3)_{4-n}Cl_n^{2-n}$, on a Hg electrode dealt specially with the influence of the structure of the reacting complex on its electrochemical behavior and the applicability of the theories on anion reduction to the case of Pt(II) complexes. The results showed the easier reduction of the complexes containing chloride ligands, at a positively or slightly negatively charged Hg surface, and a different polarographic behavior for the two $Pt(NH_3)_2Cl_2$ isomers. The cis-configuration produced polarograms with a pronounced minimum which was not observed for the trans-isomer. The presence of surface-active agents suppressed the current in cis-compound solutions with no effect on those of trans-isomer.

The author explained the results by assuming a direct interaction between the unhydrolyzed species of the Pt complex and the electrode. The presence of chloride in the inner sphere of the complex would greatly facilitate the electrochemical reduction by the chloride ligands bridging from the complex to the electrode. These chloride ligands will act as a route for electron transfer between the electrode and the complex in the region before the zero charge potential of the mercury.

The difference in electrochemical behavior of the cis- and trans-$Pt(NH_3)_2Cl_2$ can be ascribed to the fact that the former isomer is a dipole with the chloride ligands at the negative end of the molecule while the trans-configuration has no dipole moment (see Section 3.1).

At a positively charged electrode surface the cis-isomer is adsorbed with the negative end toward the electrode, and the electron transfer occurs via chloride bridges. At increasing cathodic potentials the negative end of the molecule is repelled and the dipole has to reorient itself with the positive end pointing toward the electrode so that the electron transfer becomes hindered and a decrease in current is observed. On the other hand, the behavior of the trans-isomer can be explained if we take into account that the vacant p_z orbital of square planar Pt(II) complexes can accommodate two electrons donated by the electrode if the orientation of the molecule is favorable. Thus an alternate path for electron transfer from the electrode to the complex is offered in the reduction process by the vacant p_z orbital of the central Pt atom. Since the trans-isomer has no dipole moment, the planar molecule can be situated in a plane parallel to the positively charged electrode surface, and the electron transfer can occur both via chloride bridges and via the vacant orbital. As the electrode becomes more cathodic, the chloride ligands are repelled but electron transfer to the p_z orbital can proceed with no significant variation, and no decrease in current is recorded.

Further, the author tried to obtain some evidence on the mechanism controlling the reduction of Pt(II) complexes by interpretation of the transfer coefficient α in the light of the Marcus' theory of electron transfer. Measurements of the kinetic parameters yield low values of α (see Table 3.1.1) which, according to this theory, indicate an inner sphere mechanism for electron transfer as could be expected when bridging between the complex ligands and the electrode occurs. The author thus concluded that the low α-values observed for the Pt(II) complexes support the reduction mechanism via chloride bridges at a positively charged electrode surface.

Sundholm's view is similar to that developed by Hubbard, Lau, and Cushing [45, 52] for oxi-reduction of Pt complexes at Pt electrodes, and it could be further substantiated by application of the molecular orbital theory in the line attempted by Hubbard and Lai [276] for halide-bridged electrode reactions.

At higher negative potentials, beyond -1.3 V vs SCE, all the Pt(II) complexes studied are reduced, with a relative independence of the structure of the complex. As suggested by Frumkin and Nikolaeva [31], this could be explained by assuming that the planar structure of the complexes is distorted and a direct transfer of the electron to the central platinum atom occurs.

Finally, in the field of transition metal-nitrosyl compounds, polarography has been applied as a complementary technique to determine the structure of the probably hexacoordinated Pt(II) complex $[Pt(A)_4NOCl]^{2+}$, where A stands for different amine ligands [277]. Cathodic polarographs of this cation [29] in neutral or slightly alkaline media (system 15) yield irreversible diffusion waves which have been proved to correspond to reduction of the HNO_2 molecules formed by hydrolysis of the complex and not to reduction of Pt^{2+}, as reported in an earlier work [277]. The Pt(II) complexes of these ligands give no cathodic waves.

TABLE 2.1.2. Polarographic Characteristics in Nonaqueous Solvents

Substance	Product	Conditions	$E_{1/2}$	I	$E_{3/4} - E_{1/4}$	Remarks	Refs.
1. $Pt[S_2C_2(CN)_2]_2^-$	$Pt[S_2C_2(CN)_2]_2^{2-}$	Dimethylformamide + 0.1 M $LiClO_4$	0.243		rev	25°C; reference electrode, aq Ag/AgCl + 0.1 M LiCl	[34], [35]
2. $Pt[S_2C_2(CF_2)_2]_2$	$Pt[S_2C_2(CF_3)_2]_2^-$	Dimethylformamide + 0.1 M $LiClO_4$	0.852		rev	25°C reference electrode, aq Ag/AgCl + 0.1 M LiCl	[34], [35]
$Pt[S_2C_2(CF_3)_2]_2^-$	$Pt[S_2C_2(CF_3)_2]^{2-}$	Dimethylformamide + 0.1 M $LiClO_4$	-0.234		rev	25°C; reference electrode, aq Ag/AgCl + 0.1 M LiCl	[34], [35]
3. $Pt[S_2C_2(C_6H_5)_2]_2$	$Pt[S_2C_2(C_6H_5)_2]_2^-$	Dimethylformamide + 0.1 M $LiClO_4$	0.090		rev	$D = 2.27 \times 10^{-6}$	[34]
$Pt[S_2C_2(C_6H_5)_2]_2^-$	$Pt[S_2C_2(C_6H_5)_2]_2^{2-}$	Dimethylformamide + 0.1 M $LiClO_4$	-0.844		rev		
4. $Pt[S_2C_2(p\text{-}C_6H_4CH_3)_2]_2$	$Pt[S_2C_2(p\text{-}C_6H_4CH_3)_2]_2^-$		0.043		rev		[34]
$Pt[S_2C_2(p\text{-}C_6H_4CH_3)_2]_2^-$	$Pt[S_2C_2(p\text{-}C_6H_4CH_3)_2]_2^{2-}$		-0.900		rev		

5. $Pt[S_2C_2(p\text{-}C_6H_4OCH_3)_2]_2$	$Pt[S_2C_2(p\text{-}C_6H_4OCH_3)_2]_2^-$		0.004		rev	D = 5.3 × 10^{-6}	[34]
$Pt[S_2C_2(p\text{-}C_6H_4OCH_3)_2]_2^-$	$Pt[S_2C_2(p\text{-}C_6H_4OCH_3)_2]_2^{2-}$		-0.919		rev		
6. $Pt[S_2C_2(CH_3)_2]_2$	$Pt[S_2C_2(CH_3)_2]_2^-$		-0.133		rev		[34]
$Pt[S_2C_2(CH_3)_2]_2^-$	$Pt[S_2C_2(CH_3)_2]_2^{2-}$		-1.069		rev		
7. $Pt[C_6H_4(NH)_2]_2^{2+}$	$Pt[C_6H_4(NH)_2]_2^+$	DMSO + 0.05 M $(n\text{-}C_3H_7)_4NClO_4$	+0.78		rev	SCE	[36]
$Pt[C_6H_4(NH)_2]_2^+$	$Pt[C_6H_4(NH)_2]_2$		+0.23		rev		
$Pt[C_6H_4(NH)_2]_2$	$Pt[C_6H_4(NH)_2]_2^-$		-1.01	1	rev		
$Pt[C_6H_4(NH)_2]_2^-$	$Pt[C_6H_4(NH)_2]_2^{2-}$		-1.07	0.92	rev		
8. $(Ph_3P)_2Pt\cdot PhC{\equiv}CPh$	?	Dimethoxyethane + Bu_4NClO_4	-3.4 i		rev	Ag/10^{-3} M $AgClO_4$	[37]
9. $Pt(SacSac)_2$[a]	Pt(I)	Acetone + 0.1 M Et_4NClO_4	-1.0		52	Ag/AgCl + LiCl	[39]
$Pt(SacSac)_2$	Pt(0)		-1.50		72	Ag/AgCl + LiCl	

[a]SacSac = dithioacetylacetonate.

Few data on polarographic characteristics of the reduction of Pt(IV) complexes have been published. Reduction of ammonia and ethylenediamine complexes with a DME [27] shows irreversible diffusion waves (systems 9 to 12) with formation of Pt(II) complexes, no further reduction being observed. The authors discuss the small inflection appearing in the wave as corresponding to formation of a short-lived Pt^{3+} complex. Substitution of one ammonia molecule by a halide ion (Br^-, Cl^-) in the $Pt(NH_3)_6{}^{4+}$ complex shifts the half-wave potential to more positive values [28], whereas this potential becomes more negative if the ammonia ligand is replaced by the electronegative group OH^- (systems 13 and 14). This is in accordance with the results discussed above and with those obtained by voltammetry of Pt complexes [45, 52, 276].

2.1.2. Nonaqueous Solvents

In nonaqueous solvents, polarography is a very useful method to obtain a series of data that contributes to the knowledge of coordination compounds. The study of transition metal complexes with π-acceptor ligands includes some Pt compounds that share with that group the problems of ambiguous oxidation states. Thus polarographic reduction of planar $Pt(S_2C_2R_2)_2$ complexes in dimethylformamide [34] and acetonitrile [35] solutions (where R = CN^-, $CF_3{}^-$, C_6H_5, p-$C_6H_4CH_3$, and -CH_3) displays two waves that correspond respectively to one-electron transfers. These waves are reversible, diffusion controlled, and comply with the Heyrovsky equation. The $E_{1/2}$ values for both waves show a linear correlation with the inductive substituent constants.

Analogous investigations have been carried out to detect and isolate, if possible, the members of the series of compounds $Pt[o\text{-}C_6H_4(NH)_2]_2$ [36] which belong to the same type of electron-transfer complexes. The results show a five-membered series of bis chelate complexes in which the net charge ranges from -2 to +2; reversible polarographic waves were obtained in DMSO + 0.05 M $(n\text{-}C_3H_7)_4NClO_4$ solvents whose characteristics are given in Table 2.1.2 (system 7).

In the field of σ- or π-complexed organometallic derivatives, very few Pt compounds have been studied. Dessy, King, and Waldrop [37] describe the reduction of a π-arene complex, system 8, that exhibits an ill-defined wave in dimethoxyethane solvent.

Lastly, Sundholm [38] has repeated the polarographic reduction of $Pt(NH_3)_2Cl_2$ isomers and K_2PtCl_4 complexes in dimethylformamide to avoid the hydrolysis of the compounds. The waves for cis-$Pt(NH_3)_2Cl_2$ are very similar to those obtained in aqueous solutions. At potentials more positive than that of the electrocapillary maximum, the reduction is controlled by slow electron transfer; at more negative potentials, desorption occurs and the current decreases. Reduction of the trans-isomers shows no minimum and, therefore, no adsorption, as was observed in aqueous media. Conversely, K_2PtCl_4 complexes exhibit a kinetic wave at variance with the diffusion-controlled reduction recorded in water solutions. These results fall within the interpretation frame proposed above for polarographic reduction in aqueous solutions.

TABLE 2.2.1. Voltammetric Characteristics

Substance	Product	Conditions	Electrode	Technique	Potential (V vs SCE)	Rev	Remarks	Refs.
1. $Pt(NH_3)_4^{2+}$	trans-$Pt(NH_3)_4Br_2^{2+}$	0.1 M NaBr	rot Pt dsk	E swp	$E_{1/2}$ = (0.45)	irr	i_a depends on concn of the supplied electrolyte and electrode pretreatment	[40]
		0.1 M HBr			(0.50)	irr		
	trans-$Pt(NH_3)_4Cl_2^{2+}$	0.1 M NaCl			(0.49)	irr		
		0.1 M HCl			(0.56)	irr		
2. trans-$Pt(NH_3)_4Br_2^{2+}$	$Pt(NH_3)_4^{2+}$	0.1 M NaBr	rot Pt dsk	E swp	$E_{1/2}$ = 0.37	irr		[40]
		0.1 M NaBr (pH = 12)			0.20	irr		
		0.1 M $NaNO_3$			0.36	irr		
3. trans-$Pt(NH_3)_4Cl_2^{2+}$	$Pt(NH_3)_4^{2+}$	0.1 M NaCl	rot Pt dsk	E swp	$E_{1/2}$ = 0.31	irr		[40]
		0.1 M NaCl (pH = 12)			0.13	irr		
		0.1 M $NaNO_3$			0.22	irr		
4. $PtCl_6^{2-}$	Pt	0.1 M NaCl	Pt wr	E swp	$E_{1/2}$ = -0.41	irr		[44]
5. $PtCl_6^{2-}$	$PtCl_4^{2-}$	0.1 M Na_2SO_4	Pt wr	E swp	$E_{1/2}$ = -0.46	irr		[44]
6. $PtCl_6^{2-}$	Pt	0.1 M Na_2SO_4	Pt wr	E swp	$E_{1/2}$ = -0.66	irr		[44]
7. $PtCl_6^{2-}$	$PtCl_5^{2-}$	1 M HCl	black-Pt-wr	i stp	-	irr	$i\tau^{1/2}/Ac$ = not const	[41]
8. $PtCl_2$	Pt	Molten KCl-LiCl	Pt wr	Intermittent E swp		rev	450°C; D = 1.49 × 10^{-5} vs Ag/AgCl-KCl-LiCl	[42]
9. $PtCl_2$	Pt	Molten $MgCl_2$-NaCl-KCl	Pt wr	V stp		rev	475°C; vs Pt/Pt^{2+}	[43]

2.2. VOLTAMMETRIC CHARACTERISTICS

Some of the platinum complexes have been studied by voltammetric methods in order to avoid adsorption phenomena and other interferences inherent to conventional polarography. Thus Mason and Johnson [40] have investigated the oxidation of $Pt(NH_3)_4{}^{2+}$ complexes in neutral and acid solutions of Cl^- and Br^- with rotating disk electrodes to obtain irreversible and irreproducible waves that correspond to formation of trans-$Pt(NH_3)_4Br_2{}^{2+}$ and trans-$Pt(NH_3)_4Cl_2{}^{2+}$ isomers, no wave being observed in alkaline media (system 1). Reduction of these trans-halide isomers reproduced the Pt(II) ammonia complex in neutral and basic solutions (systems 2 and 3). Voltammetry in acid solutions produced complex curves that the authors ascribed to interferences from platinum oxide formation and dissolution. Discussion of the results underlines the absence of aquo ions, which suggests a previous adsorption of chloride ions on the Pt surface providing bridges for electron transfer during the oxidation process. This cannot happen in alkaline solutions and no wave is obtained. This interpretation is confirmed by further research on the kinetics of the electrooxidation and reduction of this complex (as will be seen later) and concurs with the explanation published by Sundholm on reduction of the halogeno-amine Pt(II) complexes. Voltammetric data are given in Table 2.2.1.

In an earlier paper, Lingane [41] studied the reduction of $PtCl_6{}^{2-}$ in HCl solution with Pt microelectrodes and could only obtain polarographic waves with Pt black electrodes. The oxidation and reduction of $PtCl_4{}^{2-}$ in the same medium proved to be irreversible and to correspond to a two-electron transfer process.

In molten salt baths, voltammetry [42] of $PtCl_2$ solutions in KCl-LiCl eutectic melts at 450°C with intermittently polarized Pt electrodes shows polarograms that are better described by the Kolthoff-Lingane equation, half-wave potentials changing linearly with the log of concentration. The diffusion coefficient is of the same order as those observed for $PtCl_4{}^{2-}$ in aqueous solutions. Similar results were obtained for solutions in a $MgCl_2$-NaCl-KCl eutectic at 475°C [43].

3. KINETIC PARAMETERS AND DOUBLE-LAYER PROPERTIES

3.1. KINETIC PARAMETERS

Little work has been reported on the kinetic parameters of electrochemical processes with Pt compounds. The few data published refer mainly to reduction and oxidation of halide and chloro-amine complexes. The electroreduction of chloro-complexes of Pt(II) and Pt(IV) was studied by Hubbard and Lau [45] applying the thin-layer voltammetry method. Kinetic parameters are collected in Table 3.1.1 (systems 1 and 2). The data obtained on Pt(IV) compounds suggest a ligand bridged charge transfer, whereas reduction of $PtCl_4{}^{2-}$ appears to require a direct interaction between the electrode and the entire reactant complex.

TABLE 3.1.1. Kinetic Parameters

Reaction	Conditions	Electrode	Technique	Rate constant	α	Remarks	Refs.
1. $PtCl_6^{2-} + 2e \rightarrow PtCl_4^{2-} + 2Cl^-$	0.01 M $HClO_4$ + $NaClO_4$ (ionic force = 1.01)	Pt rd	Thin layer V	$k° = 6.1 \times 10^{-6}$ cm/sec	$(\alpha n) = 0.29$	25°C	[45]
	0.01 M $HClO_4$ + $NaClO_4$ (ionic force = 1.01) +						
	0.01 M NaCl			$k° = 3.5 \times 10^{-6}$ cm/sec	$(\alpha n) = 0.37$		
	0.1 M NaCl			$k° = 1.5 \times 10^{-6}$ cm/sec	$(\alpha n) = 0.37$		
	0.3 M NaCl			$k° = 5.9 \times 10^{-7}$ cm/sec	$(\alpha n) = 0.45$		
	1 M NaCl			$k° = 5.7 \times 10^{-7}$ cm/sec	$(\alpha n) = 0.44$		
2. $PtCl_6^{2-} + 2e \rightleftarrows PtCl_4^{2-} + 2Cl^-$	3 M H_2SO_4 + 1 M NaCl	Pt, Rh, Ir wr	E stp	$j_0 = (2 \pm 0.1) \times 10^{-7}$ at $C_{Pt^{2+}} = C_{Pt^{4+}} = 5 \times 10^{-3}$ M	$\alpha_{Pt} = 0.43$ to 0.38 $(1 - \alpha)_{Pt} = 0.46$ to 0.55 $\alpha_{Rh} = 0.33$ $(1 - \alpha)_{Rh} = 0.43$ to 0.55 $\alpha_{Ir} = 0.33$ $(1 - \alpha)_{Ir} = 0.47$ to 0.59	5 to 80°C	[46], [47]

(continued)

TABLE 3.1.1 (continued)

Reaction	Conditions	Electrode	Technique	Rate constant	α	Remarks	Refs.
3. $PtCl_4^{2-} + 2e \rightarrow Pt + 4Cl^-$	0.01 M $HClO_4$ + $NaClO_4$ (ionic force = 1.01) +	Pt rd	Thin layer V	$k° = 8.1 \times 10^{-6}$ cm/sec	$(\alpha n) = 0.38$	25°C	[45]
	0.01 M $HClO_4$ + $NaClO_4$ (ionic force = 1.01) +						
	0.01 M NaCl			$k° = 3.4 \times 10^{-6}$ cm/sec	$(\alpha n) = 0.32$		
	0.1 M NaCl			$k° = 1.6 \times 10^{-7}$ cm/sec	$(\alpha n) = 0.32$		
	0.3 M NaCl			$k° \leq 10^{-9}$ cm/sec			
	1 M NaCl			$k° \leq 10^{-9}$ cm/sec			
4. $PtCl_4^{2-} + 2e \rightarrow Pt + 4Cl^-$	1 M $NaClO_4$	Hg hng drp	E swp	$k° = 6.5 \times 10^{-6}$ cm/sec at $C_{PtCl_4^{2-}} = 10^{-4}$ M	0.23		[48]
	1 M $NaClO_4$			$k° = 9.2 \times 10^{-6}$ cm/sec at $C_{PtCl_4^{2-}} = 5 \times 10^{-5}$ M	0.24		

	1 M $NaClO_4$ + 10^{-3} M NaOH			$k° = 4.7 \times 10^{-6}$ cm/sec at $C_{PtCl_4^{2-}} = 10^{-4}$ M	0.18		
5. $PtCl_4^{2-} \rightarrow PtCl_4(H_2O)_2 + 2e$	1 M $HClO_4$	Pt wr	Thin layer V	$k° = 1.45 \times 10^{-6}$ cm/sec	$(1 - \alpha)n = 0.5$		[52]
6. $PtBr_6^{2-} + 2e \rightleftarrows PtBr_4^{2-} + 2Br^-$	3 M H_2SO_4 + 1 M NaBr	Pt wr	E stp	$j_0 = (1.8 \pm 0.3) \times 10^{-6}$ at $C_{Pt^{2+}} = C_{Pt^{4+}} = 5 \times 10^{-3}$ M	0.40 $(1 - \alpha) = 0.43$	25°C; $D_{PtBr_4^{2-}} = 4.92 \times 10^{-6}$ $D_{PtBr_6^{2-}} = 3.42 \times 10^{-6}$	[11]
7. $PtI_6^{2-} + 2e \rightleftarrows PtI_4^{2-} + 2I^-$	3 M H_2SO_4 + 1 M NaI	Pt wr	E stp	$j_0 = 4.3 \times 10^{-4}$ at $C_{Pt^{2+}} = C_{Pt^{4+}} = 5 \times 10^{-3}$ M		25°C; $D_{PtI_4^{2-}} = 3.03 \times 10^{-6}$	[10]
8. trans-$Pt(NH_3)_2Cl_2$ + $2e \rightarrow Pt + 2NH_3 + 2Cl^-$	1 M $NaClO_4$	Hg hng drp	E swp	$k° = 5.3 \times 10^{-4}$ cm/sec at $C = 5 \times 10^{-4}$ M	0.12 ± 0.02		[32]
				$k° = 3.5 \times 10^{-4}$ cm/sec at $C = 10^{-3}$ M	0.11 ± 0.01		
				$k° = 1.6 \times 10^{-4}$ cm/sec at $C = 1.5 \times 10^{-3}$ M	0.13 ± 0.01		

(continued)

TABLE 3.1.1 (continued)

Reaction	Conditions	Electrode	Technique	Rate constant	α	Remarks	Refs.
9. cis-$Pt(NH_3)_2Cl_2$ + 2e → Pt + $2NH_3$ + $2Cl^-$	1 M $NaClO_4$	Hg hng drp	E swp		0.14		[32]
10. $Pt(NH_3)_4^{2+}$ → $Pt(NH_3)_4Cl_2^{2+}$ + 2e	1 M $HClO_4$ + 0.01 M Cl^-	Pt wr	Thin layer V	$k° \geq 10^{-4}$ cm/sec		a	[52]
11. $Pt(dien)NO_2^+$ → $Pt(dien)NO_2Cl_2^+$ = 2e[b]	1 M $HClO_4$ + 0.01 M Cl^-	Pt wr	Thin layer V	$k° = 2 \times 10^{-5}$ cm/sec	$(1 - \alpha)n = 0.4$		[52]
12. cis-$Pt(NH_3)_2(NO_2)_2$ → $Pt(NH_3)_2(NO_2)_2Cl_2$ + 2e	1 M $HClO_4$ + 1 M Cl^-	Pt wr		$k° = 2 \times 10^{-6}$ cm/sec	$(1 - \alpha)n = 0.39$		[52]
13. cis-$Pt(NH_3)_2Cl_2$ → $Pt(NH_3)_2(H_2O)_2Cl_2^{2+}$ + 2e	1 M $HClO_4$	Pt wr	Thin layer V	$k° = 1.3 \times 10^{-5}$ cm/sec	$(1 - \alpha)n = 0.37$		[52]

	cis-$Pt(NH_3)_2Cl_2 \rightarrow Pt(NH_3)_2(H_2O)Cl_3^+ + 2e$	1 M $HClO_4$ + 2 × 10^{-3} M Cl^-	$k° \geq 10^{-3}$ cm/sec		a	
	cis-$Pt(NH_3)_2Cl_2 \rightarrow Pt(NH_3)_2Cl_4 + 2e$	1 M $HClO_4$ + 1 M Cl^-	$k° \geq 10^{-4}$ cm/sec		a	
14.	trans-$Pt(NH_3)_2Cl_2 \rightarrow Pt(NH_3)_2(H_2O)Cl_3^+ + 2e$	1 M $HClO_4$ + 2 × 10^{-3} M Cl^-	$k° \geq 10^{-3}$ cm/sec		a	[52]
	trans-$Pt(NH_3)_2Cl_2 \rightarrow$ trans-$Pt(NH_3)_2Cl_4 + 2e$	1 M $HClO_4$ + 1 M Cl^-	$k° \geq 10^{-5}$ cm/sec		a	
15.	$Pt(NO_2)_4^{2-} \rightarrow Pt(NO_2)_4Cl_2^{2-} + 2e$	1 M HClO + 1 M Cl^-	$k = 1.2 \times 10^{-6}$ cm/sec	$(1 - \alpha)n = 0.66$		[52]

[a]Too large rates to determine kinetic parameters by this method.

[b]dien = diethylenetriamine.

The system $PtCl_6^{2-}/PtCl_4^{2-}$ has also been studied by Kravtsov and Simakov [46, 47] with Pt, Ir, and Rh electrodes; the kinetic parameters obtained are given in Table 3.1.1 for system 3. It was observed that Cl^- ions were adsorbed at the electrode, hindering the adsorption of oxygen, this phenomenon being stronger at Pt and weaker at Ir. This adsorption facilitates the electron discharge through bridging by the adsorbed Cl^-. The authors propose a mechanism with $PtCl_5^-$ participation in the determining step. However, Hubbard and Lau [45] contend that previous $PtCl_5^-$ formation, in the absence of intense UV irradiation, just by hydrolysis or by dissociation of one Cl^-, is too slow a reaction.

The $PtCl_4^{2-}$ reduction was also studied by Sundholm [48] with the hanging mercury drop technique. It was shown that this anion is not strongly adsorbed at positive potentials and seems also to be reduced via chloride bridges.

Earlier papers reported on reduction and oxidation of bromide [49] and iodide [9] platinum complexes in halide solutions by thin-layer voltammetry. The results revealed that there is a reduction to Pt^{2+} and oxidation of this ion to Pt^{4+} by discharge of two electrons. Further reduction of Pt^{2+} to Pt metal was also observed in both solutions. These results were confirmed by Kravtsov and Simakov [10, 11] who determined the kinetic parameters of the redox process (systems 5 and 6).

Recently [50], Kravtsov and Simakov in additional communications discussed the adsorption of these complexes at the electrode, considering that adsorption through the ligands facilitates the reduction to the metal and that the adsorption strength depends upon the bond formed between the ion and the electrode metal. This manifests itself through increases of the reduction rate. Thus, in the $PtCl_6^{2-}/PtCl_4^{2-}$ system studied with Pt, Ir, and Rh electrodes, the mechanism depends upon the specific adsorption of the ligand.

Kinetic parameters for the electrochemical reduction and oxidation of $Pt(NH_3)_{4-n}X_n^{2-n}$ complexes, where X = NO_2^- or Cl^- and n = 0 to 4, have been studied by several workers. Sundholm [32] investigated the reduction of trans- and cis-$Pt(NH_3)_2Cl_2$ complexes with a stationary mercury electrode (systems 7 and 8); the trans-isomer appeared very weakly adsorbed whereas the cis-compound was strongly adsorbed from a freshly prepared solution. In hydrolyzed solution the interpretation of results underlines the different dipole moments of the isomers. The Cl^- ligands, which form the negative pole in cis-isomers, are adsorbed at the positively charged electrode surface (<-0.5 V) and the electron transfer occurs through the bridged ligands. In the trans-complex the planar molecule is placed parallel to the electrode, and the electron transfer takes place by filling the p_z orbital of the platinum ion. The difference in polarographic behavior of cis- and trans-$Pt(NH_3)_2Cl_2$ should thus be due to a reorientation effect of the dipole moment of the cis-isomer on increasing the potential to more negative values.

The rate of reduction of the anion $Pt(NH_3)Cl_3^-$ has been studied by conventional polarography as a case of anion reduction [51].

Additional work has been published on electrooxidation of Pt(II) complexes of this type, using thin-layer voltammetry [52]. The authors pointed out that in the oxidation of $PtCl_4^{2-}$ ions in chloride solutions, not only $PtCl_6^{2-}$ is produced, as reported by Simakov and Kravtsov [46], but a mixture of this ion and trans-$Pt(H_2O)_2Cl_4$.

The study of kinetic parameters for the electrooxidation of the Pt complexes mentioned above in chloride-free and chloride solutions has shed some light on the type of mechanism taking place during the electrochemical process [45, 52]. The results obtained in this way (systems 10 to 16) show that the electrooxidation is sharply accelerated by free chloride and that inclusion of the Cl^- ion in Pt(IV) complex occurs. In halide-free solutions the reactivity of the complex decreases greatly but is higher with the increase of the number of chloride ligands in the coordination sphere. It seems that the two entering ligands have the trans-configuration and the other four ligands retain their original position. Oxidation in chloride media of positively charged Pt(II) complexes leads to inclusion of two chlorides, whereas oxidation of negative complexes tends to inclusion of one chloride and one water molecule. The authors can only conclude that the electrooxidation proceeds through an intermediate complex adsorbed at the electrode surface through a bridging halogen, but not which is the halogen involved. A rate-limiting charge transfer of one electron followed by a rapid transfer of a second electron is suggested for the oxidation process on the basis of the kinetic results.

Kinetic parameters of the reaction $Pt \rightleftharpoons Pt^{2+} + 2e$ in molten eutectic KCl-LiCl at 400°C were studied by the impedance method [53]. The exchange current obtained was higher (system 1) than that evaluated by Laitinen et al. [54] (system 2) for the same electrochemical system at 450°C by applying the double pulse galvanostatic method. The latter authors stress the fact that certain features of the results are not in accordance with the simple model of a one-step charge transfer mechanism. Further research on specific adsorption or prior dissociation as a step in the main reaction was suggested by the authors in order to reach definite conclusions on the mechanism. Application of a refined galvanostatic current step technique [55] to the same electrochemical system at 450°C (system 3) allows measurements to be made at potentials other than the equilibrium redox potential; preliminary results are shown in Table 3.1.2.

TABLE 3.1.2. Kinetic Parameters in Fused Salts

Reaction	Conditions	Electrode	Technique	Rate constant	α	Remarks	Refs.
1. $Pt^{2+} + 2e \rightleftharpoons Pt$	LiCl-KCl	Pt wr	farad. imp	$j_0 = 37$ at $C_{Pt^{2+}} = 10^{-2}$ M		400°C	[53]
2. $Pt^{2+} + 2e \rightleftharpoons Pt$	LiCl-KCl	Pt wr	double i stp	$k° = 0.03$ cm/sec	0.27	450°C	[54]
3. $Pt^{2+} + 2e \rightleftharpoons Pt$	LiCl-KCl	Pt wr	i stp	$j_0 = 0.14$ at $C_{Pt^{2+}} = 1.1 \times 10^{-3}$ M	0.35	450°C $D = 9.1 \times 10^{-6}$	[55]
				$j_0 = 0.19$ at $C_{Pt^{2+}} = 1.85 \times 10^{-3}$ M	0.35	$D = 6.2 \times 10^{-6}$	
				$j_0 = 0.26$ at $C_{Pt^{2+}} = 3.02 \times 10^{-3}$ M	0.35	$D = 5.7 \times 10^{-6}$	

3.2. DOUBLE-LAYER PROPERTIES AND ADSORPTION PHENOMENA

3.2.1. Double-Layer Capacity and Potential of Zero Charge

No detailed picture of the double-layer structure on Pt has so far been elaborated comparable to that on the Hg electrode.

The surface heterogeneity of solid metals and the nonelectrostatic interactions between the metal and the constituents of the solution introduce new aspects in the treatment of the Pt double layer with respect to that of Hg. The problem is complicated by oxygen and hydrogen adsorption and the leakage of electric charges in the so-called double-layer region which make the Pt a nonideally polarized electrode in aqueous solutions.

The experimental data needed to characterize the double layer on the Pt electrode (double-layer capacity, potential of zero charge, amount of specifically adsorbed species, and interfacial tension) have been investigated to varying extents. However, lack of agreement among the values of the double-layer capacity and potential of zero charge has introduced a large degree of uncertainty about the structure of the Pt double layer. Values of these parameters differ widely according to the medium in which they are determined and are often affected by the experimental method of measurement. Thus capacitance values range from 70 $\mu F/cm^2$, obtained by charging curves, to 20 $\mu F/cm^2$ by impedance measurements in acidified solutions, or even lower values in the presence of I^- ion. Values near 40 $\mu F/cm^2$ are yielded by charging curves in 1 $\underline{N}$ HCl [56] and by impedance measurements in 1 $\underline{N}$ $HClO_4$ [64]. As for the zero charge potential (p.z.c.), we may quote for the same solutions, $HClO_4 + NaClO_4$, and by the same experimental method, the diverging values 0.1 V [63] and 0.56 V (NHE) [65] (see Table 3.2.1).

A review of the methods employed for determination of p.z.c. on solid metals is given by Andersen and Perkins [62], and an evaluation of these methods is done by Bockris et al. [72]. These authors select as more suitable the methods of capacitance, static friction, immersion, and tension vibration, while Frumkin et al. [59] favor the adsorption method for Pt and Trasatti [66] considers the open-circuit scrape as more reliable.

A broad controversy on the validity of these values for Pt in aqueous solutions, especially for those obtained by capacitance-potential measurements in dilute solutions, has been carried out among the Russian school, Bockris et al., and other investigators without reaching a final consensus [59, 62-66, 71-73]. As Frumkin noted in earlier work [56, 285], an unusual difficulty in measurements of double-layer capacitance by the impedance method is the frequency dependence of the measured capacitance which yields lower values, between 14 and 20 $\mu F/cm^2$, at higher frequencies. This effect can be minimized by using electrodes with smooth surfaces or single crystals but is never completely eliminated. The residual frequency dependence and the strong influence of the solution pH on the capacity-potential curves for Pt were associated by the author to the inclusion in the electrode capacitance of a faradaic component that depends on the exchange current of the reaction $H \rightleftharpoons H^+ + e$ of the hydrogen adsorbed on the electrode surface.

Frumkin et al., in their thermodynamic treatment of the Pt/H_2 electrode [56, 59], consider that the point of zero charge for Pt in aqueous solutions does not coincide with the maximum of the electrocapillary curve since the interface potential depends not only on the metallic surface charge but also on the amount of hydrogen adsorbed. Therefore, discrepancies among p.z.c. estimated from minimum capacity values and those from other methods can be attributed to intrusion of a pseudocapacitance arising from the adsorption and subsequent partial ionization of hydrogen or oxygen. Following Frumkin's theory, a relation between the p.z.c. and the solution pH should be expected owing to the polarity of the H—Pt bond. Consequently, the Russian school chose as more dependable the values produced mainly by the adsorption method and took the value 0.16 to 0.19 V (NHE) as the more correct p.z.c. for Pt in aqueous solutions [59]. This potential shifts to more negative or positive values by anion or cation adsorption, respectively [286, 288, 289].

However, these results are contested by Bockris et al. [65] who found by the impedance method a higher p.z.c. for Pt (0.565 ± 0.025 V) and a linear relationship between p.z.c. and solution pH (p.z.c. $= 0.56 - 2.3 \frac{RT}{F}$pH) which is not accounted for by Frumkin's theory. Moreover, they showed that the p.z.c. becomes more negative when hydrogen is dissolved in the electrode metal. Hence, they claimed, the p.z.c. must be very sensitive to electrode pretreatment and the low p.z.c. values are due to "hydrogen-loaded" electrodes. Confirmation of these results by application of the friction and organic adsorption methods to Pt and other metals [72] led Bockris et al. to the conclusion that no pseudocapacity was involved in their capacitance measurements and that the p.z.c.-pH linear relationship is independent of hydrogen adsorption.

These conclusions have been strongly criticized by the Russian school [59, 286] on both theoretical and experimental grounds. Theoretically, the p.z.c. is too high and cannot be explained by any theory. As to the experimental work, most probably the pretreatment technique to remove hydrogen completely from the electrode had little effect on eliminating oxygen from inside the metal; hence the results were likely obtained with an oxidized electrode. Some other results can be advanced to support Frumkin's views: Balashova et al. [73] found a p.z.c. value of 0.18 ± 0.025 V by the ionic adsorption method for dilute Na_2SO_4 and Cs_2SO_4 solutions and no pH dependence in the interval from 2 to 5. The difference between Pt and Hg p.z.c. in HCl and H_2SO_4 solutions (0.29 and 0.37 V, respectively) should be equal to the difference in work function of the metals, but experimental results for Pt (4.65 V) and Hg (4.5) show this difference to be much lower [63, 67]; this discrepancy can be ascribed to interference of the hydrogen adsorption in the potential measurements. Impedance measurements of the zero charge point of Pt in dilute solutions of $NaClO_4 + HClO_4$ [63] give a capacity minimum of 20 $\mu F/cm^2$ at 0.1 V (NHE), and no change of the potential is observed in the pH range from 2 to 3. Schuldiner et al. [287] have obtained potential-capacity curves for Pt in H_2SO_4 solutions at concentration 1 to 3.6×10^{-4} $\underline{M}$ by the charge-step method. The curves exhibit a capacity minimum of 10 $\mu F/cm^2$ at 0.23 V (NHE) and concentrations lower than 10^{-3} $\underline{M}$, which the authors assign to the potential of maximum

TABLE 3.2.1. Zero Charge Potentials for Pt in Aqueous Inorganic Salt Solutions with Respect to NHE

p.z.c.	Solution	Method	Refs.
1. 0.11 (PtH)	0.1 N Na_2SO_4 + 0.01 N H_2SO_4	Adsorption	[62],[71]
2. 0.11 (platinized)	1 N Na_2SO_4 (acidified solutions)	Adsorption	[61]
3. 0.18 (platinized)	0.001 N Na_2SO_4 + 0.001 N H_2SO_4	Adsorption	[62]
4. 0.28	1 N Na_2SO_4 (acid)	Contact angle	[62]
5. 0.48 (PtH)	0.02 N Na_2SO_4 (pH = 2)	Capacity	[62]
6. -0.21 (PtH)	0.02 N Na_2SO_4 (pH = 7)	Capacity	[62]
7. 0.22	0.1 N Na_2SO_4	Capacity	[62]
8. 0.05 (smooth)	0.1 N K_2SO_4 (pH = 7)	Open circuit scrape	[62]
0.02	1 N $NaClO_4$ or 1 N K_2SO_4 (pH = 7)	Open circuit scrape	[70]
9. 0.12 (PtH)	0.000025 M H_2SO_4	Electrode displacement	[62]
10. 0.35	0.01 N H_2SO_4; 0.005 N H_2SO_4	Capacity	[62]
11. +0.4 to +1.0 (platinized)	Varying electrolytes	Capacity	[61]
12. 0.21 ± 0.01	0.001 N H_2SO_4	Acoustical electrochemical	[61]
13. 0.18	0.01 N H_2SO_4; 0.001 N H_2SO_4	Capacity	[62]
14. 0.3	0.1 N H_2SO_4	Friction	[62]
15. 0.14 (platinized)	0.001 N NaCl + 0.001 N HCl	Adsorption	[62]
16. 0.10	0.001 N NaCl + 0.001 N HCl	Adsorption	[59]
17. 0.06	1 N NaCl + 0.01 N HCl	Changes in H^+ by adsorption	[59]
18. 0.19	0.00002 N HCl	Electrode displacement	[59]
19. 0.47	1 N HCl	Capacity	[62]
20. -0.40 ± 0.05	0.1 M KCl (pH = 12)	Very high-speed rotary scraper	[61]
21. 0.2 (smooth)	0.01, 0.001 N KCl; 0.00001, 0.0003, 0.001 N $MgSO_4$; 0.001 N H_2SO_4; 0.001 N $ZnCl_2$	Double-layer repulsion	[62]

(continued)

TABLE 3.2.1 (continued)

p.z.c.	Solution	Method	Refs.
22. 0.20	Varying electrolytes	Crossed polarized metallic threads	[61]
23. 0.04 (platinized)	0.001 N NaBr + 0.001 N H_2SO_4	Adsorption	[62]
24. -0.26 (platinized)	1 N NaBr + 0.05 N NaOH	Adsorption	[62]
25. -0.02 (platinized)	1 N NaBr + 0.01 N HBr	Adsorption	[62]
26. -0.5 (platinized)	0.005 N NaI + 0.005 N H_2SO_4	Adsorption	[62]
27. -0.51 (smooth)	0.01 N KI	Open circuit scrape	[62]
28. 0.18 (platinized)	0.001 N NaF + 0.001 N H_2SO_4	Adsorption	[62]
29. 0.18 (platinized)	0.005 N Cs_2SO_4 (pH = 2.3)	Adsorption	[62]
30. 0.17 (platinized)	0.01 N Cs_2SO_4 + 0.01 N H_2SO_4	Adsorption	[62]
31. -0.58	0.01 N CsI + 0.01 N CsOH	Adsorption	[59]
32. 0.65 (platinized)	0.01 N $CdSO_4$ + 0.1 N H_2SO_4	Adsorption	[62]
33. 0.45 (platinized)	0.01 N $CdBr_2$ + 0.1 N H_2SO_4	Adsorption	[62]
34. 0.01 (platinized)	0.01 N CdI_2 + 0.1 N H_2SO_4	Adsorption	[62]
35. 0.41 (smooth)	0.003 M ($HClO_4$ + NaOH) (pH = 2.8)	Capacity	[62]
36. 0.072 (smooth)	0.003 M ($NaClO_4$ + NaOH) (pH = 8.1)	Capacity	[62]
37. -0.111	0.003 M ($NaClO_4$ + NaOH) (pH = 11.2)	Capacity	[62]
38. 0.1	$HClO_4$ + $NaClO_4$	Capacity	[61]
39. 0.565 ± 0.025	$HClO_4$ + $NaClO_4$; NaOH + $NaClO_4$	Capacity	[61]
40. 0.04 (platinized)	1 N KCl + 0.01 N HCl	Adsorption	[71]
41. -0.03 (platinized)	1 N KBr + 0.01 N HBr	Adsorption	[71], [288]
42. -0.39 (platinized)	1 N KBr + 0.01 N KOH	Adsorption	[71], [288]
43. 0.56 (± 0.07) - 0.067 pH (smooth)	0.003 M ($NaClO_4$ + $HClO_4$)	Adsorption	[72]
	($NaClO_4$ + NaOH)	Capacitance and friction	
44. -0.83 smooth vs $Pt/PtCl_2$ (1 M)	KCl-LiCl eutectic (450°C)	Capacity	[68], [69], [55]

diffusivity of the double layer. This p.z.c. is near to that proposed by Frumkin and shows no variation with solution pH. The value of the minimum capacity increases with acid concentration.

On the other hand, other investigators [62, 64] have related the variations of the differential capacity to the impurity content of solutions and to pretreatment and heterogeneity of the electrode. Thus Formaro and Trasatti [64] contend from their impedance measurements that Pt electrodes are free of adsorbed hydrogen and oxygen only in the narrow potential range of 400 to 500 mV (RHE) where they found a double-layer capacitance of about 34 $\mu F/cm^2$ (real surface area) in 1 N $HClO_4$. This value decreases to 18 $\mu F/cm^2$ with increasing contamination of the solution.

We may finish this debate with the conclusion that the question of the real values of double-layer capacity and p.z.c. is still open although the discussion has clearly proved that pretreatment of the electrode and impurities in the solution strongly affect the capacity values. As the authors emphasize, the p.z.c. discrepancies can be attributed in part to different pretreatments with different investigation aims [72, 286], since the Russian school tried to obtain a hydrogen-equilibrated Pt electrode whereas Bockris et al. attempted to prepare a non-activated Pt electrode non-equilibrated with hydrogen. A more detailed discussion of the problem will be found in the quoted references. Zero charge potentials for Pt in different solutions and several experimental methods collected from references [59, 61, 62] and other original communications are given in Table 3.2.1.

The more consistent view of the double-layer structure on Pt has been advanced by Frumkin et al. and is derived from their theory of the Pt/H_2 electrode [56-59, 284-286]. These authors consider the surface on the Pt/hydrogen electrode in acid solution as partially covered by atomic hydrogen which is in equilibrium with H^+. In this way this surface can be envisaged as an equilibrium system amenable to thermodynamic treatment (on the other hand, the exchange current of hydrogen ionization is low enough for the electrode to be considered a quasi-ideally polarized electrode [284]).

This approach assumes that there is no chemisorption of water molecules or specific adsorption of H^+ on the electrode; the remaining ions of the solution retain their charges when adsorbed, and their concentration in the bulk solution largely predominates over that of H^+. The electrode electronic charge is identical to the H^+ ion surface density. A fundamental relation in this theory is $\Gamma_H = A_H - \varepsilon$, where Γ_H is, in the Gibbs sense, the amount of hydrogen to be added to the system for the composition of the bulk phase to remain constant when the surface area undergoes one unit increase; A_H is the fraction of this hydrogen that remains adsorbed on the Pt surface; and ε is the free electronic charge per unit surface that is given up by the fraction of ionized hydrogen. Then $-\Gamma_H$ is the electricity stored per unit electrode surface and represents the total charge of this unit surface; that is, the electronic charge equivalent to the density charge in the Lippman equation. Therefore, two potentials of zero charge can be observed on Pt, that of zero-free charge $\varepsilon = 0$ and that of $\Gamma_H = 0$; i.e., when no hydrogen remains adsorbed on the electrode surface [57, 284]. Thus thermodynamic relationships have been developed that describe the contributions of the adsorbed H and H^+ to the potential difference at the Pt/solution interface and explain, at least qualitatively, the experimental results on the double-layer effect of ion adsorption on hydrogen adsorption and kinetic behavior on Pt group metals [58]. Similar equations have been deduced for alkaline solutions and Pt electrodes that adsorb oxygen [56-59].

Frumkin's theory allows calculation of the p.z.c. dependence on the solution pH, and fair agreement between theoretical and experimental results has been observed for solutions with a large excess of neutral salt with respect to the acid or alkali concentration [71, 286, 288]. The results show the p.z.c. to shift linearly to more negative values with increasing pH (from 0 to 5), the slope of the lines depending on the solution composition (H_2SO_4 + Na_2SO_4, HCl + KCl, HBr + KBr) and the nature of the metal (Pt, Rh). The p.z.c. shift per pH unit is small and much lower than that reported by Bockris et al., 67 mV [72], except for Rh in sulfate solutions where it is about 45 mV [286, 288]. The authors stress that for oxygen- or hydrogen-adsorbing metals and those of the Pt group, the potential of zero total charge ($\Gamma_H = 0$) is more fundamentally characteristic than the potential of zero free charge ($\varepsilon = 0$), since the former does not depend on the structural model of the metal-solution interface and corresponds to the maximum of the electrocapillary curve on the electrode. Besides, in alkaline solutions where Γ_H^+ does not become zero at any potential, the point of zero free charge ($\varepsilon = 0$) cannot be measured and is meaningless [71, 284]. Both potentials of zero charge vary with the pH of the solution, but in solutions with anions that are weakly adsorbed on the electrode, the change of the potential of zero total charge is more notable. Values of both potentials for different solutions are given in Table 3.2.2.

TABLE 3.2.2. Potentials of Zero Charge at Platinized Pt [286]

Potential of zero free charge ($\varepsilon = 0$)	Potential of zero total charge ($\Gamma_H = 0$)	Solutions
0.16	0.23	1 N Na_2SO_4 + 0.01 N H_2SO_4
-	-0.25	1 N Na_2SO_4 + 0.01 N NaOH
-	-0.30	1 N KCl + 0.01 N KOH
0.04	0.14	1 N KCl + 0.01 N HCl
-0.03	0.06	1 N KBr + 0.01 N HBr
-0.39	-0.33	1 N KBr + 0.01 N KOH

An alternative nonqualitative model of the double layer on Pt derived from results on water adsorption on Pt in the gas phase is proposed by Müller [60]. The splitting of the water molecule into H^+ and hydroxide radicals occurs when it is adsorbed by the electrode. The author assumes a first layer is formed by chemisorption of the dissociated water molecule with partially delocalized bonds to the electrode metal, according to Schultze's results [125]. A special second layer would be made up of water molecules linked to the first layer by hydrogen bonds; this layer would include the specifically adsorbed ions if present (inner Helmholtz plane). Chemisorbed anions would be directly bound to the metal along with the water radicals. Cations not adsorbed would build up the diffusive part of the double layer. As indicated above, the chemisorbed layer is supposed to be polarized with a certain mobility of the electrical charges. With this model, the author tried to explain qualitatively the anodic and cathodic behavior of the Pt electrode together with the adsorption phenomena of ions and organic molecules.

A few papers have been published on p.z.c. values at Pt in molten salts [55, 62, 68, 69]. Measurements of capacity values vs potential in fused KCl-LiCl eutectic show a minimum which is attributed to the p.z.c. [68, 69]. As observed for double-layer capacities in molten salts at liquid metals, the capacity increases with the temperature [68]. These phenomena may be interpreted by the theory suggested by Frumkin et al. [74] who visualize the double layer in terms of alternate layers of positive and negative charges several ionic layers deep. This structure avoids the ideas of Helmholtz planes and a diffuse double layer which are not in agreement with the experimental facts. Inclusion in the double layer structure of cation and anion vacancies as a consequence of the quasi-crystalline structure of the melt accounts for the rise in capacity value at increasing temperature as due to a more compact distribution of ions near the Pt electrode.

3.2.2. Adsorption of Hydrogen and Oxygen on Pt Electrodes

The double-layer effect of anion adsorption on the Pt electrode with adsorbed hydrogen, oxygen, or organic molecules and their kinetic behavior has been widely studied [56, 59, 75-78]. In the case of hydrogen adsorption, interpretation of results is based on the idea of the formation of a Pt—H dipolar bond where the negative end is aimed at the solution for small and intermediate hydrogen coverages of the electrode surface.

It has been observed that simultaneous coverage of the electrode surface by significant quantities of anions and a hydrogen monolayer occurs which gives rise to a weakening of the bond strength between hydrogen and the electrode metal. If a monolayer of I^- is previously adsorbed on the electrode, no hydrogen adsorption occurs; at Br^- concentrations higher than 10^{-2} N, hydrogen adsorption is decreased [289]. The metal-anion bond increases in ionic character in the order I < Cl < Br < HSO_4^-, and thus a rise in repulsion of the adsorbed anions and a lesser θ (fraction of surface coverage) is observed in the same order.

In contrast, the influence of anion adsorption on that of oxygen and organic molecules on Pt electrodes appears different. Oxygen adsorption takes place only on that fraction of the surface free of adsorbed anions, as was observed in Br^- and other anion solutions. Similar phenomena are noted for organic substances, for instance, CH_3OH. No influence of the crystal geometry on oxygen adsorption has been observed [79-81].

Few investigations have been devoted to the study of the effect of inorganic cation adsorption on the hydrogen Pt electrode [56, 82-86]. Results obtained with specifically adsorbed cations [56] showed that, at relatively high concentration, cations act upon hydrogen adsorption by occupying the active centers of the electrode surface and not only through electrostatic influence. Research on adsorption of alkali and alkaline-earth cations on bright and platinized Pt electrodes [86] in acidic solutions by charging curves, potentiodynamic, and impedance measurements revealed that the cations are weakly adsorbed and do not affect the total amount of hydrogen adsorbed, although a decrease in the fraction of weakly bonded hydrogen balanced by a rise in that strongly attached to the electrode metal is observed. This effect increases from Li^+ to Cs^+ in accordance with the specific adsorptivity of the alkali metals on Pt. At low coverages of electrode surface by adsorbed hydrogen, introduction of alkali cations in the double layer must increase the strength of the Pt—H bond dipole with the negative end on hydrogen. At higher hydrogen coverages, Pt—H dipoles appear with opposed orientation, the number of which is lowered by the presence of cations.

Adsorption of oxygen species on Pt is favored by the alkali metal cations in the solution which, on the other hand, hinders its reduction. Adsorbed oxygen also forms a dipole with the electrode metal with the negative end pointed to the solution. At small coverages the effect of adsorbed cations on those dipoles is similar to that described for hydrogen adsorption. A weaker influence on hydrogen and oxygen adsorption on Pt is exerted by the alkaline-earth cations.

Specifically adsorbed Cd^{2+}, Pb^{2+}, and Tl^+ ions decrease the amount of hydrogen adsorbed in the same order as their attachment strength to the electrode metal. Low concentrations of Tl^+ reduce preferentially the amount of loosely bonded hydrogen, perhaps through an electrostatic effect. At high concentrations these cations displace the adsorbed hydrogen, irrespective of the strength of the hydrogen bonding, by blocking the adsorption active centers [289]. The Zn^{2+} ions are nonspecifically adsorbed on Pt electrodes.

3.2.3. Adsorption of Anions on Pt Electrodes

It has been shown [87] that halides, HSO_4^-, ClO_4^-, and acetate go through a two-step adsorption with a strong dependence on concentration and potential for freshly prepared electrodes. This dependence may be expressed as a product of two independent functions of concentration and potential, respectively. Adsorption of S^{2-} and SO_2 takes place in amounts larger than a monolayer with formation of a poorly reducible oxide; these adsorption products are more easily reduced if formed on oxidized Pt electrodes [88]. Specific adsorption of SO_4^{2-} ions has been revealed by radiotracer and capacity [89] methods, the adsorption being different if individual faces of a single crystal, bright, or platinized Pt electrode are used; however, no differences in adsorption of HSO_4^- was observed by Veselovskii [79] with the pretreatment of the bright or oxidized Pt or with crystal lattice defects [80] in the electrode. The differences reported by other papers are attributed by the author to differences in the true surface area of the electrode.

3.2.4. Adsorption of Inorganic Cations

The study of inorganic cation adsorption by charging curves, radiotracers, and adsorption potentials in acid solutions with platinized Pt electrodes shows the adsorbability of cations to increase in the order $H^+ < Li^+ < Na^+ < Mg^{2+} < Ca^{2+} \simeq Cs^+ < Sr^{2+} < Ba^{2+} < Zn^{2+} < Cd^{2+}$ at low concentrations of cations [90]. At a higher ratio of cation to acid concentration, cations displace the H^+ from the double layer but the amount of hydrogen adsorbed remains constant. Generally, the cation adsorption goes through a maximum as a function of concentration, except for Na^+ and Cs^+ which are specifically adsorbed and undergo a partial dehydration. Adsorption of alkaline earth cations [91] is small and has little influence on the H adsorbed on the electrode. Zn^{2+} adsorption in acid media is slow, irreversible, and is possibly controlled by dehydration of the Zn^{2+} ions. The adsorption of Cu^{2+} [92, 93] takes place at $E > E_{Cu/Cu^{2+}}$ through formation of a monolayer and is described by a Temkin isotherm for middle values of θ. The Cu^{2+} ions penetrate into the double layer but are not completely discharged; under these conditions the process is reversible.

3.2.5. Adsorption of Organic Substances on Pt Group Metals

The adsorption of organic compounds on electrodes of metals of the Pt group has recently received wider attention as experimental methods have been improved and more results are available [75, 94-96] on the oxidation of organic substances. Several specific features have been pointed out for adsorption on these metals: chemisorption with dissociation of the species, uniform distribution of the heterogeneities of the surface, and a special potential dependence of the adsorption. Thus dehydrogenation of the adsorbed species and formation of chemisorbed products have been proved to happen simultaneously with the adsorption process of organic compounds; for instance, formic acid on Pd and methanol on Pt electrodes. Studies on the adsorption of methanol, n-hexanol, and isobutyl alcohol as a function of potential, temperature, and pH, on bright Pt, reveal that the phenomenon is described by an isotherm of the Temkin-Frumkin type. At the same time, the rate of adsorption on this metal depends in an exponential way on the coverage of the electrode surface and the energy

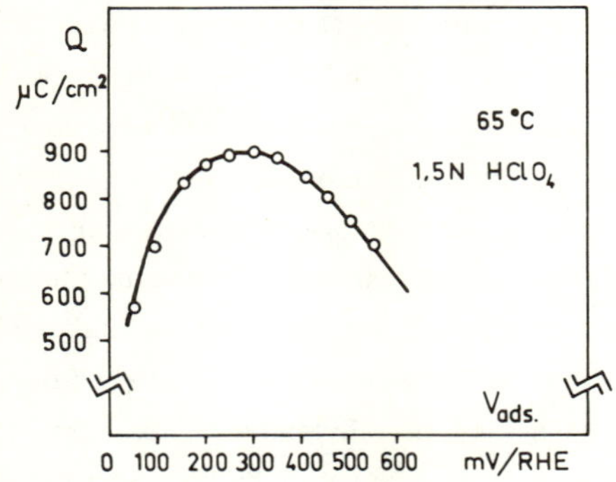

FIG. 3.2.1. Propylene adsorption on bright Pt as a function of potential. (From J. Llopis and M. Sánchez, to be published.)

of activation is a linear function of that coverage. These facts are, to a first approximation, quite general for many organic substances (methanol, ethyleneglycol, etc.) and comply with the characteristics defined for adsorption on a uniformly heterogeneous surface [95].

The potential dependence of the coverage of these metals during the adsorption of organic substances exhibits special characteristics due to their capacity to adsorb hydrogen or oxygen to some extent. The θ-E plot assumes the well-known bell-shaped form with the maximum coverage placed in the region of potentials between that of hydrogen and oxygen adsorption, respectively e.g., Fig. 3.2.1. In the latter region the decrease in adsorption of organic substances at more positive potentials may be produced by oxidation of the adsorbate, but it is mainly due to competitive adsorption of oxygen as is shown in solutions of long-chain alcohols [95] which do not oxidize at potentials lower than 1 V (RHE). One may deduce from these facts that adsorption of organic compounds will be affected by solution pH. Experimental results show a shift of the potential for a maximum coverage of the electrode with the solution pH; the extent of such adsorption also varies with pH. On the other hand, the chemisorption of organic substances on Pt as a function of potential does not depend practically on the surface charge of the electrode, chemisorption of hydrogen and oxygen being the most important influences.

A summary of the experimental results on organic adsorption on Pt group metal electrodes is outlined here [75, 94, 95]. Data on formic acid yield curves of the fractional coverage on Pt anodes as a function of potential with the initial portion being independent of potential; however, the independent portion varies considerably [94]. Brunner and Makrides contend that a Langmuir-type isotherm adequately describes adsorption of formic acid, but no agreement has been reached among researchers on the nature of the species adsorbed from these solutions.

In the group of alcohols, the main body of research deals with adsorption of methanol [75, 95] which has been found to follow a Temkin-type isotherm with a rapid decrease at higher potential as oxide formation starts. However, it has been argued that these results should be considered very qualitative because of the limitations and uncertainties inherent in the fast potential-sweep technique applied [94]. In a similar way, long-chain alcohols are adsorbed according to a Temkin isotherm and exhibit maximum coverage in the region of 0.35 to 0.55 V (RHE). At higher potentials, some of the higher alcohols show a second maximum [95] which may be due to different forms of adsorbed oxygen.

In the field of hydrocarbons [94], an ethylene coverage of $\theta = 0.77$ at 25°C on Pt black electrodes with little or no dissociative adsorption has been reported by Niedrach. This author has also investigated a series of hydrocarbons which he divides, according to the results obtained, into species adsorbed to a small extent, i.e., $\theta < 0.1$ (methane), higher saturated hydrocarbons that are adsorbed to a greater extent $\theta \simeq 0.2$ to 0.3 (ethane, propane), and unsaturated hydrocarbons with much higher coverages. In these experiments the solvent affects in a marked way the adsorption behavior of saturated species but shows no effect on unsaturated compounds. Lower ethylene coverages, $\theta \simeq 0.4$, were observed by Bockris et al. who applied his theory of "competitive adsorption with water" to explain the results. The Langmuir-type isotherm described the ethylene adsorption fairly well. The usual bell shape of the coverage-potential curves was obtained for solutions of ethane, propane, and butane by radiotracer, anodic galvanostatic charging, and other techniques.

The adsorption of some nitrogen-containing organic compounds (methylamines and pyridine homologs) on Pt electrodes has been studied in alkaline solutions [97]. The inductive effect of the methyl substituent on the electron densities of amino groups and the steric effect of the molecules appear to be the controlling influences on the phenomenon; thus the extent of adsorption decreases from mono- to dimethylamine, being negligible for the tri-substituted compounds.

A study of the surface effect on adsorption by exposing single crystal and polycrystalline Pt to methanol solutions reveals only a slight influence of the face orientation [98]. In every case the surface coverage is governed by a Temkin isotherm and attains approximately the same maximum value; the adsorption rate is a linear relationship with the logarithm of time. However, in all the measurements, slightly higher values are observed for the more densely packed face <111>, but if these values are referred to the atomic surface density, the differences fade out. The results demonstrate these surfaces to be uniformly heterogeneous.

A comparison of the adsorption characteristics of the metals of the Pt group allows one to draw some interesting conclusions about the influence of the metal nature on adsorption phenomena on solid electrodes. The analogies and differences in the behavior of Pt, Ir, Pd, and Rh metals are illustrated by Table 3.2.3 [96] where the adsorption of methanol in 1 N H_2SO_4 solutions is reported. As can be seen, adsorption properties vary with the metal but they have been proved to be very similar for the different substances (hydrogen, methanol)

TABLE 3.2.3. Analogies and Differences in the Behavior of Pt, Ir, Pd, and Rh Metals [96]

Experimental data	Pt	Ir	Pd	Rh
Stationary adsorption isotherms	$\theta = \frac{1}{f} \ln c + a$	$\theta = KC^{1/n}$	$\frac{\theta}{1-\theta} \exp(f\theta) = KC$	$\theta = \frac{1}{f} \ln c + a$
Heterogeneity coefficient	$f = 14$	$n = 5.5$ to 6.5	$f = 9$	$f = 25 - \theta < 0.25$ $f = 6 - \theta > 0.25$
Kinetic isotherm	$\theta = A + \frac{1}{\alpha f} \ln \tau$ $\alpha \approx 0.5$	$\log \theta = A - \frac{1}{\alpha n} \ln \tau$ $\alpha = 0.46$		$\theta = A + \frac{1}{\alpha f} \ln \tau$ $\alpha = 0.5$ to 0.6
Adsorption rates	$V = KC \exp(-\alpha f\theta)$	$V = KC\theta^{(1-\alpha n)}$	$V = KC(1-\theta)\exp(-\alpha f\theta)$ $\alpha \approx 0.5$	
Variation of E_A with θ	$E_A = E_A^\circ + \alpha fRT\theta$	$E_A = E_A^\circ + \alpha n \ln \theta$	$E \approx E_A^\circ + \alpha fRT\theta$	
No. of H atoms lost by dehydrogenation	3	2		3
No. of electrons given up by oxidation of (C_nH_{3-n}—OH) particles)	3	2		3
No. of hydrogen places taken by the adsorbate	3	2		2.5

on the same metal, which suggests that the adsorption of neutral substances is determined by the heterogeneity of the metal surface with small influence of the substance nature.

On all of the metals, a breaking of the C—H bond occurs with the formation of a species strongly bonded to the electrode surface; the surface coverage, θ, reaches the higher value in the region of lesser hydrogen and oxygen adsorptivity. The extent of dehydrogenation and deep oxidation of the adsorbate depends on the nature of the metal and the solution pH.

Capacitance data [75] measured under the same voltammetric conditions on these metals as a function of methanol bulk concentration (10^{-3} to 1 M) show that capacities tend toward a limiting value at increasing concentration on Pt and Ir. This value corresponds to a monolayer on that part of the surface where carbonaceous species are adsorbed, i.e., about 75% of surface. Adsorption on Rh and Pd is noticeable only at concentrations higher than 10^{-2} M, and the amount adsorbed is much less than a monolayer even at the highest concentration studied.

4. ELECTROCHEMICAL STUDIES

4.1. SURFACE OXIDATION AND ANODIZATION

4.1.1. Aqueous Solutions

Electrochemical characteristics of the oxygen-platinum interaction and its further influence on corrosion, passivation, and other electrode processes have been broadly reviewed recently [75, 99-102]. Two main theories have been developed for the interpretation of the adsorbed and oxide film formed on Pt electrodes. One line of thought explains the film as constituted by hydrated PtO or PtO_2 molecules; the other theory favors the concept of a hydrated layer of adsorbed oxygen atoms such as Pt—O (although, for the sake of simplicity we use the term oxide and the formula PtO and PtO_2 for the products of Pt surface oxidation, it becomes apparent from the following survey that there is no direct relationship between these surface compounds and the bulk oxides of the same formula).

The surface oxidation process of a Pt electrode immersed in O_2-saturated acidic solutions manifests itself, under open circuit conditions, by a potential change up to a steady-state rest potential of about 1.06 V (NHE) [103]. As widely discussed by Hoare [100], formation of an oxide film, the presence of H_2O_2, or the mixed potential of a local cell has been diversely held responsible for the rest potential value. Experimental work suggests that it corresponds to a mixed potential, but there is no agreement about the nature of the anodic reaction of the local cell. It has been shown that under open circuit conditions the amount of adsorbed oxygen changes with the time of exposure to O_2 and the rest potential is a linear function of the coverage. This never exceeds 30% at the steady-state rest potential of 1.06 V, for both bright and platinized Pt [104, 105], irrespective of the nature of the anion. The local cell reactions:

$$O_2 + 4H^+ + 4e \rightarrow 2H_2O$$

and

$$Pt + H_2O \rightarrow Pt{-}O + 2H^+ + 2e$$

in acid, or

$$Pt + 2OH^- \rightarrow Pt{-}O + H_2O + e$$

in alkaline solutions have been proposed to account for this behavior [104] along with the argument that a complete layer of Pt—O is never attained because oxygen is continually sorbed into the Pt lattice. No surface adsorbed oxygen is observed at potentials lower than 700 mV. Treatment of the electrode with HNO_3 seems to make the metal more suitable to oxygen sorption, and a Pt—O alloy is formed with a complete layer of adsorbed Pt—O on the surface (oxygen coming from the environment, not from the HNO_3 acid). This electrode exhibits the reversible oxygen potential in O_2-saturated 2 N H_2SO_4 solutions.

Results obtained by stripping cathodic pulses, galvanostatic curves, and open circuit measurements on anodized Pt electrodes in 2 N H_2SO_4 solutions saturated with oxygen allowed Hoare [100, 104] to identify different types of Pt—O alloys as a function of the electrode potential. Below 800 mV (NHE) no sorbed oxygen was detected; as the potential is raised, oxygen is adsorbed and becomes dermasorbed (oxygen adsorption sites located in the first one or two atomic layers of the metal surface) at values greater than 1 V. A complete layer of adsorbed oxygen (Pt—O) is reached at about 1.6 V, where passivation of the metal sets in. This film is electronically conducting and may grow in thickness, with simultaneous evolution of oxygen, to a monolayer of PtO_2 on both smooth and platinized Pt [106] by converting the Pt—O sites to PtO_2 at potentials higher than 2 V. Under open circuit conditions and potentials lower than 1.55 V, PtO_2 decomposes to PtO in the presence of Pt. An amount of oxygen equivalent to many complete layers may dissolve into the Pt lattice by anodic polarization; this adsorbed oxygen may replace any dermasorbed oxygen removed. Oxygen evolution takes place on the electronically conducting monolayer of Pt—O, possibly containing some PtO_2, where the adsorbed oxygen does not participate in the electrode reactions as was proved by tracer techniques.

Tiurin et al. [107, 108] have used charging curves and triangular potential sweeps in acid and alkaline solutions to confirm the formation of PtO_2 islands at potentials higher than 1.6 V (RHE), which according to the authors develop to a porous monolayer around 1.9 V. In the 1.9 to 2.2 V region, this monolayer recrystallizes and becomes activated for O_2 evolution. Schultze [109] stated that the rates of simultaneous oxide formation and oxygen evolution in 1 N H_2SO_4 solutions change according to the potential in the 1.5 to 2 V (NHE) range and the span of time measured (from 10^{-4} to 10^{-3} sec). The oxide layer thickness varies from 1.3 to 9 Å under the mentioned conditions; at low oxide quantities (<0.4 mC/cm^2), oxide layer formation prevails, and at electric quantities larger than 0.6 mC/cm^2, oxygen evolution is predominant. Because the oxygen evolution current decreases with the increasing thickness of the oxide layer, Schultze discussed the possibility of a potential drop through the oxide layer caused by limitation of the electronic conductivity in the oxide, which would explain many of the puzzles of oxygen evolution. In this respect, the paper by Ord and Ho [110] must be quoted. Working with Pt single crystals, the open circuit transient technique, and ellipsometry measurements for layer growth under galvanostatic conditions, the authors found that the layer growth must be controlled by ionic conductivity in the oxide and that the electric field determines the layer thickness, the oxygen discharge occurring through an electronic tunneling process.

On the other hand, later work by Vetter and Schultze [111] on the formation and reduction of Pt oxides in 0.5 M H_2SO_4 solutions, applying galvanostatic and potentiostatic techniques, pointed out that oxide layer properties depend on formation conditions, i.e., growth rate, layer aging, and the method of obtaining the coverage (anodic oxidation or reduction of thicker oxides), and show no differential characteristics about the Pt—O alloys formed (oxygen adsorbed on Pt, PtO, or PtO_2) in the 1 to 2 V (NHE) range. At the same time, the current density involved in oxide formation decreases in an exponential way with coverage, i.e., the Temkin isotherm does not hold for the oxide formation, whereas at coverages larger than 1.5, oxygen evolution is the prevailing process.

In the potential range of 1.1 to 2.2 V (RHE), the presence of Cl^- ions in acid solutions affects adsorption of oxygen on Pt in a marked way. Triangular potential sweeps and stationary galvanostatic curves [112] disclose that at low potentials in this range [$\simeq$ 1 V (RHE)], Cl^- ions displace the oxygen adsorbed to a greater extent the higher the Cl^- concentration in the solution, such that over the interval 0.1 to 1 N Cl^-, the Pt surface is free of oxygen. Conversely, oxygen displaces adsorbed Cl^- ions at higher potentials, although it always turns out that a rise in halide concentration at constant potential results in a decrease of oxygen coverage. An increase of temperature does not significantly influence these phenomena except for the larger amounts of oxygen adsorbed. At 1.5 V (RHE), a monolayer of Pt—O—Cl is formed in which the oxygen to chloride ratio depends on the halide concentration. On increasing the potential to 2.1 to 2.2 V, a second layer builds up along with Cl_2 evolution or formation of other anodic products in accordance with the composition of the layer oxides; for instance, ClO_4^- is produced in 0.1 N KCl solutions. Anodic potential sweeps at Pt in 0.5 to 11.5 N $HClO_4$ solutions [113] show that the oxygen evolution begins at 1.85 V (NHE) and formation of a PtO—OH or Pt—O layer occurs at 2.2 V; this oxide layer prevents chloride evolution and favors oxygen and ClO_3^- production.

At potentials above 2.1 V (RHE), Tiurin and Volodin [114] found that oxide formation in acidic and alkaline solutions does not depend on the oxidation potential, and at 2.4 V (RHE) two types of oxide occur, the ratio of which does not change with time (up to 36 days), the total coverage adding up to 2.5. The authors assume that one oxide consists of a monolayer of PtO_2 and the other of a Pt(IV) compound that would include a steady-state concentration of intermediates of oxygen evolution. Shibata and Sumino [115] also report the formation of two types of oxide on Pt at potentials between 2.07 and 2.22 V (NHE) in the temperature range 20 to 35°C in 0.5 M H_2SO_4 solutions, but the interpretation of results differs completely. One oxide is described as a rapidly forming monolayer on top of the anodization layer and the second oxide as a slowly growing multilayer that penetrates into the metal with a maximum growth rate at 2.13 V. At potentials greater than 2.17 V (NHE) or temperatures below 22°C (passivation temperature), no multilayer is formed. These points are strongly contested by James [116], who obtained a multilayer at potentials above 2.17 V by longer anodization and found no passivation temperature. This author emphasized that the oxidizability of Pt is controlled by the metal stress, and suggested a composition of PtO for the monolayer and hydrated PtO_2 for the multilayer.

Research by impedance measurements on the anodic behavior of Pt in $HClO_4$, H_2SO_4, and H_3PO_4 solutions at high concentrations (5 to 12 M) at potentials above 2.5 V [117] revealed the oxide layer on the electrode to be partly made up of anions. On this oxide a mixed layer of adsorbed anion radicals and oxygen is formed which takes part in the final anodic products.

Finally, Bagotskii et al. [118] found no influence of neutron irradiation, face orientation of single crystals, or structural crystals defects on the surface oxidation of Pt electrodes in acid media.

The anodic activation of Pt electrodes by repetitive formation and reduction of the oxygen layer [75, 99] or its pretreatment by fast cyclic oxidation-reduction methods in order to obtain reproducible surface conditions has led to some research on the surface state of the metal after such a test. Deterioration of the Pt surface was noticed by Bruckenstein et al. [119] by scanning the electrode potential several times in the range -0.3 to 1.5 V (SCE) at rates from 1 to 10 V/min in 1 M H_2SO_4 and 0.1 M $HClO_4$ solutions and using ring-disk electrode techniques. Among the soluble species produced after the potential sweeps, Pt(II) was detected by spectrophotometry during the cathodic scanning of the electrode at potentials corresponding to thermodynamical reduction of PtO_2, whereas a nonidentified species was formed during oxidation. Costa et al. [120] verified this deterioration by ellipsometry measurements on an electrode submitted to cyclic treatment between -800 and +1200 mV (SCE) in 0.1 M Na_2SO_4 solutions. The authors assume that after reduction of the dermasorbed oxygen, the Pt atoms go back to the lattice sites of lower energy through surface diffusion. An increase of roughness and oxidizability was also reported by James [116].

Studies [100] on the kinetics of the oxygen coverage growth reveal that the extent of coverage increases logarithmically with time; in the mechanism proposed for this reaction, oxygen adsorption would be the determining step. There is an almost general consensus on the mechanism of oxide formation in acid media: the theory sets forth, at potentials below 1.25 V (NHE), a two-step process:

$$H_2O + Pt \rightarrow PtOH + H^+ + e$$

$$2PtOH \rightarrow PtO + Pt + H_2O$$

in which the first step is rate determining. Results obtained by cyclic voltammetry [121, 122] and ring-disk electrode techniques support this view. Bonnemay et al. [123] have dealt mathematically with the anodization of Pt in acid solutions as a function of potential and time under potentiostatic conditions. The equations reveal a change of mechanism at about 990 mV (NHE); below this potential, discharge of H_2O occurs by one-electron transfer and Pt—OH formation; at higher potential, PtO is formed by two-electron transfer. The transference coefficient, α, calculated in this way agrees fairly well with the experimental values obtained in 0.01 N H_2SO_4 solutions. Tarasievitch and Vilinskaya [124] gave experimental proof that the oxygen discharged in acidic and alkaline solutions comes from the electrolyte and not from oxygen dissolved in the solution. On the other hand, Vetter and Schultze [111] advance a mechanism in which oxygen ions are adsorbed in equilibrium with the electrolyte: $H_2O \rightleftharpoons O^{2-}_{ads} + 2H^+$, covering the electrode surface less than 10%. Then, a site exchange between Pt^{2+} and oxygen, which is rate controlling and depends on the electric field, occurs inside the inner Helmholtz layer, giving rise to an epitaxial surface oxide. The more dense oxide layers (coverage from 1 to 2.5) are formed in this way at the metal/oxide and oxide/electrolyte interfaces, followed by migration of oxygen and Pt ions through vacancies and interstitial sites in a high electric field. The site exchange mechanism was formally put forward by Bockris et al. [102] who assumed a Temkin isotherm adsorption for water molecules (later proved experimentally by Schultze [125]) and derived an equation that fits the experimental relationship between the amount of oxide and the potential or the logarithm of time.

In alkaline solutions [100], the hydrogen and oxygen adsorption regions overlap and a greater double-layer capacity value is observed, possibly due to OH^- adsorption in this region in comparison to that in acidic solutions. According to the few results reported on the Pt—O_2 system in this medium, the oxidation mechanism is essentially the same as that described above except for a higher contribution of the OH^- adsorption to the process. Charging curves and triangular potential sweep techniques show two types of oxygen adsorption, with the process starting at potentials more negative than those in acidic media [121]. Oxygen adsorption at potentials lower than 700 mV (RHE) is quite reversible as long as that at higher potential takes place irreversibly. A monolayer of adsorbed oxygen is reached in the 1.1 to 1.25 V potential range [126]. Calculation of kinetic parameters demonstrates that experimental results agree with the oxidation mechanism advanced for acid solutions, the reversible adsorption corresponding to that of OH^- ion. Reduction of oxygen previously adsorbed at 0.7 to 0.75 V (RHE) on Pt in alkaline solutions [127] discloses a different bonding strength for oxygen depending upon the adsorption time; this must be related to the adsorbed and dermasorbed kind of oxygen referred to in acidic solutions.

Impedance measurements carried out by Yakovleva and Vaselovskii [128] in 0.1 to 10 N KOH solutions in the 0 to 25°C temperature interval reveal that oxygen is adsorbed in the potential span of 0.5 to 1 V (NHE), forming dipoles as in acidic solutions. At potentials of about 2.4 V, a rise in the capacity value is observed that the authors ascribe to oxygen adsorption on Pt oxide; at potentials over 2.5 V the capacity decreases, which is supposed to be due to adsorption of oxidized intermediates that lead to the final anodic products. Thus addition of Na_2SO_4 inhibits oxygen evolution, especially at low temperatures, and decreases the formation of ozone which evolves at high potentials. The results suggest that at 2.8 V (NHE) the slow step is the anion discharge:

$$PtO(OH)_{ads} + OH^- \rightarrow PtO + \frac{1}{2}O_2 + H_2O + e$$

4.1.2. Molten Salts

Open circuit measurements and anodization of Pt electrodes in melts containing oxyanions demonstrate the formation of passivating layers, the composition of which has not generally been studied, although they are supposed to consist of Pt oxides. Thus gray-green films that become thicker at increasing temperatures in the range 790 to 980°C have been produced in molten Li_2CO_3 [129]. Similar nonidentified oxides have been detected in nitrate [130-132] and alkaline silicate melts [133, 134]. However, Grinsley and Every report formation of a thick layer of brown oxide on Pt at 400°C in fused KNO_3, the composition of which was PtO and PtO_2 in the ratio 1:3 [135]. The extensive corrosion of the Pt electrode seems to be due to the introduction of impurities such as H_2O in the melt [278]. A semiconducting layer is formed by anodization of Pt in molten aluminum-calcium silicates (CaO 40%, SiO_2 40%, Al_2O_3 20%), in the 1350 to 1500°C temperature range, the thickness of which amounts to 100 Å. The concentration of Pt ions in this layer decreased in an exponential way with the distance to the Pt/oxide interface [136]. Oxide formation on Pt in alkali hydroxide melts at 200 to 500°C allows the metal to perform as a reversible oxygen electrode [137] in this bath. The oxide composition has been identified as PtO or PtO_2, depending on the water content of the melt [20].

In chloride melts a passivating film of unknown nature is detected by anodic polarization in $AlCl_3$-NaCl at 200°C [138], while addition of alkaline oxides to the NaCl-KCl mixture at 750°C helps to form a Pt oxide under potentiostatic conditions [139, 140]. In the molten KCl-LiCl eutectic at temperatures between 450 and 600°C, a passivating layer grows at potentials lower than 1 V (vs the Pt/Pt^{2+} electrode) [140-142], the composition of which is said to consist of oxides [143] or K_2PtCl_6 [144].

5. APPLIED ELECTROCHEMISTRY

5.1. CORROSION

In the theoretical corrosion diagram, the domain of Pt stability overlaps most of the water stability region (Fig. 1.1.1). Except for strongly oxidizing conditions, the Pt metal is stable in aqueous solutions at all pH's. The best known attacking reagent is aqua regia which combines the oxidizing action of HNO_3 with the complexing ability of HCl to dissolve Pt and produce H_2PtCl_6. Similarly, acidic solutions that contain Cl_2 can corrode Pt metal.

The electrochemical corrosion of Pt has been studied mainly at potentials of the oxygen layer and in the range of oxygen evolution or in the presence of complexing ions. Under other conditions, experimental difficulties arise from the extremely small rates of the metal corrosion although sensitive techniques have been developed in recent years [99] that will allow the extension of knowledge of this phenomenon.

In HCl solutions at 25°C [99], Pt dissolves at the anode to produce $PtCl_6^{2-}$ at potentials prior to passivation. The latter shows up in the galvanostatic polarization curves as a potential shift, after which Cl_2 evolution is the main process at the electrode. To achieve passivation, higher current densities are necessary as the Cl^- concentration, temperature, and acidity increase. Pt anodic corrosion in HCl solutions over the concentration range 0.1 to 8 N displays a linear relationship between log i and the potential at low current densities, reaching the passivation state at about 1.1 to 1.2 V (NHE) at low HCl concentrations [99, 145] and at 1.4 V [146] at high HCl concentrations. The plot of log i vs 1/T at a given potential (0.860 V vs SCE) yields an activation energy of 20 kcal/mole [99]. This value indicates that Pt anodic corrosion under these experimental conditions is a highly irreversible process. A quantitative study of corrosion reveals that it is controlled by the potential so that Pt goes into solution as $PtCl_6^{2-}$ with a current efficiency of 100% at potentials below that of passivation. Dissolution of Pt decreases almost to nil as the electrode passivates or is polarized at high potentials [99]. A linear dependence of corrosion current density, at constant potential, on Cl^- concentration has been found for 0.0031 to 0.28 N HCl solutions in 15 N H_2SO_4 [145], and a similar relationship for passivation current density in HCl solutions at concentrations over the range 2 to 8 N [146]. Above 8 N HCl, no passivation did occur however high the applied current might be.

Under passivating conditions, radiochemical techniques show that the corrosion rate falls very slowly to a steady-state value of the order of 2 to 5×10^{-9} A/cm^2 [99, 145]. This value is highly dependent upon experimental conditions and is mostly affected by the potential applied. At conditions above the passivation point, Cl_2 and O_2 evolution occurs and approximate Tafel lines are again obtained. Polarographic analysis identified the existence of Pt(II) at these potentials, probably as $PtCl_4^{2-}$, with a current efficiency of dissolution less than 10^{-3}% [146]. In this region the Cl^- concentration does not influence Pt corrosion, and even in 3 N ($HClO_4$ + HCl) solutions the rate of attack decreases at rising HCl concentration [145], with the current efficiency becoming almost independent of the potential.

Similar phenomena have been observed for 1 to 5 N NaCl and NaCl + HCl mixed solutions but with lower corrosion rates, an effect undoubtedly associated with the solution pH [146].

As a result of these studies it can be concluded that dissolution of the Pt electrode takes place in the passive region through a mechanism different from the one occurring in the active state, where Cl^- ions participate directly in the first one-electron step of the anodic dissolution of Pt in HCl. It is likely that under these conditions Pt chloro-complexes are not formed directly in some electrochemical step but rather through a secondary chemical reaction. In order to slow down the dissolution rate of the metal, inclusion of inhibiting ions into the passivating layer must take place. This inhibiting effect at high potentials has been verified for corrosion in mixed solutions of Cl^- + $HClO_4$ and in Cl_2 saturated 3 N $HClO_4$ at -15 and 25°C [145, 147, 148]. By contrast, addition of alkali bromide or iodide to $HClO_4$ solutions hardly affects the deterioration rate of the metal and NaF does increase it. Pt corrosion in 3 N HBr displays the same rate values and passivating potentials as in HCl solutions, no active state of the electrode being observed in 3 N HI or in 3 N $HClO_4$ + 0.1 N NaF solutions at potentials lower than 1 V (NHE) [145]. In solutions of anhydrous HF + 0.01 M NaF at -20°C, the polarization curve for Pt electrodes exhibits a passivation region above which, at potentials of F_2 evolution ($\sim$ 5 V vs Hg/Hg_2F_2), the anodic dissolution yield is 14.3%, assuming PtF_2 formation; addition of H_2O to the solution favors the passivation reactions [149].

Anodic corrosion of Pt is almost undetectable in solutions of oxy-acids or NaOH. Still, the sensitivity of the radiochemical method allowed the study of this corrosion in solutions of acids unable to form complexes with Pt [99, 150]. Thus corrosion rates of bright Pt in 1 N H_2SO_4 have been detected at all positive potentials although the rate decays with time to a final value for a period that depends on the polarization time. The final rate increases with the potential, going through a region of constant value, and is independent of the previous hydrogen and double-layer potential applied to the electrode and of the concentration of the solution (similar results are obtained in 15 N H_2SO_4). The relationship between steady corrosion rate and potential becomes linear in the range 1.2 to 2 V (NHE) of Pt oxidation. Corrosion rate-potential curves in 1 N HNO_3 and 3 N $HClO_4$ show the same passivation region and dissolution rates (10^{-10} to 10^{-6} A/cm^2) as those for H_2SO_4 solutions, the current density for passivation becoming higher with increasing temperature. It is characteristic for all these cases that the dissolution yield at high anodic potentials is constant and independent of the potential. These results suggest that the nature of the anion plays a minor role in the corrosion of Pt in oxy-acids. Kolotyrkin et al. [150] concluded that the corrosion of Pt in acidic solutions that do not contain Cl^- ions is determined by dissolution of oxides formed on the metal surface. Therefore, deterioration increases in the region of Pt oxidation potentials and depends on the polarization time, being independent of the electrolyte anion. The increase of corrosion under passivated conditions may be considered as due to a sort of transpassivation process.

The formation and dissolution of surface oxides may also account for the anodic corrosion of Pt in the solutions used for the Kolbe synthesis. In acetate solutions at 30°C [151], the radiochemical method reveals that the corrosion rate increases markedly at potentials higher than 1.8 V (NHE) (from 10^{-10} to 10^{-6} A/cm^2) and the current efficiency of metal dissolution is almost constant and about 2×10^{-3}% in the potential region of the Kolbe synthesis (2.4 to 3.2 V). The use of methanol solutions of acetate instead of aqueous solutions [152] decreases the corrosion rate; the same effect is obtained by addition of other alcohols, the monomethyl ester of adipic and glutamic acid, or aromatic and cyclic compounds that are adsorbed on the electrode at high potentials (benzene, naphthalene, cyclohexene, isoprene, etc.); in every case the dissolution rate amounts to about the same value, $\simeq 10^{-6}$ A/cm^2. However, the addition of water to the solutions notably increases the corrosion of the metal.

Pt dissolves anodically as Pt(II) in molten KCl-LiCl eutectic [14] before reaching a passivation state [139-142] as reported above (see Section 4.1.2).

Superimposition of ac produces a remarkable enhancement of anodic corrosion with results that are summarized next [99]. Earlier research by Margules, Stichlik, Ruer, and other investigators established that an intense corrosion of Pt is obtained in H_2SO_4 solutions by superimposing an ac on the polarizing dc. It has been found that at a given value of the dc density, corrosion starts only when a minimum value of the ac is reached, this particular value being proportional to the dc density. A maximum value of corrosion is attained above the threshold value of ac as this variable increases. Careful analysis of the anodic products disclosed them to consist of $PtO_2 . nH_2O$.

Application of this method to Pt in KCN solutions leads to considerable corrosion of the metal with formation of $Pt(CN)_4^{2-}$.

The effect of ac on the anodic corrosion in HCl solutions has been studied by Rius, Llopis, and Colom who found the attack to be stronger as the ac density increases and its frequency decreases. They also showed that the dc density required for passivation is higher the lower the frequency of the superimposed ac or at increasing ac densities. As reported for H_2SO_4 solutions, at current densities high enough to passivate the electrode, corrosion occurs only when the ac superimposed is much higher than the dc density. At values of ac slightly higher than the critical one, the rate of corrosion rapidly reaches the theoretical value corresponding to the dc applied, assuming the formation of $PtCl_6^{2-}$ and taking into account Faraday's law (Fig. 5.1.1). Subsequently, Llopis and Sancho observed that Pt corrosion is very notable for a frequency of 50 Hz and becomes very intense for 10 Hz square-wave ac, the effects being much more pronounced as the acidity and temperature of the electrolyte are increased.

Analogous results can be obtained with 50 and 100 Hz ac superimposed on anodic currents for Pt and platinum-plated Ti in NaCl solutions [153].

Experiments performed with only ac reveal an intense corroding action on the Pt electrode in HCl at frequencies lower than 100 Hz [99]. Using 50 Hz ac and concentrated solutions of $CaCl_2$, Brouillet and Epelboin demonstrated the feasibility of electrolytic polishing of small Pt electrodes. In a similar manner the etching required for metallographic observation of Pt has been produced in dilute HCl solutions, and corrosion in H_2SO_4, HNO_3, and their mixtures. In all cases, corrosion becomes nil at high frequencies. The influence of the cations on this type of Pt corrosion has been studied at a current density of 1 A/cm^2 for alkali and alkaline-earth metal ions with Cl^-, PO_4^{3-}, SO_4^{2-}, NO_3^-, OH^-, and I^- solutions [154]. The highest corrosion effects were observed in hydroxide and chloride solutions, and

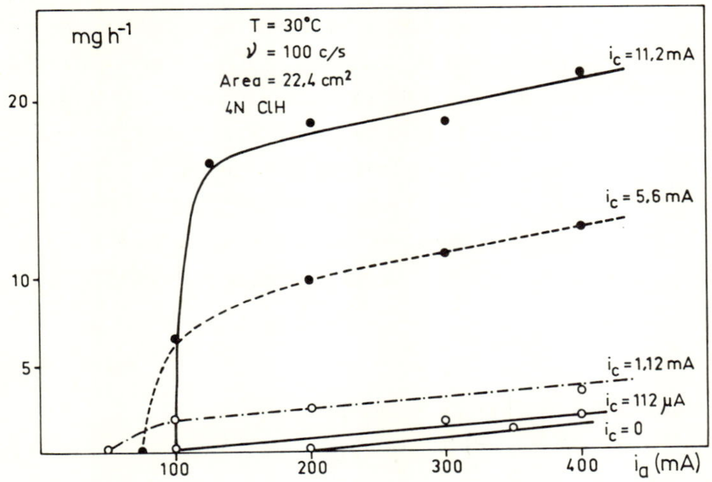

FIG. 5.1.1. Corrosion rates of a Pt electrode in HCl solutions using superimposed currents. (From Ref. 99.)

in the latter the rates of attack decrease, according to the cation, from alkaline-earth to alkali and hydrogen ion solutions. A rise in temperature lowers the corrosion rate which, on the other hand, goes through a maximum as a function of the ac frequency. This maximum lies between 10 and 80 Hz for the different cations, and the rates tend to zero at frequencies higher than 200 Hz.

The method of electrolytic dissolution of powdered metal electrodes in HCl solutions by ac has been employed with noble metals [155], among which Pt shows a corrosion at 50 Hz ac to be proportional to time with the formation of $PtCl_6^{2-}$. The attack of the metal increases with higher temperature and current density.

5.2. ELECTROPOLISHING

The electrochemical polishing of Pt has been performed with success in molten baths of KCl or NaCl at 1020°C by anodization at 0.15 A/cm^2 against a Ni cathode [156], in fused mixtures of KCl (48 mole %) and NaCl at 700°C, and KF-NaF (50 mole %) at 800°C at 2.5 and 3 A/cm^2, respectively, with the formation in all cases of Pt^{2+} [157]. Another melt reported as an electrolyte bath is the 1:10 mixture KCN + KCNS over the range 200 to 300°C [158], in which Pt dissolves anodically to the Pt^{2+} state with 100% current efficiency against Ta, Rh, or Ni cathodes. The polishing rate was $\simeq$ 0.3 mil/min at 0.4 A/cm^2. At lower temperatures (60 to 70°C) the noble metals have been polished with 20 to 60 V ac of 50 Hz frequency in electrolytes containing $HClO_4$ (5 to 30 g/l) and water (95 to 70 g/l) [159].

Application of ac to Pt polishing at room temperatures has given good results for small electrodes in slightly acidic solutions (1% concentrated HCl) of $CaCl_2$ (120 wt%) at $\simeq 3\ A/cm^2$ with 50 Hz ac and for large electrodes in the same electrolyte with 10 Hz square-wave ac at $1\ A/cm^2$ for 2 hr [157]. A similar solution cooled to -35 and -40°C has been used with 60 Hz ac at $3\ A/cm^2$, the recommended bath composition being 100 ml of $CaCl_2$ (saturated solution, density 1.46 g/cc) + 73 ml H_2O + 6 ml of concentrated HCl [160]. Recently, Toušek [161] reported the polishing of Pt with a dissolution rate of $3 \times 10^{-6}\ g/cm^2$ at $0.5\ A/cm^2$ of 50 Hz ac at room temperature in a mixed 1:1:1 volume solution of H_2SO_4 (96%), H_3PO_4 (80%), and HNO_3 (65%).

5.3. ELECTRODEPOSITION; ELECTROPLATING

Platinization is widely used to produce surfaces for hydrogen reference electrodes, conductivity cells, and electrocatalytic electrodes. The electrodeposition of the metal is usually carried out from chloroplatinic acid solutions and involves three types of reactions:

$$PtCl_6^{2-} + 2e \rightleftharpoons PtCl_4^{2-} + 2Cl^-$$

$$PtCl_4^{2-} + 2e \rightleftharpoons Pt + 4Cl^-$$

$$PtCl_6^{2-} + 4e \rightleftharpoons Pt + 6Cl^-$$

Normal potentials and kinetic parameters are given for a variety of conditions in Sections 1 and 2, respectively. From early electrochemical studies of Pt deposition it is known that better coatings and adherence of deposits to metal substrate are obtained if a small quantity of lead acetate is added to the plating solution. The presence of small amounts of lead acetate in the bath affects the deposit appearance and growth considerably, lead metal becoming incorporated into the coating.

In spite of the long and broad use of the platinized Pt black electrode, very little is understood about the deposition process and considerably less about electrodeposition from lead-containing solutions. Feltham and Spiro [162], after reviewing the electrochemical data and plating conditions for Pt platinization, recommend the following recipe for routine platinizing: a solution of 0.072 M (3.5%) chloroplatinic acid plus 1.3×10^{-4} M (0.005%) lead acetate with a current density of $30\ mA/cm^2$ for up to 10 min. A deposition time of 5 min is thought to be adequate for hydrogen electrode and conductance electrodes, since a smaller amount of deposit speeds equilibration and reduces adsorption. Good stirring and no gas evolution at the Pt cathode are considered essential so as to prevent the interaction of the cathode with the chlorine evolved at the anode.

For electrocatalytic research, where the properties of the electrode play an important part and depend on the deposition potential [163], potentiostatic platinization is advised: suitable electrodes are obtained by plating at +50 mV (NHE) from a 0.041 M (2%) chloroplatinic acid solution in 1 M HCl. For systems sensitive to trace impurities, platinization without lead additive and at low current densities (10 to 20 mA/cm^2) has been recommended. Application of ultrasonic vibration during the electrolysis was found to produce nonporous and thicker coatings [164].

Bright and active Pt electrodes can also be prepared by plating from chloroplatinous acid solutions [163]. The electrodeposition rate of Pt^{2+} is increased by the acidity of the solution and by adding hydrazine to the bath, but it is decreased by a rise in Cl^- concentration [165-167].

Pt coatings for industrial use have been produced from a number of freshly prepared electrolytes [168-170a]. Commonly employed baths listed include: 7.5 g/l $PtCl_4$ + 20 g/l NH_4Cl + 100 g/l Na_2HPO_4 + 20 g/l $(NH_2)_2HPO_4$, with which deposits are obtained at low current densities (0.2 to 0.5 A/dm^2) and with low current efficiency (10 to 50%); bright and coherent coatings up to 25 μm in thickness are deposited with 100% efficiency at 1 A/dm^2 from 18.5 g/l $Na_2Pt(OH)_6$ + 5.1 g/l Na(OH) at 65 to 80°C.

However, this bath is subject to deterioration through carbonation which accelerates the sodium salt decomposition and leads to formation of a yellow precipitate in operation or even in storage of the solution. A more suitable solution is formed with potassium instead of sodium salts.

The more conventional plating electrolytes are based on the platinum "P" salt $Pt(NH_3)_2(NO_2)_2$ and are the more widely used. In alkaline solutions of composition 16.5 g/l "P" salt + 100 g/l $(NH_4)NO_2$ + 10 g/l $NaNO_3$ + 50 g/l NH_3 (20%), electrodeposition requires operation at 90 to 95°C and a current density of 1 to 20 A/dm^2; even then it shows a low cathodic efficiency (15 to 20%). Better results are produced by applying a periodic reverse current plating of a 5-sec cathodic, 2-sec anodic cycle with a 5 to 6 A/dm^2 current density. Crack- and porous-free Pt coatings up to 50 μm thick are obtained in this way, but reservations are noted about the stability of this solution [169].

Acidic solutions of the platinum "P" salt make more satisfactory plating electrolytes, e.g., 20 g/l "P" salt + 50 to 100 g/l HBF_4 + 90 to 120 g/l $NaBF_4$, in operation at 70 to 90°C and 2.5 A/dm^2 with a current efficiency of about 14 to 18% and coatings 4 to 8 μm thick; or the sulfamic acid solution 10 g/l "P" salt + 20 g/l NH_2SO_3H at 80 to 90°C and a current density of 0.2 to 2 A/dm^2, though the latter bath produces lower yields (6 to 9%).

5.4. FUEL CELL ELECTRODES

The platinum metals are widely used as catalysts for both anodes and cathodes in fuel cells. The extensive literature in this field is discussed in several monographs [170b].

Chapter VI-5

IRIDIUM

J. F. LLOPIS* and F. COLOM

Department of Surface Chemistry and Electrochemistry
Institute of Physical Chemistry, C.S.I.C.
Madrid, Spain

Even though the chemistry of iridium involves a variety of species in which the metal takes up oxidation states ranging from 0 to +VI, only the I, III, and IV compounds are commonly found and studied. Aqueous solutions of III and IV compounds mainly produce complex species with coordination number 6 while the I state forms square complexes with π-bonding ligands (CO, tertiary phosphines, olefins, etc.).

---

*Deceased.

The References for Chapters VI-4 through VI-7 appear on pages 267-275. A General Bibliography appears on pages 328 and 329.

1. STANDARD AND FORMAL POTENTIALS

According to the potential-pH diagram constructed by Van Muylder and Pourbaix [171] for the iridium-water system at 25°C (Fig. 1.1), the stability of the metal almost overlaps the domain of water stability provided that complexing agents are absent and insoluble salts are not formed. The thermodynamic equilibrium relations represented in the diagram are:

1. $IrO_4^{2-} + 8H^+ + 3e = Ir^{3+} + 4H_2O$ $\qquad E_0 = 1.448 - 0.1576\ pH + 0.0197 \log \frac{IrO_4^{2-}}{Ir^{3+}}$

2. $Ir_2O_3 + 6H^+ + 6e = 2Ir + 3H_2O$ $\qquad E_0 = 0.926 - 0.0591\ pH$

3. $IrO_2 + 4H^+ + 4e = Ir + 2H_2O$ $\qquad E_0 = 0.926 - 0.0591\ pH$

4. $2IrO_2 + 2H^+ + 2e = Ir_2O_3 + H_2O$ $\qquad E_0 = 0.926 - 0.0591\ pH$

5. $Ir_2O_3 + 6H^+ = 2Ir^{3+} + 3H_2O$ $\qquad \log(Ir^{3+}) = -3.79 - 3\ pH$

6. $Ir^{3+} + 3e = Ir$ $\qquad E_0 = 1.156 + 0.0197 \log(Ir^{3+})$

7. $IrO_2 + 4H^+ + e = Ir^{3+} + 2H_2O$ $\qquad E_0 = 0.233 - 0.2364\ pH - 0.0591 \log(Ir^{3+})$

8. $2IrO_4^{2-} + 10H^+ + 6e = Ir_2O_3 + 5H_2O$ $\qquad E_0 = 1.680 - 0.0985\ pH + 0.0197 \log(IrO_4^{2-})$

9. $IrO_4^{2-} + 4H^+ + 2e = IrO_2 + 2H_2O$ $\qquad E_0 = 2.057 - 0.1182\ pH + 0.0295 \log(IrO_4^{2-})$

A large number of Ir(III) and (IV) complexes are known, but only the halo-compounds have been studied from an electrochemical point of view. The Ir(IV) halo-complexes have octahedral geometry and are easily reduced to the III state in basic, neutral, or slightly acidic solutions, with retention of their configuration; in acidic media they seem to be quite stable and kinetically inert. In contrast, the IrX_6^{3-} (X = Cl, Br, I) ions are thermodynamically stable but kinetically labile, forming aquo-halo-complexes in aqueous solutions. The aquo-complexes are easily converted into hydroxy-aquo complexes in basic solutions.

A list of data on standard and formal potentials of iridium systems in aqueous solutions referred to the NHE at 25°C is presented in Table 1.1. Also, the only known normal potential for Ir in fused salts is included, no data for nonaqueous solutions being available.

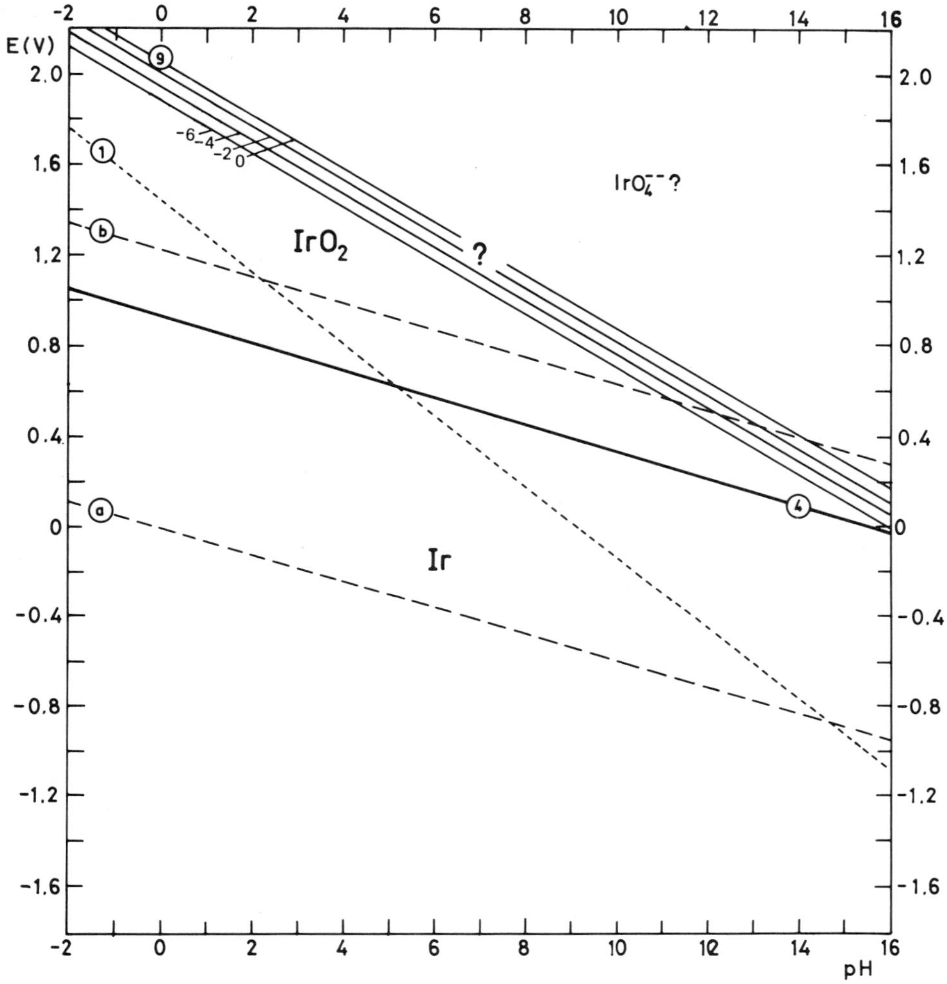

FIG. 1.1. Potential-pH diagram of the Ir-H_2O system at 25°C. (From Ref. 171, p. 375, by courtesy of the Centre Belge de la Corrosion.)

Goldberg and Hepler [172] calculated thermodynamic data for Ir compounds from the published standard potentials, and conversely they used these data to estimate the formal potentials for reduction of the III and IV chloro-complexes to metal. The results are gathered in Table 1.1. Redox potentials were usually measured with an Ir wire electrode at 10^{-3} M by potentiometric methods.

These potential values show that the stability of the Ir(IV) complexes rises with an increasing volume of the halogen ligand and decreases with a growing number of H_2O molecules in the aquo-complex.

TABLE 1.1. Standard and Formal Potentials in Aqueous Solutions

Half-reaction	Standard or formal potential (V)	Conditions	Refs.
1. $IrO + 2H^+ + 2e = Ir + H_2O$	+0.870 ± 0.020	2 N H_2SO_4	[172]
2. $IrO_2 + 4H^+ + 4e = Ir + 2H_2O$	+0.935 ± 0.005	2 N H_2SO_4	[172]
3. $IrCl_6^{2-} + 4e = Ir + 6Cl^-$	+0.86	calcd	[172]
4. $IrCl_6^{3-} + 3e = Ir + 6Cl^-$	+0.86	calcd	[172]
5. $IrCl_6^{2-} + e = IrCl_6^{3-}$	+0.867	0.3 M HCl (Rh electrodes)	[172]
6. $IrCl_6^{2-} + e = IrCl_6^{3-}$	+0.933	1 M HCl; 1 M NaCl	[173]
$IrCl_6^{2-} + e = IrCl_6^{3-}$	+1.05	0.2 M HCl (DME)	[174]
6. $IrCl_5(H_2O)^- + e = IrCl_5(H_2O)^{2-}$	+1.0	0.2 M HNO_3	[172]
7. trans-$IrCl_4(H_2O)_2 + e =$ trans-$IrCl_4(H_2O)_2^-$	+1.22 ± 0.02	0.4 M HNO_3	
		0.4 M $HClO_4$ + 0.2 M NaH_2PO_4	[172]
8. $IrCl_3(H_2O)_3^+ + e = IrCl_3(H_2O)_3$	+1.30 ± 0.01	0.4 M $HClO_4$	[172]
9. $IrBr_6^{2-} + e = IrBr_6^{3-}$	+0.883 ± 0.003	1 M $HClO_4$	[175]
10. $IrI_6^{2-} + e = IrI_6^{3-}$	+0.49	1 M KI	[172]
11. Ir^{3+} (0.2 to 0.006 M) + 3e = Ir	−0.092 vs Pt/1 M $PtCl_2$	KCl-LiCl at 450°C (41:49 M%)	[176]

2. VOLTAMMETRIC CHARACTERISTICS

2.1. POLAROGRAPHIC AND VOLTAMMETRIC CHARACTERISTICS

Some work has been carried out on the direct polarography of Ir in aqueous solutions, but positive results have been obtained only in HCl solutions. No polarographic waves [177] were observed for Ir(IV) and (III) compounds in bromide, cyanide, ethylenediamine, pyridine, and thiocyanate electrolytes and for Ir(III) in chloride, tartrate, and tetraethylammonium iodide solutions. A very poorly defined reduction wave was recorded in NaF solutions, the results discouraging further investigation. In 0.2 M HCl [174] supporting electrolyte, the reduction of Ir(IV) to Ir(III) takes place at the DME with half-wave potentials more positive than the oxidation potential of the Hg but with well-defined waves. The limited current density is proportional to the Ir concentration, thus making this solution suitable for polarographic analysis. No consistent results have been found in $HClO_4$ and H_2SO_4 solutions.

In basic solutions, Page and Van Loon [178] showed the reduction of Ir(IV) to Ir(III) in NaCl solutions to be reversible and diffusion controlled at constant ionic strength and between pH 11.3 and 13. The linear dependence of half-wave potentials on pH, as well as the results of coulometric and spectrophotometric analysis, led to a description of the reaction as $Ir(OH)_6^{2-} + H_2O + e — Ir(OH)_5H_2O^{2-} + OH^-$. The polarographic oxidation of the Ir(III) complex is also reversible but not suitable for quantitative work since the complex is easily oxidized by air in basic solutions.

The reduction of $IrCl_6^{2-}$ to Ir(III) in acidic NaCl solutions at a Pt rotating electrode [177] and the reoxidation of Ir(III) to Ir(IV) at slightly more positive potentials has been demonstrated by voltammetry. The limiting diffusion current was proportional to concentrations in the range 2×10^{-3} to 5×10^{-6} M, which can be of practical use in analysis. The galvanostatic study of the $IrCl_6^{2-}/IrCl_6^{3-}$ system in a 1 M HCl electrolyte [179] revealed the redox process to be reversible and controlled by diffusion of the reacting species with no slow chemical reaction preceding the electrode process.

In the field of organometallic electrochemistry, polarography has been applied to study the influence of the Ir coordination on the reactivity of π-bonded organic ligands in quinone complexes [180]. Thus it was found that the reduction of π-duroquinone-Ir-π-cyclopentadienyl, 1 in aqueous and aquo-alcoholic solutions is reversible and the half-wave potential independent of concentration while the cathodic wave is irreversible in acetonitrile solvent. Comparison with polarographic data for noncoordinated quinones led to the conclusion that a change of the electronic density in the quinone part of the complex takes place during the electrode reduction, metal coordination having the effect of increasing the energy of the quinones lowest vacant orbital. In the same way, the structure of the neutral chelate $(\pi\text{-}C_5H_5)IrS_2C_4F_6$ has been studied by polarography in dimethoxyethane solutions [37], and a reversible cathodic wave corresponding to one-electron exchange was obtained.

Tables 2.1.1 and 2.1.2 give a list of the polarographic and voltammetric data on Ir systems published up to 1971.

TABLE 2.1.1. Polarographic Characteristics in Aqueous Solutions

Substance	Product	Conditions	$E_{1/2}$ (V vs SCE)	I	$E_{3/4} - E_{1/4}$	Remarks	Refs.
$IrCl_6^{2-}$	$IrCl_6^{3-}$	0.5 M NaF, pH = 5	-1.4i	-	-	0.01% Gelatin	[177]
$IrCl_6^{2-}$	$IrCl_6^{3-}$	0.2 M HCl		1.49	rev		[174]
$Ir(OH)_6^{2-}$	$Ir(OH)_5(H_2O)^{2-}$	NaOH + NaCl; μ = 0.5 M; pH = 12	-0.6w		rev	$E_{1/2}$ depending on pH	[178]
π-Cyclopentadienyl—Ir—duroquinone	?	0.1 N KCl	-1.67		rev		[180]
		Nonaqueous solvents					
π-Cyclopentadienyl—Ir—duroquinone	?	CH_3—CN + 0.1 N $(C_2H_5)_4NClO_4$	-1.99 (SCE)		irr		[180]
$(\pi\text{-}C_5H_5)$—Ir—$S_2C_4F_6$		Dimethoxyethane + Bu_4NClO_4	-1.7 vs (Ag/10^{-3} M $AgNO_3$)		rev		[37]

TABLE 2.1.2. Voltammetric Characteristics in Aqueous Solutions

Substance	Product	Conditions	Electrode	Tech-nique	Potential (V vs SCE)	Rev	Remarks	Refs.
$IrCl_6^{2-}$	$IrCl_6^{3-}$	Acid; 0.1 M NaCl	rot Pt dsk	E swp	$E_{1/2}$ = +0.73	sl irr	$\frac{i_e}{CA}$ = const	[177]
$IrCl_6^{2-}$	$IrCl_6^{3-}$	1 M HCl	Pt wr	I stp	$E_{1/2}$ = +0.930 (NHE)	rev	$\frac{i\tau^{1/2}}{AC} = 244 \pm 6$; D = $(8.2 \pm 0.4) \times 10^{-6}$	[179]
$IrCl_6^{3-}$	$IrCl_6^{2-}$	1 M HCl	Pt wr	I stp	$E_{1/2}$ = +0.930 (NHE)	rev	$\frac{i\tau^{1/2}}{AC} = 239 \pm 5$; D = $(7.8 \pm 0.3) \times 10^{-6}$	[179]

3. KINETIC PARAMETERS AND DOUBLE-LAYER PROPERTIES

3.1. KINETIC PARAMETERS

Research on the kinetics of electrochemical reactions with Ir compounds is almost nonexistent. Mehl and **Lohman** [181] have found the standard exchange current density to be $j_0 = 1.7$ to 7×10^{-3} A/cm^2 for the redox system $IrCl_6^{2-}/IrCl_6^{3-}$ at 0.1 M of the species in 0.5 M HCl solutions with anthracene electrodes. Their purpose was to study the behavior of insulator semiconductors in a redox electrolyte solution. No considerations of the kinetics of the redox process were made.

3.2. DOUBLE-LAYER PROPERTIES

The few available data on the zero charge potential (p.z.c.) on Ir electrodes by direct measurements were obtained by Andersen et al. through application of the "open circuit scrape" method [62, 70] in K_2SO_4 and $NaClO_4$ solutions at different pH values. The presence in the solution of anions that may be specifically adsorbed causes a p.z.c. shift to more negative potentials which in the Br^- case amounts to about 200 mV. The authors observed that the p.z.c. of Ir becomes more negative with increasing pH value. Although the p.z.c. values reported by Andersen are quite imprecise, they are close to those derived by Petrii and Nguen Van Tue [182, 183] from hydrogen adsorption studies on iridized Pt by the isoelectric potential shift method and different from the early value recorded through capacitance measurements [184]. An outline of the reported results is given in Table 3.2.1.

3.2.1. Hydrogen and Oxygen Adsorption on Ir Electrodes

The adsorption properties of Ir with respect to hydrogen and oxygen ions have been the subject of several investigations. Earlier work by Breiter and Knorr [75] showed a considerable surface coverage with adsorbed atomic hydrogen for Ir; at the reversible potential in 8 N H_2SO_4 it was estimated to be ~ 45%. The coverage of the electrode, calculated from integration of the anodic potential sweeps or directly from anodic charging curves in 2.3 M H_2SO_4 solutions at 30°C [75], reveals a rapid reduction of the coverage from 80 to 30% with increasing electrode overvoltage in the range 0.05 to 0.1 V, followed by a more gradual decrease at higher potentials. The coverage is smaller than on Pt and Pd, but similar to that on Rh.

Charging curves and isoelectric potential shift measurements [75, 182, 183] at iridized and smooth Ir electrodes in acidic and alkaline solutions indicate overlapping of the

TABLE 3.2.1. Potential of Zero Charge for Ir (NHE)

p.z.c.	Conditions	Method	Refs.
-0.04 ± 0.05 (pH = 7)	1 N $NaClO_4$ or 1 N K_2SO_4, 25°C	Open circuit scraper	[62],[70]
(-0.17 ± 0.05) (pH = 11)	1 N $NaClO_4$ or 1 N K_2SO_4, 25°C	Open circuit scraper	[62],[70]
-0.12 (iridized)	1 N KBr + 10^{-2} N HBr, 20°C	Isoelectric potential shift	[182]
-0.13 (iridized)	1 N KCl + 10^{-2} N HCl, 20°C	Isoelectric potential shift	[183]
-0.06 (iridized) (pH = 2.4)	1 N Na_2SO_4 + 10^{-2} N H_2SO_4	Isoelectric potential shift	[183]
+0.55 (pH = 0.5)	2 N NaCl	Capacitance	[184]

hydrogen and oxygen regions even in the presence of high concentrations of Cl^-. It has been observed that hydrogen is adsorbed reversibly at small polarizations in both types of solutions although the adsorbed amount is smaller than for Pt [75, 182, 183, 185]. The heat of hydrogen adsorption on Ir shows a rapid decrease with increasing coverage from 0 to 28% [75], and a slower diminution at higher coverages. This variation suggests two types of hydrogen adsorption, as in the case of Pt, with strongly and weakly bonded H atoms to the metal. This suggestion is supported by the appearance of two peaks in the potential region 0.4 to 0.15 V and 0.15 to 0 V (RHE) of the cathodic potentiodynamic (i-E) curves at Ir in H_2SO_4 solutions and the corresponding humps in the charging curves [75, 185]. Bagotzkii et al. [185] claimed that the Freundlich equation governs the isothermal adsorption of hydrogen for the middle values of θ, which would indicate an exponential distribution of the surface heterogeneity, whereas Breiter found that the Langmuir isotherm works fairly well with the weakly bonded H atoms on Ir [75].

The anion effect at a bright Ir electrode has been established by independent measurements in acidic and alkaline solutions [75, 182, 183], and it appears to be similar to that observed on Pt. This effect is more pronounced the higher the adsorbability of the anion, and it manifests itself through a decrease in the differential heat of hydrogen adsorption due to competitive adsorption of anion and hydrogen atoms. Comparison of these results and of the dependence of hydrogen adsorption on potential and solution pH with those reported for Pt leads to the conclusion that for the hydrogen and oxygen atoms bonded to the metal the net negative end of the dipole points toward the electrolyte.

Hydrogen adsorption on Ir deposits has proved to be very sensitive to poisoning by Hg and As atoms [186]. At Hg coverages of 0.1 % in 0.1 N H_2SO_4 solutions, one Hg atom displaces 74 H atoms on the electrode surface. This effect decreases with higher Hg coverage although the extension covered by hydrogen and, to a lesser degree, the energy of the H-metal bond are also reduced. On the other hand, small quantities of As (coverages from 0.01 to 20%) showed no influence on hydrogen adsorption but sharply affected the results at coverages larger than 50%. Ir electrode poisoning by As increases the energy of the H bonding to the metal and displays insignificant desorption activity. The decrease in hydrogen adsorption is rather due to a lowering of the ionization rate of hydrogen atoms.

These electrode poisoning effects on hydrogen adsorption are in good agreement with those observed for other Pt metals (Os [218], Pt [279], Rh [280], and Ru [281]) which can be summarized as follows. Progressive poisoning of the electrode reduces the hydrogen coverage, the desorption activity being much more pronounced for small amounts of contaminant and, under the same conditions, more notable for Hg than for As. The bonding energy of the adsorbed hydrogen is either fundamentally unaffected by the contamination or slightly decreased by Hg adsorption or increased by As poisoning. The anodic charging curves for hydrogen show no distortion except for strongly poisoned electrodes. Finally, As-contaminated electrodes can be partly regenerated by anodic polarization, but the treatment is not effective for Hg poisoning since the Hg adsorbed is not completely removed from the electrode even at high anodic potentials [279]. On the other hand, As poisoning favors the adsorption of oxygen and the oxidation of the metals.

These results could be explained in part by postulating an initial adsorption of As and Hg atoms on the active sites of the metal, which would account for enhanced hydrogen desorption at low levels of poisoning, followed by accumulation of the contaminant on the blocked sites which should exert a weaker influence upon hydrogen adsorption. This has been proved for Pt poisoning where Hg "islands" on the metal and formation of β-Hg_4Pt alloy have been detected [282].

More information is needed on the interaction between the Hg and As atoms adsorbed and the hydrogen-metal dipoles in order to explain the null or contrary effects of these contaminants on the hydrogen bonding energy.

Fundamental information on oxygen adsorption and formation of oxides on Ir has been furnished by the work of Hoare [187, 100] and Breiter [75] in H_2SO_4 solutions. From galvanostatic, potentiostatic, capacitance, and voltammetric measurements they concluded that atomic oxygen (or, according to Breiter, OH^- ions adsorbed initially and later converted to atomic oxygen) is reversibly chemisorbed in the potential region 1.18 to 1.38 V (NHE) with a surface coverage smaller than 100% and with possible formation of oxide patches. The latter do not exceed 20% of the total oxygen adsorbed. The adsorbed layer is a good electronic conductor that obeys the Langmuir adsorption isotherm. At potentials higher than 1.4 V, the Ir—O layer begins to oxidize to IrO_2, forming a reversible and stable oxide no more than about two monolayers thick, with good electronic conductivity and catalytic properties on which oxygen evolution takes place. It has been found that adsorption of carbonaceous species has a negligible influence on the oxygen adsorption on Ir.

4. ELECTROCHEMICAL STUDIES

4.1. SURFACE OXIDATION AND ANODIZATION

As mentioned earlier, surface oxidation of the Ir electrode in H_2SO_4 solutions starts at about 1.1 V (NHE) with the formation of a chemisorbed layer of oxygen which is converted to a layer of IrO_2 as the electrode is anodized at higher potentials. Veselovskii et al. [188], in their studies on the production of peroxodisulfuric acid in H_2SO_4 at Ir electrodes at low temperatures and high potentials (1.9 to 11.0 V, RHE), recorded three successive Tafel lines that the authors ascribed to different processes on the electrode surface. They assume that at a potential of 2.3 to 2.4 V, sulfate anions begin to be adsorbed, and at potentials higher than 3.4 V and up to 11 V the electrode goes through a series of passivation states that correspond to changes of composition and the formation of high-order iridium oxides. However, the authors did not identify such oxides or the type of intermediate built up on the electrode surface. In addition, the Russian authors observed that the anodic behavior of the metal depended on the previous treatment of the electrode; thus a fresh electrode prepared by mechanical polishing and then dipped successively in HNO_3 and alkali solutions dissolves anodically in 2 to 3 $\underline{N}$ H_2SO_4 solutions and produces no $H_2S_2O_8$ acid at higher potentials. A similar dissolution of the metal was noticed as the electrode was anodized for a long time at a current density greater than 1 A/cm^2 in 10 and 12 $\underline{N}$ H_2SO_4 solutions at 3.3 V (RHE). With a H_2SO_4 concentration higher than 14 $\underline{N}$, the iridium anode undergoes such a deep passivation that it can be used later in dilute solutions without any anodic corrosion.

In concentrated solutions of HCl [99], iridium dissolves at the anode as Ir(III) and Ir(IV), the formation of Ir(III) being more significant at less positive potentials. This process requires a high overvoltage, and Ir passivates readily even at extremely low current densities. At low HCl concentration, anodic formation of a chemisorbed layer of oxygen causes the passivation of the electrode; the fraction of oxidized surface decreases with an increasing concentration of the acid.

Passivation of the iridium electrode has also been detected in the eutectic melt $AlCl_3$-NaCl-KCl at 110°C [189] where cathodic and anodic charging curves and impedance measurements yielded inconclusive results on the composition of the passivating layer. It seems that a nonconducting, insoluble, porous compound about 100 monolayers thick is formed on the metal surface, although the passivation layer can be cathodically reduced.

5. APPLIED ELECTROCHEMISTRY

5.1. ELECTROPLATING AND ELECTRODEPOSITION

Owing to the mechanical properties of the metal, iridium coatings should be expected to show good erosion resistance and wear qualities, which along with the high melting point and work function should make them highly commendable for the electronic industry. However, commercial electrolytes for deposition of iridium are generally scarce, and the high cost and low availability of the metal are likely to restrict its practical applications. From a commercial view, the more promising plating baths are those developed from bromide and from chloride-sulfamate solutions [169]. The former bath is prepared by dissolving hydrated IrO_2 in boiling HBr, with the most advantageous concentration found to be IrO_2 5 g/l; cathodic efficiency decreases sharply with small deviations from the critical current density, 0.15 A/dm^2, at 75°C. Deposits under these conditions appear bright and adherent but somewhat porous and subject to cracking at thickness larger than 1 μm. The cathodic efficiency amounts to 65% on base metals (Cu, brass, Ni, mild steel, etc.) but only 45% on Ti, and the plating rate is limited to about 1 μm/hr.

In the case of chloride-sulfamate solutions, the composition of the electrolyte is $IrCl_3.4H_2O$ (5 to 15 g/l) and NH_2SO_3H (25 to 50 g/l). The current efficiency again shows a decline with increasing current density and with time which is more marked at high temperature. Nevertheless, the efficiency can be maintained constant by using auxiliary anodes of Ir with high ac or dc current densities (10 to 50 A/dm^2). This fact was interpreted as being due to a shift in the equilibrium of the Ir complexes in solution toward a more favorable one for deposition. The operating parameters recommended for use are: T, 50 to 80°C; cathode current density, 2 to 6 A/ft^2; anode current density, 100 to 200 A/ft^2; ac current density, 100 to 200 A/ft^2; and ac current plated surface, 350 to 700 A/ft^2. The maximum coating thickness reported was 25 μm, although heavy deposits appear to be highly stressed and cracked.

Compact Ir deposits were obtained by Nidola [190] from chloride solutions at 70°C and high current density. The plating bath was a solution of $(NH_4)_2IrCl_6$ (24 g/l) and H_3BO_3 (40 g/l) at pH 5. It was observed that current efficiency increased with increasing boric acid concentration, but after several hours of electrolysis the NH_4Cl concentration in the solutions rose and the iridium deposition declined. A rotating Ir wire at 140 rpm was used as anode with a current density of 35 A/dm^2. A similar electrolyte is recommended by Hänsel [168], the composition of which is $(NH_4)_2IrCl_6$ (2.5 g/l) and Na_2HPO_4 (100 g/l) at pH 8.5. By electrolysis at 0.1 to 0.5 A/dm^2 current density and 12 to 25°C, bright coatings are obtained on mechanically polished brass electrodes after 1 to 3 hr of operation. At higher temperatures and in more acidic solutions, the deposits become black and deteriorate. On the other hand, higher current densities and operation times longer than 3 hr produce non-adherent and cracked coatings due to adsorption of hydrogen and the building of stresses in the deposit.

Krasikov and Krivonos [184] studied the electrodeposition of Ir at 5×10^{-3} to 6×10^{-2} M $IrCl_6^{2-}$ concentration in 2 M NaCl solutions at pH 5 and 25°C. The authors noticed a simultaneous coating of the electrode and hydrogen evolution, with a Tafel line describing the relation between the fraction of current density used in metal deposition and the potential. Current efficiencies were very small (0.08%) at -0.136 V (NHE) in dilute solutions and 0.3% at 6×10^{-2} M; hydrogen evolution increased and metal deposition decreased with rising current density. The discussion of results suggests that the determining step in the Ir deposition is the addition of one electron to an intermediate anion containing Ir(III), probably $(IrCl_5H_2O)^{2-}$. Reference 184 is related to the preliminary experiments reported by MacNamara [169a] on Ir electrodeposition from H_2IrCl_6 solutions, where plating conditions were notably dependent on temperature. The optimum operating parameters were 10 g/l of metal, pH 2, 60°C, and cathodic current density of 6.6 A/dm^2. Current efficiency under these conditions was ~ 6%; better yields (14%) were obtained at a current density of 2.2 A/dm^2 but the coatings showed lower quality.

Attempts have been made to obtain Ir platings from molten baths. Andrews, Kenahan, and Schlain [191] produced a uniform, gray, 1.8-mil thick deposit on Mo from a fused solution of 2% Ir in NaCN at 600°C after 4 hr electrolysis at 50 mA/cm^2. Coatings up to 15 mil thick can be deposited on W and Mo substrates by using higher Ir concentrations and longer polarizations in the current density range from 10 to 100 mA/cm^2.

6. CORROSION

The anodic corrosion of Ir has been investigated to a lesser extent than that of Pt. Extant data on electrochemical corrosion of the metal indicate a high resistance to anodic attack. Although Ir metal dissolves at the anode with the production of $IrCl_6^{3-}$ and $IrCl_6^{2-}$ complexes in solutions containing a high concentration of Cl^- ion, [99], the dissolution rate is very small. Llopis et al. [192] have shown by radiochemical methods that Ir at 50°C corrodes anodically in HCl solutions only at HCl concentrations higher than 6 N; at constant concentration, corrosion rates increased with rising anodic potential. In saturated solutions of NaCl a very small rate of attack (0.024 $\mu g/cm^2 hr$) was detected. Under these conditions, Ir(III), Ir(IV), and Cl_2 are formed at the anode, the formation of Ir(III) tending to be more significant as the potential becomes less positive but Ir(IV) always remaining the main product. This behavior is easily explained if we take into account the standard potentials of the Cl_2/Cl^- and $IrCl_6^{3-}/IrCl_6^{2-}$ systems according to which, in aqueous solutions containing

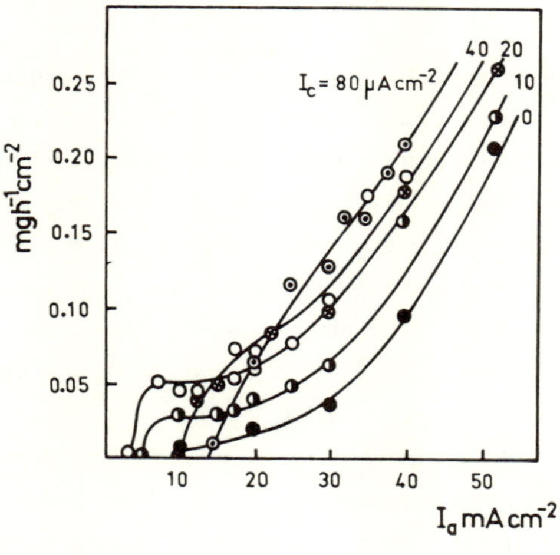

FIG. 6.1. Corrosion rates of an Ir electrode in 6 N HCl solutions by superimposition of 50 Hz ac on different current densities of direct current. Temperature, 19°C. (From Ref. 99 by courtesy of the Journal of the Electrochemical Society.)

Cl_2 at 1 atm, $IrCl_6^{3-}$ ions must be completely oxidized to $IrCl_6^{2-}$. From these results it can be concluded that anodic dissolution of Ir by dc in chloride solutions is very small but finite if the Cl^- concentration is high enough, the corrosion rates depending upon potential and the H_3O^+ and Cl^- concentrations.

As reported above, Veselovskii et al. [188] noticed the anodic dissolution of Ir in 3 to 5 N H_2SO_4 solutions at -30°C after successive treatment of the electrode in alkaline and nitric acid solutions. A similar deterioration of the electrode was observed in 10 and 12 N H_2SO_4 solutions after prolonged anodic polarization with current densities greater than 1 A/cm^2 at 3.3 V (RHE).

A notable rise in the corrosion rate of the metal is achieved if an alternating current is superimposed on the polarizing direct current [99, 192]. As seen in Fig. 6.1, in 6 N HCl solutions the superimposition of a 50-Hz sine wave ac does not affect the passivity of the electrode until its current density, i_a, is much higher than the direct current current density, i_d, applied. This effect is more pronounced as the ac frequency is lowered or the temperature increased. The decrease in HCl concentration reduces the corrosion rates if the other operating parameters are held constant. Under these conditions, the metal goes into the solution as Ir(III) and Ir(IV) complexes. The relative proportion of Ir(III) decreases as the potential becomes more positive, and it is enhanced as i_d decreases and/or the i_a/i_d ratio increases. Absorption spectra of corrosion products in 2 N HCl solutions obtained by polarization of Ir with direct and 50 Hz alternating currents reveal the formation of aquo- and diaquo-Ir(III) complexes at low current densities of direct current and aquo-pentachloro Ir(IV) complexes at high current densities.

Attack of the metal has also been observed when only ac has been applied to a sheet or powdered Ir electrode in aqueous solutions. In 0.25 to 1 N HCl solutions [192] at 50 mA/cm^2 (50 Hz ac), the corrosion rate of an iridium foil was about 250 μg/cm^2/hr, and did not change significantly with the acid concentration, Ir(III) and (IV) chloro-complexes being formed by the metal dissolution. Absorption spectrophotometry analysis showed that the Ir(III)/Ir(IV) ratio increased as the HCl concentration was lowered and that at the same time the proportion of aquo-chloro complexes in the solution also increased. These phenomena have been confirmed by Kryshchenko and Styrkas [193] for powdered electrodes with 50 Hz ac at high current densities. At high HCl concentration and 3.5 A/cm^2, the corrosion rate increased with time up to a constant value (5 mg/cm^{2}hr) after 15 min of electrolysis. In dilute HCl solutions the corrosion rate at a current density of 3 A/cm^2 goes through a maximum as a function of time; these rates are lower as the temperature of the electrolyte is raised. Similar trends are reported by Kadaner et al. [194] for corrosion of Ir electrodes with ac in HCl and H_2SO_4 solutions; current efficiencies, however, with respect to the anodic component of the ac, were very low, 3 and 10% at the most, respectively. On the other hand, the yields went through a maximum as the current density increased and fell off at higher temperatures and ac frequencies.

Studies on corrosion of Ir in nonaqueous solutions are almost nonexistent. Tiurin et al. [195] report that the metal dissolves at a rate of 11-12 mg/mole of product, when used as anode with a current density of 0.16 A/cm^2, for the electrochemical synthesis of dimethyl esters from some dicarboxylic acids in methanol and ethylene glycol solvents at 60 to 65°C and 50 to 55°C, respectively. The metal goes into solution as ionic species which are in part reduced to organometallic sols.

In the field of molten salts, Laitinen and Plambeck [176] carried out the anodic dissolution of Ir in molten eutectic KCl-LiCl at 450°C and low current densities, assuming the formation of Ir(III) ions.

Chapter VI-6

OSMIUM

J. F. LLOPIS* and F. COLOM

Department of Surface Chemistry and Electrochemistry
Institute of Physical Chemistry, C.S.I.C.
Madrid, Spain

Osmium is the most chemically reactive of the platinum metals, and no less than nine oxidation states (from +VIII to 0) have been identified for the metal compounds. As is well known, platinum metals show a remarkable tendency to form complexes with π-bonding ligands, which in the case of Os help generally to stabilize higher or lower oxidation states according to whether they are good π-donor or π-acceptor ligands, respectively. Some of these systems have been studied electrochemically, as shown below.

With the exception of a few important compounds (the tetroxide, osmiamate, and pentacarbonyl), all osmium complexes in aqueous solutions are octahedrally coordinated.

---

*Deceased.

The References for Chapters VI-4 through VI-7 appear on pages 267-275. A General Bibliography appears on pages 328 and 329.

1. STANDARD AND FORMAL POTENTIALS

The equilibrium diagram represented by Pourbaix and Zoubov [196] for the osmium-water system at 25°C shows the metal to be thermodynamically stable in aqueous solutions at any pH except in the presence of strong oxidizing substances or complexing agents (Fig. 1.1).

The drawing of the diagram was based on:

$H_2OsO_5 = HOsO_5^- + H^+$ $\qquad \log \frac{(HOsO_5^-)}{(H_2OsO_5)} = -10.0 + pH$

$HOsO_5^- = OsO_5^{2-} + H^+$ $\qquad \log \frac{(OsO_5^{2-})}{(HOsO_5^-)} = -14.50 + pH$

$H_2OsO_5 + 2e = OsO_4^{2-} + H_2O$ $\qquad E_0 = 0.418 + 0.0295 \log \frac{(H_2OsO_5)}{(OsO_4^{2-})}$

$HOsO_5^- + H^+ + 2e = OsO_4^{2-} + H_2O$ $\qquad E_0 = 0.714 - 0.0295\ pH + 0.0295 \log \frac{(HOsO_5^-)}{(OsO_4^{2-})}$

$OsO_5^{2-} + 2H^+ + 2e = OsO_4^{2-} + H_2O$ $\qquad E_0 = 1.142 - 0.0591\ pH + 0.0295 \log \frac{(OsO_5^{2-})}{(OsO_4^{2-})}$

The limits of the domain where these species occur are:

$H_2OsO_5/HOsO_5^-$ $\qquad pH = 10.0$

$HOsO_5^-/OsO_5^{2-}$ $\qquad pH = 14.50$

OsO_4^{2-}/H_2OsO_5 $\qquad E_0 = 0.418$

OsO_4^{2-}/OsO_5^{2-} $\qquad E_0 = 1.142 - 0.0591\ pH$

$OsO_4^{2-}/HOsO_5^-$ $\qquad E_0 = 0.714 - 0.0295\ pH$

$OsO_2 + 4H^+ + 4e = Os + 2H_2O$ $\qquad E_0 = 0.687 - 0.0591\ pH$

$OsO_4 + 4H^+ + 4e = OsO_2 + 2H_2O$ $\qquad E_0 = 1.005 - 0.0591\ pH$

$OsO_4 + H_2O = H_2OsO_5$ $\qquad \log(H_2OsO_5) = -0.55$

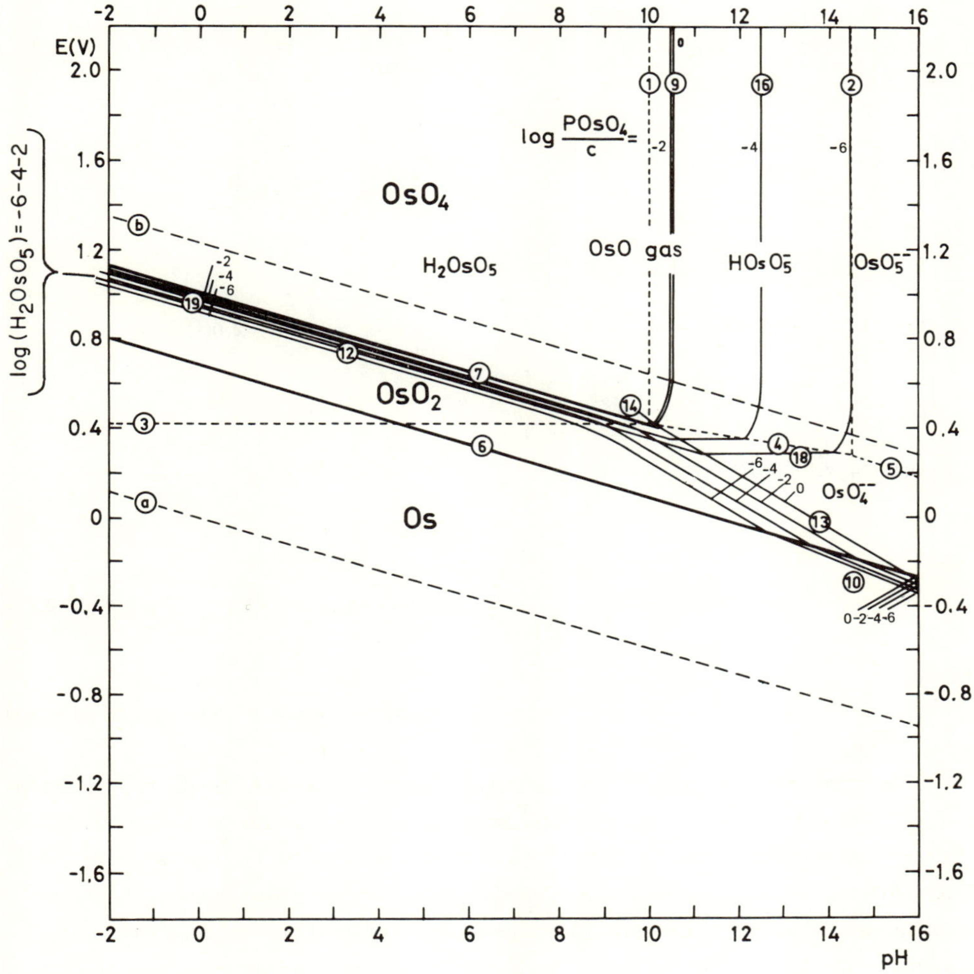

FIG. 1.1. Equilibrium diagram for the Os-H_2O system. (From Ref. 196, p. 367, by courtesy of the Centre Belge de la Corrosion.)

$OsO_4 + H_2O = HOsO_5 + H^+$	$\log(HOsO_5^-) = -10.55 + pH$
$OsO_4^{2-} + 8H^+ + 6e = Os + 4H_2O$	$E_0 = 0.994 - 0.0788\ pH + 0.0098 \log(OsO_4^{2-})$
$H_2OsO_5 + 8H^+ + 8e = Os + 5H_2O$	$E_0 = 0.850 - 0.0591\ pH + 0.0074 \log(H_2OsO_5)$
$OsO_4^{2-} + 4H^+ + 2e = OsO_2 + 2H_2O$	$E_0 = 1.607 - 0.1182 + 0.0295 \log(OsO_4^{2-})$
$H_2OsO_5 + 4H^+ + 4e = OsO_2 + 3H_2O$	$E_0 = 1.013 - 0.0591\ pH + 0.0148 \log(H_2OsO_5)$
$OsO_4 + 2e = OsO_4^{2-}$	$E_0 = 0.402 - 0.0295 \log(OsO_4^{2-})$

$$OsO_4 + H_2O = H_2OsO_5 \qquad \log \frac{p(OsO_4)}{(H_2OsO_5)} = -1.50$$

$$OsO_4 + H_2O = HOsO_5^- + H^+ \qquad \log \frac{p(OsO_4)}{(OsO_5H^-)} = 8.50 - pH$$

$$OsO_4 + H_2O = OsO_5^{2-} + 2H^+ \qquad \log \frac{p(OsO_4)}{(OsO_5^{2-})} = 22.99 - 2\ pH$$

$$OsO_4 + 2e = OsO_4^{2-} \qquad E_0 = 0.463 + 0.0295 \log \frac{p(OsO_4)}{(OsO_4^{2-})}$$

$$OsO_4 + 4H^+ + 4e = OsO_2 + 2H_2O \qquad E_0 = 1.035 - 0.0591\ pH + 0.0148 \log p(OsO_4)$$

The available data on the equilibrium of osmium systems at 25°C are given in Table 1.1.

Only a few potential values have been added to those quoted by Goldberg and Hepler [172] in their comprehensive review on thermodynamic properties and oxidation potentials of Pt metals compounds. These authors have recalculated some of Latimer's values [197], taking into account more recent thermodynamic data. They have also included the potentiometric measurements of the osmium oxides system reported by Cartledge and earlier published values for the Os(IV)/Os(III) couple in bromide and chloride solutions, although it is probable that in the latter solutions the halide species are hydrolyzed. Dwyer, Buckingham, and Sargeson [199] have measured the potential of 20 different cells involving Os(II) and Os(III) complexes with π-ligands vs the SCE in order to determine the influence of the overall charge, conjugation, coordinated halide, and substitution in the ligand on the stability of the complex cation. The results show that a higher stabilization of the Os(III) species is achieved by lowering the overall charge and/or decreasing the ligand conjugation, as well as by halide coordination, particularly the Cl^- ligand. Substitution in the ligand also seems to increase the stability of Os(III), but the effect on the redox potential is too small to draw reliable conclusions.

Publications on standard state potentials of Os compounds in nonaqueous electrolytes are completely lacking. Calculations of emf by Hamer, Malmberg, and Rubin [200] for cells containing a single molten fluoride, bromide, and iodide gave values of 0.094 and 0.162 V at 800 and 1000°C, respectively, for the $Os/OsBr_2/Br_2$ system.

TABLE 1.1. Standard and Formal Potentials in Aqueous Solutions

Half-reaction[a]	Standard or formal potential (V)	Conditions	Refs.
1. $OsO_4 + 8H^+ + 8e = Os + 4H_2O$	0.84	calcd	[172], [197]
2. $HOsO_5^- + 4H_2O + 8e = Os + 9OH^-$	0.00	calcd for 1 M (OH^-)	[172], [197]
3. $OsO_2.2H_2O + 4e = Os + 4OH^-$	-0.12	calcd	[172], [197]
4. $OsO_2.2H_2O + 4H^+ + 4e = Os + 4H_2O$	0.72	calcd	[172], [197]
5. $HOsO_5^- + 4H_2O + 4e = OsO_2.2H_2O + 5OH^-$	0.1	calcd	[172], [197]
6. $OsO_4 + 4H^+ + 4e = OsO_2.2H_2O$	0.964 ± 0.002	1.2×10^{-3} M K_2SO_4 or 10^{-2} N Na_3PO_4	[172]
7. $OsBr_6^{2-} + e = OsBR_6^{3-}$	0.349	2.1 M HBr extrapolated to infinite dilution	[172]
8. $OsCl_6^{2-} + e = OsCl_6^{3-}$	0.45	May be hydrolized	[172]
9. $Os(CN)_6^{3-} + e = Os(CN)_6^{4-}$	-0.99	calcd from polarographic data	[198]
10. $[Os(trpy)(bipy)(py)](ClO_4)_3/[Os(trpy)(bipy)(py)](ClO_4)_2$	0.8713	At zero ionic strength and Au electrodes	[199]

(continued)

TABLE 1.1 (continued)

Half-reaction[a]	Standard or formal potential (V)	Conditions	Refs.
11. $[Os(trpy)(bipy)Cl]Cl_2/[Os(trpy)(bipy)Cl]Cl$	0.5628	At zero ionic strength and Au electrodes	[199]
12. $[Os(py)_2(bipy)_2](ClO_4)_3/[Os(py)_2(bipy)_2](ClO_4)_2$	0.8345	At zero ionic strength and Au electrodes	[199]
13. $[Os(bipy)_2(py)Cl](ClO_4)_2/[Os(bipy)_2(py)Cl]Cl$	0.4836	At zero ionic strength and Au electrodes	[199]
14. $[Os(bipy)(py)_4](ClO_4)_3/[Os(bipy)(py)_4](ClO_4)_2$	0.8052	At zero ionic strength and Au electrodes	[199]
15. $[Os(bipy)(py)_3Cl](ClO_4)_2/[Os(bipy)(py)_3Cl]Cl$	0.4255	At zero ionic strength and Au electrodes	[199]
16. $[Os(bipy)_3](ClO_4)_3/[Os(bipy)_3](ClO_4)_2$	0.8847	At zero ionic strength and Au electrodes	[199]
17. $[Os(bipy)_2(acac)](ClO_4)_2/[Os(bipy)_2(acac)]ClO_4$[b]	0.1530	At zero ionic strength and Au electrodes	[199]
18. $[Os(trpy)_2](ClO_4)_3/[Os(trpy)_2](ClO_4)_2$	0.9866	At zero ionic strength and Au electrodes	[199]
19. $[Os(trpy)(py)_3](ClO_4)_3/[Os(trpy)(py)_3](ClO_4)_2$	0.7999	At zero ionic strength and Au electrodes	[199]

20.	$[Os(trpy)(bipy)Br](ClO_4)_2/[Os(trpy)(bipy)Br]Cl$	0.5675	At zero ionic strength and Au electrodes	[199]
21.	$[Os(trpy)(bipy)I](ClO_4)_2/[Os(trpy)(bipy)I]Cl$	0.5666	At zero ionic strength and Au electrodes	[199]
22.	$[Os(bipy)_2(py)I](ClO_4)_2/[Os(bipy)_2(py)I]Cl$	0.4888	At zero ionic strength and Au electrodes	[199]
23.	$[Os(bipy)_2(py)Br](ClO_4)_2/[Os(bipy)_2(py)Br]Cl$	0.4871	At zero ionic strength and Au electrodes	[199]
24.	$[Os(bipy)(py)_3I](ClO_4)_2/[Os(bipy)(py)_3I]Cl$	0.4508	At zero ionic strength and Au electrodes	[199]
25.	$[Os(bipy)(py)_3Br](ClO_4)_2/[Os(bipy)(py)_3Br]Cl$	0.4437	At zero ionic strength and Au electrodes	[199]
26.	$[Os(trpy)(bipy)(3\text{-}CH_3py)](ClO_4)_3/[Os(trpy)(bipy)(3\text{-}CH_3py)](ClO_4)_2$	0.8672	At zero ionic strength and Au electrodes	[199]
27.	$[Os(trpy)(bipy)(4\text{-}C_3H_7py)](ClO_4)_3/[Os(trpy)(bipy)(4\text{-}C_3H_7py)](ClO_4)_2$	0.8609	At zero ionic strength and Au electrodes	[199]
28.	$[Os(trpy)(bipy)(4\text{-}C_2H_5py)](ClO_4)_3/[Os(trpy)(bipy)(4\text{-}C_2H_5py)](ClO_4)_2$	0.8587	At zero ionic strength and Au electrodes	[199]
29.	$[Os(trpy)(bipy)(4\text{-}CH_3py)](ClO_4)_3/[Os(trpy)(bipy)(4\text{-}CH_3py)](ClO_4)_2$	0.8506	At zero ionic strength and Au electrodes	[199]

[a] trpy = terpyridine; bipy = bipyridine; py = pyridine; acac = acetylacetone. Systems 10 to 29 are Os(III)/Os(II) couples.

[b] Os(II)/Os(I) couple.

2. VOLTAMMETRIC CHARACTERISTICS

2.1. POLAROGRAPHIC CHARACTERISTICS

The voltammetric study of osmium compounds is riddled with difficulties stemming from the several possible oxidation states that the element is able to assume in solution, depending on pH values and type of media, and from the deficient knowledge possessed on the chemistry of the element. Advantage must be taken of other physico-chemical techniques in order to identify the chemical species formed and particularly the valency and coordination number of the complexes.

The few polarographic papers published on osmium are mainly concerned with electroreduction of Os(VIII) in solutions of the metal tetroxide. The pioneering investigations of Heyrovsky et al. [201] on the reduction of OsO_4 found inconsistent and ill-defined waves in solutions of alkali metal hydroxide. Improvement in definition of polarograms was observed for alkaline earth hydroxide solutions, where in $Ca(OH)_2$ saturated media (~0.02 M) OsO_4 showed a three-step wave which was described as corresponding to reduction of Os(VIII) to (VI), (IV), and (III) states, respectively. Preliminary attempts of OsO_4 polarography in acid electrolytes were not successful because of Hg oxidation by the osmium compounds. However, Meites [202], Meites and Cover [203], and Cover and Connery [204] have investigated with some detail the polarography of osmium in basic and acidic solutions. In alkaline solutions, the results obtained in the concentration range 0.1 to 10 N NaOH may be summarized as follows: at NaOH concentrations greater than 0.25 N, the polarograms of OsO_4 consist of three well-defined, diffusion controlled reduction waves which coincide with those reported by Heyrovsky. The first wave rises from zero applied voltage, showing that Os(VIII) is spontaneously reduced by Hg in this medium; the change of the diffusion current constant with concentration is due to variation of the solution viscosity. The other two waves seem to be irreversible and show constant half-wave potentials at concentrations above 1 N, irrespective of the alkali metal ion present. The height ratio of the waves and the results of controlled potential electrolysis appeared to confirm the reduction scheme proposed by Heyrovsky, although no conclusive evidence could be furnished for the Os(IV) → Os(III) process being responsible for the third wave since a grayish solid precipitated during the electrolysis. With intermediate concentrations, a maximum suppressor must be used in order to obtain well-defined waves, in keeping with the difficulties encountered by previous researchers in their work with these systems. With concentrations lower than 0.25 N NaOH the polarograms become more complicated, for additional cathodic waves are recorded, the nature of which is hard to account for.

A similar picture is drawn from the polarographic oxidation of osmium ions. In the concentration range 0.1 to 4 N NaOH, the polarograms show a single anodic wave that corresponds to two-electron oxidation of Os(IV) to Os(VI). At OH^- concentrations outside this range, two anodic waves are recorded for the same oxidation process. A detailed study of Os(VI) and Os(IV) alkaline solutions by polarographic and coulometric techniques [203] suggested the formation of intermediate Os(V) which quickly disproportionates, giving rise to the new waves observed. Further work [204] with constant current coulometry, voltammetric techniques, and vibrating DME provided no evidence on the presence of Os(V) in the solution, and the former results were explained by the authors as due to the existence of three Os(VI) and three Os(IV) different species which are in a moderately rapid equilibrium with OH^- ions. It should be pointed out that a clearer view of the problem should be obtained if the coordination requirements of the osmium ion and the possible hydroxy-aquo complexes equilibria were taken into account. Furthermore, except for the polarographic evidence on Os(III) existence in hydroxide solutions reported here, no Os(III) complex with oxy, hydroxy, or aquo-ligands has been isolated [198].

An extensive series of experiments on polarographic reduction of osmium complexes in mineral and organic acid electrolytes in acidic solutions has been reported [202, 203]. In phosphoric, sulfuric, hydrochloric, and perchloric acids, the reduction of OsO_4 yields a single diffusion-controlled wave which rises from zero potential and corresponds to the direct reduction of Os(VIII) to the (III) state. The same result is obtained for Os(VI) in HCl and H_2SO_4 solutions, where the reduction proceeds from (VI) to the (III) state. In organic acid media the cathodic polarography of Os(VIII) has produced consistent results only in saturated solutions of tartaric acid. Polarograms of these solutions show a three-step wave which describe the Os(VIII) → (IV) and (IV) → (III) processes, the third step being due to catalytic evolution of H_2. In weakly acidic buffers the electroreduction of Os(VIII) yields four waves, in agreement with the scheme (VIII) → (VI) → (IV) → (III); the fourth wave, in the case of 0.1 M pyridine, 0.1 M pyridinium ion (pH = 5.4) solution, may be due to the possible reduction of Os(III) to (II), or in acetate buffers (pH = 4.7) to catalytic evolution of H_2. No consistent interpretation was achieved for the polarograms recorded in $NH_3 + NH_4Cl$ solutions.

More promising results for analytical purposes have been produced in cyanide media. Addition of Os(VIII) to 0.1 to 1 M KCN solutions results in almost immediate reduction to Os(VI), probably with formation of $K_2OsO_2(CN)_4$. Polarograms of these solutions show two waves that are related to the Os(VI) → Os(III) and Os(III) → Os(II) reduction. This was corroborated by coulometry of the solutions and the final production of $K_4Os(CN)_6 \cdot 2H_2O$. The solutions of Os(III) in 0.1 N KCN render a single cathodic wave for the Os(III) → Os(II) reduction and another single wave for the polarographic oxidation of Os(II) in the same solutions. The anodic wave is not produced in 1 N KCN solutions.

Lal and Christian [205] recently undertook the study of the polarographic behavior of Os(VIII) in alkali metal hydroxide solutions and in the presence of a series of buffers at varying pH values. The results showed that in almost all the supporting electrolytes a three-wave polarogram is obtained similar to the early one reported by Heyrovsky. The second and, to a greater extent, the third wave appear to be irreversible; in a general way, the

TABLE 2.1.1. Polarographic Characteristics of Os in Aqueous and Nonaqueous Solutions at 25°C

Substance	Product	Conditions	$E_{1/2}$ (V vs SCE)	I = $i_d/cm^{2/3}t^{1/6}$	$E_{3/4}$ - $E_{1/4}$ (mV)	Remarks	Refs.
OsO_4	Os(IV)	~0.02 M $Ca(OH)_2$	-0.41				[201]
Os(VIII)	Os(III)		-1.16				[201]
OsO_4	Os(VI)	1 to 5 N NaOH	w	3.19 ± 0.21			[202]
Os(VIII)	Os(IV)	0.3 to 10 N NaOH	-0.61 ± 0.01w		irr	0.005% Triton X-10 added at 0.5 to 4 N NaOH	[202]
Os(VIII)	Os(III)	0.3 to 10 N NaOH	-1.51 ± 0.02w		irr	5 X 10^{-3}% Triton X-10 added to 0.5 to 4 N NaOH concn	[202]
Os(VI)	Os(IV)	0.5 to 10 N NaOH	-0.596 ± 0.009w		irr		[203]
Os(VI)	Os(III)	0.5 to 10 N NaOH	-1.521 ± 0.015w		irr		[203]
OS(VI)	Os(III)	0.1 M KCN	w	5.3			[202]
Os(VI)	Os(II)	0.1 M KCN	-0.81 ± 0.01w	7.0 ± 0.2	-69.5 rev		[202]
Os(II)	Os(III)	0.1 M KCN	(-0.74 ± 0.02)w	-1.6 ± 0.1			[202]
Os(VI)	Os(IV)	0.1 N $NaGH_4$[a] + 1 N NaOH	-0.495w	2.44 ± 0.01	rev	Diffusion	[206]
Os(VI)	Os(III)	0.1 N $NaGH_4$[a] + 1 N NaOH	-1.465w		rev	Diffusion	[206]
Os(IV)	Os(VI)	0.1 N $NaGH_4$[a] + 1 N NaOH	(-0.447)w		rev		[206]
$Os(III)(SacSac)_3$[b]	$Os(II)(SacSac)_3^-$	Acetone + 0.1 M $[(CH_3CH_2)_4]ClO_4$	-0.05w		58 ± 2 rev	vs Ag/AgCl (LiCl + acetone)	[207]

[a] $NaGH_4$ = sodium gluconate.

[b] SacSac = $[CH_3-CS-CH-CS-CH_3]^-$.

reduction waves of Os(VIII) for acidic solutions exhibit a poor reproducibility. Further evidence as to the irreversibility of these waves is provided by cyclic voltammograms with triangular sweeps.

It may be stated from the results reported that the polarographic behavior of Os conforms to the following pattern: In strongly acid media there is a direct reduction of Os(VIII) to Os(III); in more dilute solutions a second wave appears that corresponds to intermediate formation of Os(IV); and at pH values higher than 5 a third wave describes reduction to the (VI) state. The cyanide solutions seem to be more suitable for polarographic work. This picture is confirmed by the voltammetric work on these solutions.

In the search for analytical techniques to detect the several oxidation states of osmium species in solution, the polarography of gluconate complexes of Os in alkaline solutions has been tried [206]. Sodium gluconate spontaneously and quickly reduces Os(VIII) to Os(VI), and these species, as well as Os(IV) and Os(III), are strongly complexed and stabilized by the organic anion, making them suitable for polarographic analysis. The polarograms of Os(VI) and Os(IV) in 1 N NaOH + 0.1 N $NaGH_4$ (sodium gluconate) solutions display diffusion controlled waves similar to those described above for alkaline solutions as corresponding to Os(VI) → (IV) and (IV) → (III) reduction and (IV) → (VI) oxidation. The results showed the waves to be reversible and that some OH^- ligands were lost during reduction. Spectrophotometry of Os(VI) indicates that the osmium (VI) ion is chelated by two gluconates and that the complexing equilibration is established very slowly. The linear relationship between the height of the first wave and the Os(VI) concentration over the range 3×10^{-5} to 3×10^{-3} N allows the quantitative determination of this ion by polarographic analysis.

The only known polarographic data for Os compounds in nonaqueous solvents have been published by Bond, Heath, and Martin [207], who studied the reduction and oxidation of dithioacetylacetonate complexes of several transition metals in acetone media. The reduction of Os(III)$(CH_3{-}CS{-}CH{-}CS{-}CH_3)_3$ complex against the Ag/AgCl reference electrode in solutions of acetone and 0.1 N $[(CH_3{-}CH_2)_4N]ClO_4$ as supporting electrolyte yields a reversible wave controlled by diffusion of the reacting species. The height of the wave is linearly dependent on the concentration, and the electrode process is determined by a one-electron transfer. Polarographic oxidation of the product exhibits similar characteristics. No ligands are lost during either electrode process, and the bidentate structure of the complex is maintained for reduced and oxidized species.

The more reliable characteristic data published on the polarographic behavior of Os are given in Table 2.1.1.

2.2. VOLTAMMETRIC CHARACTERISTICS

The voltammetric study of Os compounds in aqueous solutions essentially confirms the findings of the polarographic research reported above. Buvet et al. [208] investigated the redox couple Os(VIII)/Os(VI) at Pt electrodes as a possible oxygen electrode for fuel cells at OsO_4 concentrations of 5.3×10^{-4} and 1.5×10^{-3} M in 0.5 M KOH solutions (pH = 13.7). The voltammetric reduction of these solutions occurs in two waves, the half-potential values of which are very close to those obtained by polarography for the Os(VIII) → (VI) and Os(VI) → (IV) steps. The limiting diffusion current is proportional to the OsO_4 ooncentration, but in oxygenated solutions the reduction of O_2 interferes with the Os(VI) → Os(IV) process, producing a deformation of the second wave and formation of solid OsO_2. The variation of the half-potential with the solution pH reveals that Os(VI) is not detected at pH < 8.9 because of its rapid disproportionation in OsO_2 + OsO_4 so that the OsO_4 compound is reduced directly to OsO_2. The work of Bardin and Goncharenko [209] on the same solutions with rotating Pt disk electrodes corroborated these results for OsO_4 concentrations over the range 10^{-4} to 10^{-2} M in 1 N NaOH solutions, including the disturbing influence of oxygen in the solution upon the reduction process. Absorption spectra of the solutions confirmed the reduction scheme proposed for this process.

Later work on 0.12 N NaOH solutions of 10^{-3} M OsO_4 [210] with a vibrating Pt electrode confirmed these results. However, with hydroxide concentrations higher than 2 N NaOH the voltammograms disclosed a third wave that was ascribed to intermediate formation of Os(VII). This species is produced as an intermediary step in the chemical reduction of Os(VIII) to Os(VI) by alkali metal hydroxide solutions and has also been detected by potentiometric titration, spectrophotometry, and analytical techniques [210, 211]. The oxidation of Os(VI) to Os(VIII) in 1 N NaOH solutions (the real composition of the complexes being trans-$OsO_2(OH)_4^{2-}$ and -$OsO_4(OH)_2^{2-}$) with a rotating Pt wire electrode yields a single wave controlled by diffusion. The same results were obtained by Kalvoda [212] for reduction and oxidation of Os(VIII) and Os(VI), respectively, in solutions of 1 N KOH.

An attempt has been made to reduce OsO_4 in acidic media with rotating Pt and graphite electrodes for dilute and concentrated solutions [213]. In H_2SO_4 solutions the cathodic polarization curve exhibits a wave at pH 5 and two waves at pH 3 that are attributed to the reduction steps $OsO_4 \rightarrow OsO_2$ and $OsO_4 \rightarrow OsO_2^{2+} \rightarrow OsO_2$. The insoluble OsO_2 remains as a black product on the electrode and dissolves anodically into the solution, showing a peak in the current-potential curve. The current density at this anodic maximum depends on the solution pH and the concentration of Os(VIII). The authors suggest that this relationship may be applied to analytical work by recording a previous calibration curve relating Os concentration and peak current density in 0.1 N H_2SO_4 solutions.

The voltammetric method has been applied to prove the existence of the dinitrogen complex of Os(III) amine [214] which otherwise has not yet been isolated. Cyclic voltammetry on the $[Os(NH_3)_5N_2]^{2+}$ complex in H_2SO_4 + K_2SO_4 solutions (pH = 1.1 and μ = 0.31) at a Pt disk electrode shows voltammograms which are consistent with a reversible one-electron charge-transfer followed by a break-up of the product into $[Os(NH_3)_5(H_2O)]^{3+}$. The rate constant of the complex decomposition was thus calculated, and the results were confirmed by chronoamperometry and electrolytic oxidation at constant potential of the original complex.

Voltammetric data from the reported work are presented in Table 2.2.1.

TABLE 2.2.1. Voltammetric Characteristics

Substance	Product	Conditions	Electrode	Technique	Potential (V vs SCE)	rev	Remarks	Refs.
OsO_4	Os(VI)	0.5 N KOH	Pt	E swp	+0.025		pH = 13.7	[208]
Os(VIII)	Os(IV)	0.5 N KOH	Pt	E swp	-0.5			[208]
Os(VIII)	Os(VI)	1 N NaOH	rot Pt dsk	E swp	+0.08		Diffusion wave	[209]
Os(VIII)	Os(IV)	1 N NaOH	rot Pt dsk	E swp	-0.56			[209]
OsO_4	Os(VII)	5 N NaOH	vib Pt	E swp	+0.14			[210]
Os(VIII)	Os(VI)	5 N NaOH	vib Pt	E swp	-0.11			[210]
Os(VIII)	Os(IV)	5 N NaOH	vib Pt	E swp	-0.41			[210]
Os(VI)	Os(VIII)	1 N NaOH (KOH)	rot Pt wr	E swp	(+0.1)	rev	Diffusion	[211]
Os(VIII)	Os(VI)	1 N NaOH	stat Pt gauze	E swp	+0.05	rev		[212]
Os(VI)	Os(VIII)	1 N NaOH	stat Pt gauze	E swp	(+0.13)	rev		[212]
$[Os(NH_3)_5N_2]^{2+}$	$[Os(NH_3)_5N_2]^{3+}$	$H_2SO_4 + K_2SO_4$	Pt dsk	CV	(E_p = +0.135)	$E_{pc} - E_{pa}$ = 120 mV	pH = 1.1; μ = 0.3	[214]

3. KINETIC PARAMETERS AND DOUBLE-LAYER PROPERTIES

3.1. KINETIC PARAMETERS

As described above, Lal and Christian [205] studied the polarographic behavior of Os(VIII) in several electrolytes and reported the only known data on kinetic parameters of the reduction process at the dropping mercury electrode. The polarograms showed reduction of OsO_4 in the different media to proceed through a three-step reaction. An initial examination suggested the second and third steps, which correspond to Os(VI) → (IV) and Os(IV) → (III), to be irreversible. The electrochemical parameters were calculated by the Meites and Israel equation for a totally irreversible wave, and the results are given in Table 3.1.1.

TABLE 3.1.1. Kinetic Parameters at 25°C

Reaction	Conditions	Technique	Rate constant k° (cm/sec)	αn	Hg height (cm)
Os(VI) + 2e → Os(IV)	satd $Ca(OH)_2$	E swp	5.1×10^{-8}	0.73	59.1
	satd $Ca(OH)_2$	E swp	1.6×10^{-7}	0.61	49.1
Os(IV) + e → Os(III)	satd $Ca(OH)_2$	E swp	3.8×10^{-17}	0.68	59.1
Os(VI) + 2e → Os(IV)	satd LiOH	E swp	1.2×10^{-8}	0.63	59.1
	satd LiOH	E swp	8.4×10^{-7}	0.43	49.1
Os(IV) + e → Os(III)	satd LiOH	E swp	1.1×10^{-10}	0.36	59.1
	satd LiOH	E swp	9.2×10^{-11}	0.36	49.1
Os(VI) + 2e → Os(IV)	2 N KOH	E swp	1.1×10^{-8}	0.51	59.1
	8 N KOH	E swp	5×10^{-14}	1	59.1
Os(VI) + 2e → Os(IV)	0.05 M $Na_2B_4O_7$ (pH = 9.15)	E swp	8.9×10^{-7}	0.79	49.1
Os(IV) + e → Os(III)	McIlvaine buffer (pH = 2.2)	E swp	6.8×10^{-7}	0.24	49.1

The data corroborate the early analysis (rate constants are less than 3×10^{-5} cm/sec). In $Ca(OH)_2$, LiOH, and borax solutions the half-wave potentials for the second step become more positive with a decreasing head of mercury which signifies a totally irreversible process. Rate constants are higher for the second wave but decrease for the third wave when the mercury height is raised. Table 3.1.1 lists the parameters for the same mercury heights in different solutions so that the processes may be compared. The irreversibility of the waves was confirmed by current-time curves and cyclic triangular sweep voltammetry.

3.2. DOUBLE-LAYER PROPERTIES

There are no published data regarding the potential of zero charge of osmium.

Studies of hydrogen and oxygen adsorption on Os electrodes by anodic potentiodynamic [215] and charging curves [216, 217] in acidic and basic solutions show a broad overlapping of the hydrogen, double-layer, and oxygen potential region which makes the determination of the amount of hydrogen adsorbed on the electrode difficult. The hydrogen region in acidic solutions extends from 0.06 to 0.15 V (RHE), and the amount of hydrogen adsorbed increases with the pH of the solution. Hydrogen adsorption also depends on the electrolyte, decreasing in the order KOH > H_2SO_4 > HCl > HBr [215-218]. By contrast, the heat of hydrogen adsorption on Os seems unaffected by changes of electrolyte or solution pH; therefore, Breiter [75] applied with some success the Langmuir approximation to the isotherms of hydrogen adsorption on Os. The potentiodynamic curves at Os electrodes in 0.1 N H_2SO_4 and HCl solutions disclosed the adsorbed hydrogen to consist only of the very weakly bonded type [215]. Halide ions are adsorbed on the electrode in the order $I^- > Br^- > Cl^-$, hindering the adsorption of hydrogen and oxygen and acting as poison for catalytic hydrogenation reactions [217]. Vovchenko, Khomchenko, and Ul'ko [218] investigated the poisoning effect of As, Hg, and I^- on the adsorption of hydrogen on Os by charging curves in 0.1 N H_2SO_4. They showed that contamination with As facilitates the oxidation of the electrode and decreases the amount of hydrogen adsorbed but slightly affects the hydrogen-metal bonding energy. These effects are much stronger at lower concentrations of the contaminant and can be reduced by anodic polarization of the poisoned electrode. Hg adsorption does not distort the anodic charging curves for hydrogen but does decrease hydrogen adsorption to a greater extent than As. Finally, the specific adsorption of I^- decreases the hydrogen and double-layer region and restrains hydrogen adsorption; at coverages higher than 100%, I^- adsorption notably affects the hydrogen-metal bond.

The adsorption of oxygen on Os in acidic and basic solutions is irreversible and begins at less positive potentials than for the other Pt metals [217]. Anodic charging curves do not indicate the formation of a new phase at high potentials and in the region of 0.85 to 0.9 V (RHE) reveal a strong adsorption of oxygen together with dissolution of the metal. Llopis and Vazquez [99] studied the anodic behavior of Os in 2 N $HClO_4$ and HCl at 20°C and found that, at 0.83 V (RHE), the metal oxidizes, forming a dark compound even at high concentrations of HCl. From the value of the potential the authors postulated the formation of OsO_4 with likely dissolution of the oxide into the electrolyte.

4. APPLIED ELECTROCHEMISTRY

Except for the papers quoted on the adsorption of oxygen [217, 99] no work has been carried out on the anodic behavior of osmium or its electrochemical corrosion. In a test for anodic electrode materials, Alder and Fleet [219] proved that Os can be used as an inert electrode for anodic polarizations up to 0.55 V (SCE) in solutions of pH 1 to 4.2 and up to 0.4 V in pH 10.

A Russian industrial patent [220] describes the electropolishing of Os at a temperature ≤ 50°C in solutions with a composition of (in weight percentage) 30% H_2O_2 (5 to 2), H_3PO_4 (5 to 25), and H_2O (25 to 75), with ac at 20 to 40 V.

4.1. ELECTRODEPOSITION AND ELECTROPLATING

Commercial electrolytes for the deposition of Os are not yet available, and possible applications have not been evaluated to a thorough extent. Osmium should be expected to show good wear properties since it has been proved to endure a thickness loss only a quarter that of chromium in comparative abrasive wear tests [169].

Very few papers have been published on the small-scale electrolytic deposition of Os. Khomchenko et al. [217] proposed a method for the preparation of black catalyst electrodes of Os analogous to that for Ru deposition. The cathodic coating of a Pt electrode was carried out in 0.1 N KOH solution at 70°C with a current density of 0.83 mA/cm^2 during 4 hr and short periods of anodic polarization. The anode consisted of a pressured brick of Os powder. A black and adherent deposit of about 40 to 45 mg was obtained in this way.

Llopis and Vázquez [221] tried to deposit Os from Na_2OsCl_6 solutions. Exploratory experiments showed that 5 g/l concentrations of chloro-osmiate yield good coatings. At low temperatures the cathodic efficiency of electrodeposition is very low, and at high temperature a rapid hydrolysis of the bath takes place. The solution pH markedly affected the current efficiency of the electrodeposition, such that in 0.1 N HCl solutions the coating yield was almost nil but in 0.01 N HCl the cathodic efficiency amounted to 34%. The bath showed strong deterioration after prolonged electrolysis. Thus the best operating recipe for Os deposition from this solution was: temperature = 50°C, solution pH $\simeq$ 6, current density = 5 mA/cm^2, electrode surface $\simeq 2\ cm^2$, and electrolysis time = 30 to 60 min under moderate stirring. Under these conditions, bright, gray, and adherent coatings are deposited.

An osmium plating process utilizing $(NH_4)_2OsCl_6$ or K_2OsCl_6 baths for making wear-resistant electric contacts has been patented by Notley and Barr [222]. Os coatings 0.05 to

5.08 μm thick are obtained from chloro-osmiate solutions with 6.25 to 87.5 g/l KCl at pH 0 to 4. The thickest Os plates were electrodeposited on a Au-coated Cu cathode from 10 g/l K_2OsCl_6 + 6.25 g/l KCl solutions, pH 1.5, at 70°C during 90 min electrolysis.

A large-scale plating bath with experimental details protected by patents has been developed by Greenspan [223]. The electrolyte is based on an anionic complex which contains 0.8 to 1.2 g/l osmium formed by reaction of OsO_4 with sulfamic acid. The solution is highly alkaline, operating at a pH between 13.5 and 14 at 75 to 80°C. The recommended current density is 15 mA/cm^2. Cathode efficiency rises to about 70% and remains relatively constant in the 10 to 20 mA/cm^2 current density range; at higher current densities the current efficiency decreases. Deposits are bright and adherent up to 1.3 μm; above 2.5 μm coatings become dull but retain their adherence and smoothness. A coating of 2.5 μm is obtained at 15 mA/cm^2 in 30 to 35 min.

Finally, Appleby [224] reported a successful attempt to prepare osmium coatings for oxygen and redox test electrodes from molten electrolytes. The melt was formed by dissolution of 8 wt% $OsCl_3$ in fused KCN-NaCN (equal parts in weight) at 600°C under nitrogen atmosphere. Bright and adherent though granular deposits were obtained on a Au cathode at 10 to 20 mA/cm^2 with graphite anodes; at lower current densities no plating was observed, and black coatings were deposited in the current density range from 20 to 50 mA/cm^2. Cathode efficiencies amounted to 20 to 40% referred to Os^{3+}, but no attempts were made to obtain deposits thicker than 2 to 5 μm because of deterioration of the bath. The author claims the method has good practical prospects if careful control of the atmosphere on the melt and replenishment of osmium salt in the bath are provided.

Chapter VI-7

PALLADIUM

J. F. LLOPIS* and F. COLOM

Department of Surface Chemistry and Electrochemistry
Institute of Physical Chemistry, C.S.I.C.
Madrid, Spain

Solid palladium compounds occur primarily in the II and IV oxidation state (PdO, PdO_2), together with less abundant I and III species (Pd_2O, Pd_2O_3). In aqueous solutions, only the oxidation states II and IV in palladium complexes with coordination numbers 4 and 6, respectively, are known.

---

*Deceased.

The References for Chapters VI-4 through VI-7 appear on pages 267-275. A General Bibliography appears on pages 328 and 329.

A main feature of this metal that is very important in the study of its electrochemical behavior is the capacity for hydrogen absorption. This phenomenon takes place along with changes of the crystalline structure of the metal, essentially with formation of α and β solid solutions and the Pd_2H (or Pd_4H_2) hydride.

1. STANDARD AND FORMAL POTENTIALS

1.1. AQUEOUS SOLUTIONS

Standard potentials in V vs NHE for several electrochemical reactions of Pd compounds are presented in Table 1.1.1. Some of these potentials have been calculated from thermodynamic data given by Latimer [197].

As can be noticed, marked discrepancies in potential values are found for the systems I, V, and XI. Nevertheless, for the half-reaction I, extrapolation to ionic strength I = 0 of potential values measured by Templeton et al. [225] at I = 4 results in a value of E° = 0.945 V, more approximate to that obtained by Izatt et al. [226], E° = 0.915 V. As for the systems V and XI, the difference in potential values arises from the dissimilar ΔG_f° data from which they have been deduced. Thus the potential E° = 0.79 V measured by Hoare [228] is in agreement with that estimated for PdO from thermochemical data [232-234] and differs from that given by Pourbaix [231] based on thermodynamic calculations. A similar disagreement in the values rendered by Pourbaix [231] and that given by Goldberg and Hepler [172] for $Pd(OH)_4$ (system XI) calculated on a thermodynamic and thermochemical basis, respectively, has been noted. Estimations on thermochemical measurements apparently present a more reliable approach than those based on thermodynamic treatments. Standard potentials in fused salts are given in Table 1.1.2.

TABLE 1.1.1. Standard and Formal Potentials in Aqueous Solutions

System	Half-reaction	Standard or formal potential[a]	Conditions	Refs.
I	$Pd^{2+} + 2e = Pd$	+0.987	exptl measured, I = 4	[225]
		+0.915	exptl measured, I = 0	[226]

(continued)

TABLE 1.1.1 (continued)

System	Half-reaction	Standard or formal potential[a]	Conditions	Refs.
II	$[PdCl_4]^{2-} + 2e = Pd + 4Cl^-$	+0.62	1 M HCl	[225], [227]
		+0.59	equil const	[172]
III	$[PdBr_4]^{2-} + 2e = Pd + 4Br^-$	+0.49	equil const	[172]
IV	$[PdI_4]^{2-} + 2e = Pd + 4I^-$	+0.18	equil const	[172]
V	$PdO + 2H^+ + 2e = Pd + H_2O$	+0.917	Thermodynamic data	[197]
		+0.79	exptl measured	[228]
VI	$Pd(OH)_2 + 2H^+ + 2e = Pd + 2H_2O$	+0.896	equil const	[172]
VII	$[PdCl_6]^{2-} + 2e = [PdCl_4]^{2-} + 2Cl^-$	+1.29	equil const	[229], [230]
		+1.26		
VIII	$[PdBr_6]^{2-} + 2e = [PdBr_4]^{2-} + 2Br^-$	+0.99	exptl measured, 1 M KBr	[230]
IX	$[PdI_6]^{2-} + 2e = [PdI_4]^{2-} + 2I^-$	+0.48	exptl measured, 1 M KI	[230]
X	$PdO_2 + 2H^+ + 2e = PdO + H_2O$	+1.263	Thermodynamic data	[231]
XI	$Pd(OH)_4 + 2H^+ + 2e = Pd(OH)_2 + 2H_2O$	+1.283	Thermodynamic data	[231]
XII	$PdO_3 + 2H^+ + 2e = PdO_2 + H_2O$	+2.030	exptl measured, extrapolated to standard conditions	[231]
XIII	$PdO_2 + 4H^+ + 2e = Pd^{2+} + 2H_2O$	+1.194	Thermodynamic data	[231]
XIV	$Pd(OH)_4 + 4H^+ + 2e = Pd^{2+} + 4H_2O$	+1.128	Thermodynamic data	[172]
XV	$2Pd + H^+ + e = Pd_2H$	+0.048	calcd from dissociation pressure	[231]

[a]All potential values are in V vs NHE.

TABLE 1.1.2. Standard or Formal Potentials in Fused Salts

Fused salt	Electrode[a]	Potential (V)	Reference electrode	Conditions	Refs.
1. LiCl-KCl (59:41 M%)	Pd/Pd^{2+} (0.001-0.5) M	-0.214	Pt/Pt^{2+} (1 M)	450°C	[14]
	Pd/PdO, O^{2-}	-0.514	Pt/Pt^{2+} (1 M)	450°C	[270]
2. $MgCl_2$-KCl (32.5:67.5 M%)	$Pd/PdCl_2$ (1.824 to 5.193 × 10^{-3} X)	-0.422	Cl_2(graphite)/Cl^-	475°C	[17]
3. $MgCl_2$-NaCl-KCl (50:30:20 M%)	$Pd/PdCl_2$	-0.281	Cl_2(graphite)/Cl^-	475°C, mole fraction	[18]
		-0.472		calcd 475°C	[200]
4. NaCl-KCl (50:50 M%)	$Pd/PdCl_2$ (10^{-5} to 10^{-3} X)	-0.374 ± 0.008	Cl_2(graphite)/Cl^-	727°C, mole fraction	[16]
		-0.357		calcd 727°C	[200]
5. Li_2SO_4-K_2SO_4 (80:20 M%)	Pd/Pd^{2+} (10^{-1} to 10^{-3} m)	+0.541 ± 0.003	Ag/1 m Ag^+	625°C	[269]
6. Li_2SO_4-Na_2SO_4-K_2SO_4 (78:8.5:13.5 M%)	Pd/Pd^{2+} (10^{-1} to 10^{-2} M)	+0.546 ± 0.004	Ag/1 M Ag^+	580°C	[268]

[a] X = mole fraction; m = molality.

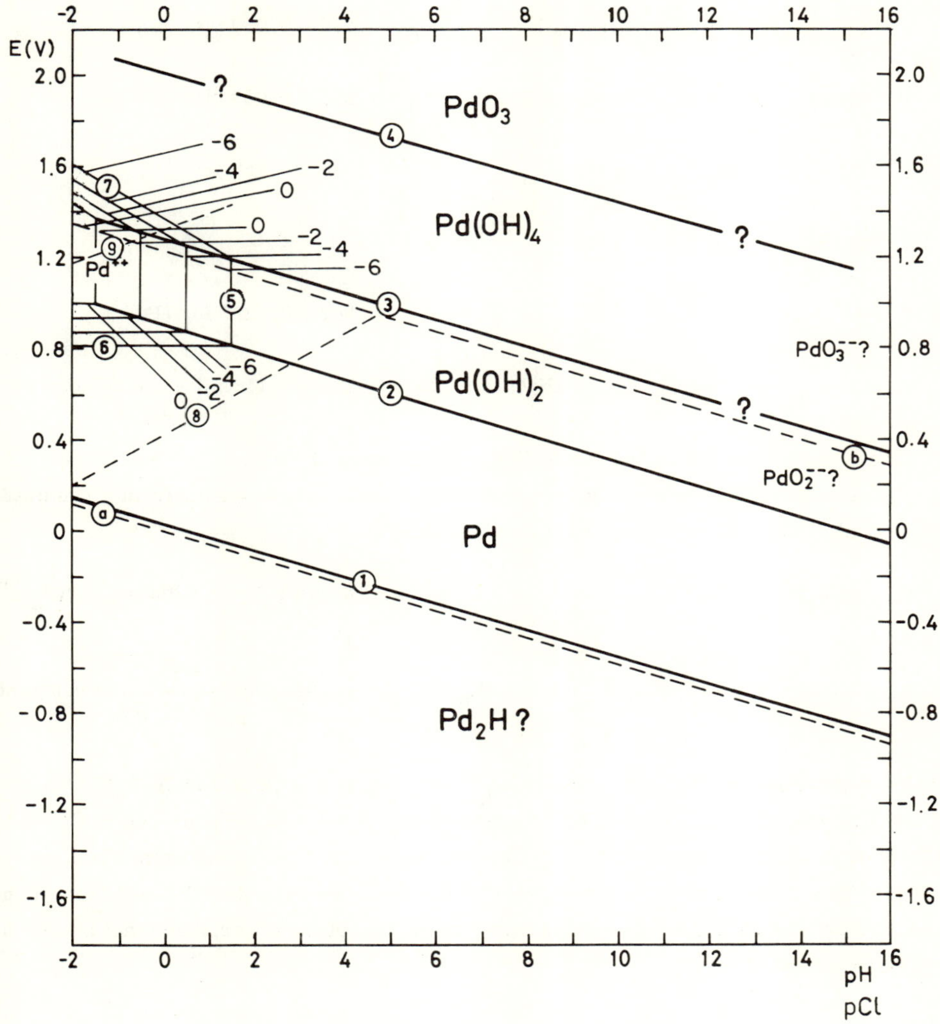

FIG. 1.2.1. Potential-pH diagram of the palladium-water system at 25°C. (From Ref. 231, p. 360, by courtesy of the Centre Belge de la Corrosion.)

1.2. POTENTIAL-pH DIAGRAMS

The potential-pH diagram of palladium at 25°C as proposed by Pourbaix [231] is plotted in Fig. 1.2.1. The "a" and "b" lines confine the stability region of the H_2O, while the numbered lines correspond to the following reactions and their equilibrium potentials, respectively:

1. $2Pd + H^+ + e = Pd_2H$ $\qquad E_0 = 0.048 - 0.059\ pH$

2. $Pd(OH)_2 + 2H^+ + 2e = Pd + 2H_2O$ $\quad E_0 = 0.896 - 0.059\ pH$

3. $Pd(OH)_4 + 2H^+ + 2e = Pd(OH)_2 + 2H_2O$ $\quad E_0 = 1.283 - 0.059\ pH$

4. $PdO + 2H^+ + 2H_2O + 2e = Pd(OH)_4$ $\quad E_0 = 2.030 - 0.059\ pH$

5. $Pd(OH)_2 + 2H^+ = Pd^{2+} + 2H_2O$ $\quad \log (Pd^{2+}) = -2.35 - 2\ pH$

6. $Pd^{2+} + 2e = Pd$ $\quad E_0 = 0.987 + 0.0295 \log (Pd^{2+})$

7. $Pd(OH)_4 + 4H^+ + 2e = Pd^{2+} + 4H_2O$ $\quad E_0 = 1.194 = 0.118\ pH - 0.0295 \log (Pd^{2+})$

Lines 5 to 7 were calculated for 10^{-6} M concentration of Pd^{2+}.

The complexing influence of Cl^- ion was taken into account in the diagram given by Kammel et al. [235]. Lines 8 and 9 relate to equilibria:

8. $(PdCl_4)^{2-} + 2e = Pd + 4Cl^-$ $\quad E_0 = 0.62 + 0.118\ pCl + 0.0295 \log (PdCl_4)^{2-}$

9. $(PdCl_6)^{2-} + 2e = (PdCl_4)^{2-} + 2Cl^-$ $\quad E_0 = 1.288 + 0.0295 \log \dfrac{(PdCl_6)^{2-}}{(PdCl_4)^{2-}} + 0.059\ pCl$

where concentrations of $(PdCl_4)^{2-}$ and $(PdCl_6)^{2-}$ are constant and equal to 10-6 M. This diagram illustrates the enlargement of the corrosion region due to the presence of Cl^- and also the change of the oxidation state of palladium inside this zone. The potentials corresponding to Reactions 3 and 5 are those given by Pourbaix and not those obtained from thermochemical data as indicated above, although introduction of the latter should not modify the course of the lines.

2. VOLTAMMETRIC CHARACTERISTICS

2.1. POLAROGRAPHIC CHARACTERISTICS

Palladium ions are easily reduced to the metallic state by contact with Hg due to their high oxidation potentials in aqueous solutions. This metal acts as a catalyst in the reduction of H^+ at the DME, as do all the others of this group. Therefore, polarographic research of palladium compounds is restricted to dissolved complex ions.

Some of these complexes (for instance, $[Pd(NH_3)_4]^{2+}$) have been studied for use in analytical determinations, but they cannot be successfully separated from similar Ir and Rh

TABLE 2.1.1. Polarographic Characteristics. Half-Wave Potentials of Palladium(II) Complexes in Various Media (maximum suppressor: 0.004% methyl res (sodium salt) for Pd(II) compounds. Temperature: 35.5 ± 0.1°C.)

Substance	Conditions	$E_{1/2}$ (V) vs SCE	Remarks	Refs.
trans-$Pd(NH_3)_2Cl_2$	0.1 M KCl (fresh)	-0.27 -0.59	Diaquo complex Chloroaquo complex	[24] [24]
Pd(en)Cl_2	0.1 M KCl (fresh)	-0.585	Chloroaquo complex	[24]
trans-$Pd(NH_3)_2Cl_2$	0.1 M KCl (aged 24 hr)	-0.265 -0.585 -1.32	Diaquo complex Chloroaquo complex Dichloro complex	[24]
Pd(en)Cl_2	0.1 M KCl (aged 24 hr)	-0.583	Chloroaquo complex	[24]
trans-$Pd(NH_3)_2Cl_2$	1 M KCl (aged 48 hr)	-0.237 -0.62 -1.37	Diaquo complex Chloroaquo complex Dichloro complex	[24]
Pd(en)Cl_2	1 M KCl (aged 48 hr)	-0.63 -1.25	Chloroaquo complex Dichloro complex	[24]
trans-$Pd(NH_3)_2(NO_2)_2$	0.1 M $NaClO_4$	-0.33		[24]
cis-$Pd(NH_3)_2(NO_2)_2$	0.1 M $NaClO_4$	-0.27		[24]
trans-$Pd(NH_3)_2(OH)_2$	0.1 M $NaClO_4$ + 0.01 M NaOH	-0.74		[24]
Pd(en)$(OH)_2$	0.1 M $NaClO_4$ + 0.01 M NaOH	-0.62		[24]
$Pd(NO_3)_2$	0.06 M HCl + acetate buffer, pH = 3	+0.2		[24]
[trans-$Pd(NH_3)_2(H_2O)_2]^{2+}$	0.1 M $NaClO_4$ + 0.01 M $HClO_4$	~-0.05		[24]
[Pd(en)$(H_2O)_2]^{2+}$	0.1 M $NaClO_4$ + 0.01 M $HClO_4$	~-0.06		[24]

compounds. Oscillographic polarography of amine, cyanide, and particularly pyridine complexes of Pd(II) characteristically yield reversible waves with heights proportional to the palladium concentration. Cyanide and sulfocyanide complexes show no reduction waves.

The cis- and trans-isomers of Pd complexes in various media (Table 2.1.1) [24] are polarographically irreversible. The polarograms disclose two or three reduction waves, depending on the age of the solution and the concentration of Cl^-, which are related to

TABLE 2.1.2.

Metal	Polarographic soln prepared by diln from	Apparent diffusion current const	Apparent calcd diffusion coefficient (cm^2/sec)
Pd(II)	Fresh, clear stock soln, 0.06 M in NaCl. Final dilution to 0.1 M acetate and 0.003 M chloride	2.5	0.4×10^{-5}
Pd(II)	Same stock soln, but final soln was made 0.1 M in acetate and 0.01 M in chloride	3.0	0.6×10^{-5}
Pd(II)	Same stock soln, but final soln was made 0.1 M in acetate and 0.1 M in chloride	3.5	0.8×10^{-5}

formation of aquochloro complexes. In general, complexes with two acidic ligands, i.e., Cl^-, NO_2^-, and OH^-, show the half-wave potentials of trans-isomers to be at least 60 mV more negative than those of cis-compounds. This fact points to a higher stability of the former. The $E_{1/2}$ for diaquo- and aquochloro complexes are very similar. All these Pd complexes display a residual current at zero potential vs SCE that is attributed to reduction of Hg(I) produced by reaction of Hg and the Pd(II) compounds. On the other hand, the shape of the polarographic waves and the low values of diffusion coefficients for solutions of simple palladium salts, i.e., $Pd(NO_3)_2$, reveal that reduction of the Pd(II) complex by Hg metal is not complete at the potential where the reduction of Hg(I) begins.

As can be seen from Table 2.1.2, low diffusion currents appear in acetate buffer (pH = 3) and acidic 0.1 N solutions. Addition of Cl^- prevents precipitation of Pd salts. In general, diffusion currents increase linearly with increasing Cl^- concentration, although uncontrolled factors also play a role in the process (Table 2.1.2).

Some data on the voltammetry of Pd in fused salts are given in Table 2.1.3.

TABLE 2.1.3. Voltammetric Characteristics in Fused Salts

Substance	Product	Conditions	Electrode	Technique	Potential (V vs Pt/ 1 M Pt^{2+})	rev	Remarks	Refs.
Pd(II)	Pd(0)	LiCl-KCl (485°C)	Pt wr	i stp	$E_{1/4} = -0.24$	rev	$i\tau/AC = 693 \pm 15$ $D = (1.68 \pm 0.05) \times 10^{-5}$	[271]
Pd(II)	Pd(0)	LiCl-KCl (450°C)	Pt wr	E swp		rev	$D = (1.39 \pm 0.06) \times 10^{-5}$	[42]

3. DOUBLE-LAYER PROPERTIES

Very few data have been published on the potential of zero charge on palladium (Table 3.1). Andersen et al. [70] measured this potential in K_2SO_4 and $NaClO_4$ solutions at pH 7 by the method of the "open circuit scrape." The value obtained is quite different from that reported by Kheifets and Krasikov [62] as corresponding to the minimum capacity of the electrode in Na_2SO_4 solutions at various pH values.

TABLE 3.1. Potential of Zero Charge for Pd (NHE)

$E_{p.z.c.}$ (V)	Conditions	Method	Refs.
-0.06	0.02 N Na_2SO_4 (pH = 3)	Capacity	[62]
-0.36	0.02 N Na_2SO_4 (pH = 7)	Capacity	[62]
-0.39	0.02 N Na_2SO_4 (pH = 11)	Capacity	[62]
0.00	1 N $NaClO_4$ or K_2SO_4 (pH = 7)	Open circuit scrape	[70]

4. ELECTROCHEMICAL STUDIES

4.1. ANODIC DISSOLUTION OF PALLADIUM

Pd metal dissolves at the anode in acidic solutions with the production of $Pd(H_2O)_4^{2+}$. However, at more positive potentials an oxide layer is formed that passivates the electrode. In solutions of 2 M HNO_3 the total passivation of the electrode is already observed at very low current densities through the formation of a thin protective coating [236]. Potentiostatic polarization curves recorded by Vetter and Berndt [237] proved that this metal corrodes anodically at potentials lower than 1.0 V vs NHE and passivates at 10^{-6} A/cm^2. In the same way, electrolysis of organic substances in acidic solutions with Pd anodes showed this metal to be less resistant to corrosion than Ir, Rh, and Pt [238].

Dissolution of Pd anodes starts at relative low positive potentials in the presence of Cl^- ions, as is also observed in Br^- and I^- solutions [236, 239]. The anodic behavior of Pd in HCl solutions was studied by Heumann and Schürmann [240] over a wide range of concentrations (Fig. 4.1.1). The results indicate that the metal goes into solution as Pd(II) complex ions at potentials from 0.35 to 0.50 V vs SCE; at a sufficiently high Cl^- concentration, the $PdCl_4^{2-}$ complex is formed. Finally, Pd dissolves as Pd(IV) at the potential of Cl_2 evolution with the formation of $[PdCl_6]^{2-}$.

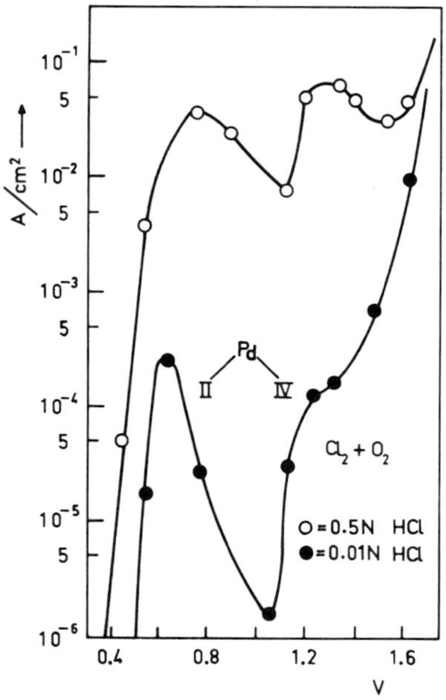

FIG. 4.1.1. Potentiostatic polarization curves for a Pd electrode in HCl solutions. (Reprinted from Ref. 240 by courtesy of Verlag Chemie GmbH.)

Passivity studies of Pd carried out by Jirsa [241] in 1 M NaOH solutions at 18°C lead the author to conclude that a cathodically prepolarized electrode shows two active regions at potentials around 0.1 V (SCE), where dissolution of the metal occurs at a rate of 10 mA/cm^2. However, Zotov et al. [242] measured a lower extent of corrosion (about 10^{-6} A/cm^2) in concentrated solutions of KOH at temperatures from 60 to 100°C. This suggests that the polarization curves observed by Jirsa included the oxidation of H_2 absorbed by the Pd electrode during the cathodic prepolarization process. Dissolution of Pd in the form of the complex $[Pd(OH)_4]^{2-}$ is assumed to occur in alkaline solutions.

4.2. CATHODIC REDUCTION

The low stability of Pd(IV) complexes in aqueous solutions causes the cathodic behavior of the metal to be relatively simple. Hence only reduction of Pd^{2+} to metal and phenomena related to hydrogen evolution on this metal should be considered.

Little work has been devoted to the problems of Pd electrodeposition. Astakhova and Krasikov [243] have given attention to this process in HCl solutions by studying the variation of the reduction rate with the potential and concentration of Cl^- and $[PdCl_4]^{2-}$ ions. In solutions of 2.4 M Cl^- and 0.094 M $PdCl_4^{2-}$ at 20°C and pH 0.5, a cathodic current of about 18 mA/cm^2 at a potential of 0.3 V (NHE) was found.

Cathodic polarization of Pd in dilute H_2SO_4 solutions produces hydrogen, which does not evolve from the electrode surface but is absorbed by the metal, distorting its surface and bulk crystalline structure [244, 245]. Two different compounds are produced in this process: a solid solution of H_2 in Pd (the α-phase) and a relatively definite compound (the β-phase) with the approximate composition Pd_4H_2 [246].

4.3. MECHANISM OF THE ELECTRODE REACTIONS

The kinetics of anodic dissolution and cathodic deposition processes for Pd in solutions of Cl^- and Br^- were investigated by Kravtsov and Zelenskii [247] who inferred from their results that the respective rates were best described by

$$i_a = k_1(X^-)^2 \exp\left(\frac{\beta FE}{RT}\right) \qquad (4.3.1)$$

$$i_c = k_2(PdX_4^{2-})(X^-)^2 \exp\left(-\frac{\alpha FE}{RT}\right) \qquad (4.3.2)$$

where X^- is Cl^- or Br^-, α and β are transfer coefficients, and E is the potential.

An underlying reaction mechanism for these equations was proposed by the authors. Adsorption of X^- ions was assumed as the first step for PdX_2 formation. Hence the $Pd(X_{ad}^-)$ species is predominant on the electrode surface and is the basis for the further building up of the $Pd(X_{ad}^-)_2$ species which act directly in the anodic process. Finally, the PdX_2 formed dissolves in a high concentration of X^-, giving rise to a definite complex. The electrode reactions may be represented by

$$Pd + X^- = PdX_{ad}^- \qquad (4.3.3)$$

$$PdX_{ad}^- + X^- = Pd(X_{ad}^-)_2 \qquad (4.3.4)$$

$$Pd(X_{ad}^-)_2 = PdX_2 + 2e^- \qquad (4.3.5)$$

$$PdX_2 + 2X^- = [PdX_4]^{2-} \qquad (4.3.6)$$

Conversely, dissociation of the complex followed by reduction of the PdX_2 is postulated for the cathodic process according to

$$PdX_4^{2-} = PdX_3^- + X^- \qquad (4.3.7)$$

$$PdX_3^- = PdX_2 + X^- \qquad (4.3.8)$$

$$PdX_2 + 2e^- = Pd + 2X^- \qquad (4.3.9)$$

Astakhova et al. [243], working in Cl^- solutions, reached the conclusion that at high Cl^- concentrations the Reaction (4.3.9) actually goes through two one-electron transfers:

$$PdCl_2 + e^- = PdCl_2^- \tag{4.3.10}$$

$$PdCl_2^- + e^- = Pd + 2Cl^- \tag{4.3.11}$$

where Reaction (4.3.10) would be the rate-determining step of the cathodic process.

4.4. PASSIVATION AND ANODIC OXIDATION

Early work on the anodic oxidation of Pd [248] took scarce notice of peroxide formation on the electrode in aqueous media, but longer experiments carried by El Wakkad and Sham El Din [249] suggested that the evolution of O_2 on Pd anodes takes place through formation of an unstable higher oxide [228]. In a similar way, Hickling et al. [250] showed by anodic oxidation of Pd at high current densities that the primary PdO formed is further oxidized to an unstable oxide of probable composition PdO_2.

In fact, the cathodic charging curves exhibit two reduction waves at potentials lower than that of H^+ ion discharge. In contrast, the anodic charging curves display no clear inflections corresponding to the two oxidation states of the electrode surface [251]. Thus it may be concluded that the oxidation states depend fundamentally on the potential at which the metal has been passivated. This view is verified by the work of Mazitov et al. [252] in which the electrode surface was covered by a PdO layer at a potential of 1.0 V in KOH solutions and another oxide, probably PdO_2, was produced as the potential of oxygen evolution, 1.5 V, was attained. At higher anodic potentials the amount of electricity required to reduce both oxides increases rapidly with the potential of oxidation [253]. Production of these oxides must originate the passivation of Pd in alkaline media and possibly also in acidic solutions in the absence of complexing agents.

In acidic solutions and in the presence of Cl^- ions, Pd displays two passivation regions, Müller [236] and recently Heumann et al. [240] explained the first passivation as due to an initial covering of the electrode surface by a precipitated salt, in the same way as for iron passivation. In the second region, at potentials of Cl_2 evolution, a compact layer of an electronically conducting oxide is formed on the electrode.

In later work [254] the two-stage theory was discarded and the first passivity region was ascribed to the formation of a layer a few molecules thick with good electronic or vacancy conductivity and a PdO or $Pd(OH)_2$ composition.

The second passivation region must be due to the formation of PdO_2 because of the potential region in which it appears.

5. APPLIED ELECTROCHEMISTRY

5.1. ELECTRODEPOSITION, ELECTROPLATING, ELECTROPOLISHING

The main application of palladium plating is in the finishing of electrical contacts for printed circuitry, where low and stable values of contact resistance are required. The physical properties of this metal make it ideal for this purpose [255]. Factors such as adhesion, protective value, physical properties of the metal, and stability are of primary importance in comparison with color and brightness of deposits, which are important in the decoration field.

One of the earlier electrolytes employed was $PdCl_2$ in a mixture of Na_2HPO_4 and $(NH_4)_2HPO_4$. Grube [256] has described the preparation of a solution of this type. A closely similar electrolyte has been reported by Ostroumov [257] (Pd, 1 to 20; Na_2HPO_4, 100; $(NH_4)_2HPO_4$, 20; NH_4Cl, 25 g/l; and aq NH_3 to pH 9), who has studied the mechanical stresses in Pd deposits and showed that they depend on the formation and decomposition of the α- and β-phases of the Pd—H system, both during and after deposition. In general, the observed stress increased when Pd and H were discharged simultaneously. Ostroumov [258] has also studied the effect of some organic additives to the above-mentioned electrolyte. Furfural brightened deposits but increased the internal stress; better results were recorded for protalbinic acid (prepared from egg albumin).

Deposition of Pd from solutions containing 20 to 200 g/l KOH and 0.3 to 3 g/l Pd (as $PdCl_2$) was described by Ostroumov [259]. However, it appears to be of no interest for general application. Above the limiting current density (2.5 mA/cm^2), β-phase formed, which after electrolysis decomposed it to form the α-phase and H_2.

Among other electrolytes that have been proposed are an iodide electrolyte [260] and a solution based on $Pd(NH_3)_2(CN)_2$. A more suitable solution is probably the acid chloride solution ($PdCl_2$, 50 and NH_4Cl, 20 to 50 g/l; pH 0 to 0.5). Other compounds used are $[Pd(NO_2)_4]^{2-}$, $[PdCl_4]^{2-}$, or $[Pd(SCN)_4]^{2-}$. Experimental data in all these cases show that deposition proceeds from the undissociated complex as a whole [261]; saccharin was used as an organic additive [262].

According to Reid [169, 263], in present day practice only three types of electrolyte are of significant interest: 1) the soluble anode solution, 2) electrolytes based on $Pd(NH_3)_2(NO_2)_2$, and electrolytes based on $[Pd(NH_3)_4]Cl_2$.

A typical soluble-anode electrolyte is based on a nitrite complex of Pd in the presence of NaCl or NH_4Cl and boric acid. With this electrolyte, Pd dissolves at the anode (100% efficiency) at a pH of 4.5 to 6, but becomes insoluble at pH 7.

The compound $Pd(NH_3)_2(NO_2)_2$, known conventionally as "P" salt, is still the basis of the most widely used electrolytes. It was originally proposed by Keitel and Zschiegner [264]. A formulation which finds commercial applications is "P" salt, 8 g/l; $(NH_4)NO_3$, 100 g/l; $NaNO_2$, 10 g/l; and NH_3 up to the desired pH. At pH 7 to 8, brighter deposits are produced, while at higher pH values deposits are of a dull appearance. An interesting modification of a "P" salt electrolyte is based on reaction of this salt with $H(SO_3NH_2)$ or $(NH_4)(SO_3NH_2)$.

A series of electrolytes is based on the complex $Pd(NH_3)_2Cl_2$. Smooth, bright, and firmly bound coatings were obtained [265] from electrolytes containing the $[Pd(NH_3)_4]^{2+}$ complex at pH 8 to 10 with $(NH_4)_2CO_3$ as a buffer. Smooth and bright deposits were also obtained from solutions of $[Pd(NH_3)_4](SO_3NH_2)_2$, the pH being 7.8 to 8.8. Deposition of Pd from $[Pd(NH_3)_4]Br_2$ solutions at pH 13 also gives deposits of good quality. Cathodic polarization during electrodeposition of Pd from ammoniacal electrolytes has little dependence on the type of anion but is determined by the NH_3 content in the solution.

5.2. CHEMICAL CORROSION

A summary of the main reactions related to the chemical corrosion of Pd is listed in Table 5.2.1. Unless experimental conditions are indicated, spontaneous reaction at room temperature is assumed.

TABLE 5.2.1. Chemical Corrosion of Pd

Reaction	Conditions	Refs.
$Pd + H_2 = Pd_4H_2$	200 to 300°C	[266]
$Pd + O_2 = PdO_2$	800°C	[266]
$Pd + F_2 = PdF_2$	400°C	[266]
$Pd + Cl_2 = PdCl_2$	480°C	[266]
$Pd + S = PdS$	Molten	[266]
$Pd + Na_2O_2 = PdO$	Molten	[266]
$Pd + KHSO_4 = PdSO_4$	Molten	[266]
$Pd + NH_4Cl + CsCl + O_2 = Cs_2PdCl_4$	500°C	[267]
$Pd + (HNO_3 + 3HCl) = H_2PdCl_4$		[266]
$Pd + (HNO_3 + 3HBr) = H_2PdBr_4$		[266]
$Pd + H_2SO_4 = PdSO_4$	Boiling concentrated acid	[266]
$Pd + HNO_3 = Pd(NO_3)_2$	Boiling concentrated acid	[266]

REFERENCES

1. M. Pourbaix, Atlas d'Equilibres Electrochimiques á 25°C, Gauthier Villars, Paris, 1963, p. 380.
2. Y. N. Kukushkin and S. Ch. Dkhara, Zh. Neorg. Khim., 14, 2816 (1969).
3. Y. N. Kukushkin and S. Ch. Dkhara, ibid., 14, 853 (1969).
4. Y. N. Kukushkin and S. Ch. Dkhara, ibid., 14, 1012 (1969).
5. Y. N. Kukushkin, S. Ch. Dkhara, and M. J. Gelfman, ibid., 13, 2530 (1968).
6. A. A. Grinberg, S. Ch. Dkhara, and M. J. Gelfman, ibid., 13, 2199 (1968).
7. D. Ginstrup and I. Leden, Acta Chem. Scand., 22, 1163 (1968).
8. H. Yamamoto, S. Tanaka, T. Nagai, and T. Takei, Denki Kagaku, 32, 43 (1964).
9. A. T. Hubbard and F. C. Anson, Anal. Chem., 38, 58 (1966).
10. V. I. Kravtsov and B. V. Simakov, Elektrokhimiya, 2, 646 (1966).
11. V. I. Kravtsov and B. V. Simakov, ibid., 2, 406 (1966).
12. Y. N. Kukushkin and S. Ch. Dkhara, Zh. Neorg. Khim., 15, 1585 (1970).
13. I. I. Chernyaev, G. S. Muraveiskaya, and L. S. Korablina, ibid., 11, 1339 (1966).
14. H. A. Laitinen and C. H. Liu, J. Amer. Chem. Soc., 80, 1015 (1958).
15. D. J. G. Ives and G. J. Janz, Reference Electrodes, Academic, New York, 1961, p. 592.
16. R. Combes, J. Vedel, and B. Tremillon, J. Electroanal. Chem., 27, 174 (1970).
17. H. C. Gaur and H. L. Jindal, Electrochim. Acta, 15, 1113 (1970).
18. H. C. Gaur and H. L. Jindal, ibid., 13, 835 (1968).
19. A. Kisza, Z. Phys. Chem. (Leipzig), 237, 97 (1968).
20. J. Goret and B. Tremillon, Electrochim. Acta, 12, 1065 (1967).
21. W. S. Hamer, M. S. Malmberg, and B. Rubin, J. Electrochem. Soc., 103, 8 (1956).
22. G. Sundholm, Soc. Sci. Fennica Comment. Phys.-Math., 34, 39 (1968).
23. J. R. Hall and R. A. Plowman, Aust. J. Chem., 9, 14 (1956).
24. R. Chakravarty and D. Banerjea, J. Inorg. Nucl. Chem., 16, 288 (1961).
25. J. R. Hall and R. A. Plowman, Aust. J. Chem., 9, 143 (1956).
26. G. Sundholm, Suomen Kemistilehti, B, 42, 401 (1969).
27. J. R. Hall and R. A. Plowman, Aust. J. Chem., 8, 158 (1955).
28. J. R. Hall and R. A. Plowman, ibid., 8, 168 (1955).
29. A. Y. Stetsenko and V. M. Kiseleva, Zh. Prikl. Khim., 15, 1385 (1970).
30. M. J. D. Brand, B. Corain, B. Fleet, and A. J. Poe, J. Electroanal. Chem., 24, 155 (1970).
31. A. N. Frumkin and N. V. Nikolaeva, J. Chem. Phys., 26, 1552 (1957).
32. G. Sundholm, Suomen Kemistilehti, B, 41, 331 (1968).
33. C. G. Macarovici, F. Manok, and C. Varhelyi, Rev. Roum. Chim., 12, 279 (1967).
34. D. C. Olson, V. P. Mayweg, and G. N. Schrauzer, J. Amer. Chem. Soc., 88, 4876 (1966).
35. A. Davison, N. Edelstein, R. H. Holm, and A. M. Maki, Inorg. Chem., 2, 1227 (1963); 3, 814 (1964).

36. A. L. Balch and R. H. Holm, J. Amer. Chem. Soc., 88, 5201 (1966).
37. R. E. Dessy, R. B. King, and M. Waldrop, ibid., 88, 5112 (1966).
38. G. Sundholm, Acta Chem. Scand., 24, 335 (1970).
39. R. L. Martin, A. M. Bond, and G. A. Heath, Inorg. Chem., 10, 2026 (1971).
40. W. R. Mason and R. C. Johnson, J. Electroanal. Chem., 14, 345 (1967).
41. J. J. Lingane, ibid., 7, 94 (1964).
42. E. Schmidt, H. Pfander, and H. Siegenthaler, Electrochim. Acta, 10, 429 (1965).
43. H. C. Gaur and W. K. Behl, J. Electroanal. Chem., 5, 261 (1963).
44. H. Yamamoto, Y. Kokubu, T. Nagai, and T. Takei, Denki Kagaku, 33, 804 (1965).
45. A. L. Y. Lau and A. T. Hubbard, J. Electroanal. Chem., 24, 237 (1970); 33, 77 (1971).
46. V. I. Kravtsov and B. Simakov, Vestn. Leningrad. Univ., Ser. Fiz. Khim., 10, 76 (1965).
47. V. I. Kravtsov and B. Simakov, Electrochim. Acta, 9, 1211 (1964).
48. G. Sundholm, Electrochim. Acta, 13, 2111 (1968).
49. F. C. Anson and A. T. Hubbard, Anal. Chem., 38, 1887 (1966).
50. V. I. Kravtsov and B. Simakov, Elektrokhimiya, 6, 275, 1243 (1970); 7, 1069 (1971).
51. N. V. Nikolaeva-Fedorovich, I. E. Barbasheva, and N. P. Berezin, ibid., 3, 836 (1967).
52. J. R. Cushing and A. T. Hubbard, J. Electroanal. Chem., 23, 183 (1969).
53. G. J. Hills and K. E. Johnson, J. Electrochem. Soc., 108, 1013 (1961).
54. H. A. Laitinen, R. P. Tischer, and D. K. Roe, ibid., 107, 546 (1960).
55. A. D. Graves, G. H. Hills, and D. Inman, Advan. Electrochem. Electrochem. Eng., 4, 117 (1966).
56. A. N. Frumkin, ibid., 3, 307 (1963).
57. A. N. Frumkin, O. Petrii, and R. Marvet, J. Electroanal. Chem., 12, 504 (1966).
58. A. N. Frumkin, O. Petrii, A. Kossaya, V. Entina, and V. Topolov, ibid., 16, 175 (1968).
59. A. N. Frumkin, N. A. Balashova, and V. E. Kazarinov, J. Electrochem. Soc., 113, 1011 (1966).
60. K. Müller, Z. Phys. Chem. (Leipzig), 243, 293 (1970).
61. L. Campanella, J. Electroanal. Chem., 28, 228 (1970).
62. T. N. Andersen and R. S. Perkins, in Modern Aspects of Electrochemistry, Vol. 5 (J. O'M. Bockris and B. E. Conway, eds.), Butterworths, London, 1969, p. 208.
63. R. Kh. Burshtein, A. G. Pschenichnikov, and L. A. Sherchark, Elektrokhimiya, 5, 332 (1969).
64. L. Formaro and S. Trasatti, Electrochim. Acta, 12, 1457 (1967).
65. E. Gileadi, S. D. Argade, and J. O'M. Bockris, J. Phys. Chem., 70, 2044 (1966).
66. S. Trasatti, J. Electroanal. Chem., 33, 351 (1971).
67. L. A. Fokina, N. A. Shumovskaya, and R. Kh. Burshtein, Elektrokhimiya, 5, 225 (1969).
68. E. A. Ukshe, N. G. Bukun, and N. S. Tkacheva, ibid., 5, 1421 (1969).
69. A. D. Graves and D. Inman, Nature, 208, 481 (1965).

70. T. N. Andersen, J. L. Anderson, and H. Eyring, J. Phys. Chem., 73, 3562 (1969).

71. A. N. Frumkin, O. A. Petrii, and Yu. G. Kotlov, Elektrokhimiya, 5, 476 (1969).

72. J. O'M. Bockris, S. D. Argade, and E. Gileadi, Electrochim. Acta, 14, 1259 (1969).

73. N. A. Balashova, N. T. Gorokhova, and M. Y. Kuleznova, Elektrokhimiya, 4, 871 (1968).

74. E. A. Ukshe, N. G. Bukun, D. I. Leikis, and A. N. Frumkin, Electrochim. Acta, 9, 431 (1964).

75. M. W. Breiter, Electrochemical Process in Fuel Cells, Springer, New York, 1969.

76. A. N. Frumkin, D. A. Petrii, and Y. G. Kotlov, Elektrokhimiya, 4, 774 (1968).

77. Y. Y. Veber, Yu. B. Vasiliev, and V. S. Bagotzkii, ibid., 5, 323 (1969).

78. Ying-Chen Chin and M. A. Genshaw, J. Phys. Chem., 73, 3571 (1969).

79. N. Y. Pospelova, A. A. Rakov, and V. I. Veselovskii, Elektrokhimiya, 6, 722 (1970).

80. I. I. Pyshnograeva, A. M. Skundin, and Y. B. Vasiliev, ibid., 5, 1469 (1969).

81. S. S. Skundin, Y. B. Vasiliev, and V. S. Bagotzkii, ibid., 6, 204 (1970).

82. A. D. Obrucheva, Dokl. Akad. Nauk SSSR, 142, 859 (1962).

83. B. G. Bowles, Electrochim. Acta, 10, 717, 731 (1965).

84. A. N. Frumkin, G. N. Mansurov, V. E. Kazarinov, and N. A. Balashova, Collect. Czech. Chem. Commun., 31, 806 (1966).

85. F. M. Toktabaeva, G. D. Zukamaeva, and D. V. Sokolskii, Dokl. Akad. Nauk SSSR, 180, 652 (1968).

86. A. N. Frumkin, O. A. Petrii, I. G. Schshigorev, and W. A. Safimov, Z. Phys. Chem. (Leipzig), 243, 261 (1970).

87. G. Horanyi, J. Solt, and F. Nagy, J. Electroanal. Chem., 31, 95 (1971).

88. T. Loucka, ibid., 31, 319 (1971).

89. S. Shuldiner and M. Rosen, J. Electrochem. Soc., 118, 1138 (1971).

90. O. A. Petrii, V. E. Kazarinov, S. Ya. Vasina, Y. G. Shigorev, and Zh. N. Malisheva, Dvoinoi Sloi Adsorbtsya Tverd. Elektrodakh, 2nd Symposium on Materials, Gosudarstvennogo University, Tartu, SSSR, 1970, p. 296.

91. O. A. Petrii, S. V. Vasina, and A. N. Frumkin, Elektrokhimiya, 7, 418 (1971).

92. J. W. Schultze, Ber. Bunsenges. Phys. Chem., 74, 705 (1970).

93. G. N. Mansurov, O. A. Petrii, F. S. Safiulina, and V. K. Pavlovich, Dvoinoi Sloi Adsorbtsya Tverd. Elektrodakh, 2nd Symposium on Material, Gosudarstvennogo University, Tartu, SSSR, 1970, p. 253.

94. E. Gileadi and B. Piersma, in Modern Aspects of Electrochemistry, Vol. 4 (J. O'M. Bockris and B. E. Conway, eds.), Butterworths, London, 1966, p. 47.

95. V. S. Bagotskii and Y. B. Vasiliev, in Uspekhi Elektrokhimii Organicheskikh Soedinenii, Izda "Nauka," Moscow, 1966, p. 38.

96. V. S. Bagotzkii, Y. B. Vasiliev, V. A. Gromiko, S. S. Sedova, and D. A. Khasova, Dvoinoi Sloi Adsorbtsya Tverd. Elektrodakh, 2nd Symposium on Material, Gosudarstvennogo University, Tartu, SSSR, 1970, p. 380.

97. Y. Matsuda and H. Tamura, Electrochim. Acta, 14, 427 (1969).

98. L. I. Pyshograeva, Y. B. Vasiliev, and V. S. Bagotzkii, Elektrokhimiya, 6, 433 (1970).

99. J. Llopis, Catal. Rev., 2, 161 (1968).

100. J. P. Hoare, The Electrochemistry of Oxygen, Wiley-Interscience, London, 1968.

101. J. P. Hoare, Advan. Electrochem. Electrochem. Eng., 6, 201 (1967).

102. A. Damjanovic, in Modern Aspects of Electrochemistry, Vol. 5 (J. O'M. Bockris and B. E. Conway, eds.), Butterworths, London, 1969, p. 369.

103. J. P. Hoare, J. Electrochem. Soc., 109, 858 (1962).

104. J. P. Hoare and R. Thacker, J. Electroanal. Chem., 30, 1, 15 (1971).

105. L. D. Burke and A. Moynihan, Trans. Faraday Soc., 67, 3550 (1971).

106. T. Biegler, D. A. J. Rand, and R. Woods, J. Electroanal. Chem., 29, 269 (1971).

107. Y. M. Tiurin, G. F. Volodin, and V. S. Paducheva, Elektrokhimiya, 7, 1309 (1971).

108. G. A. Alfonshin, G. F. Volodin, and Y. M. Tiurin, ibid., 7, 1338 (1971).

109. J. W. Schultze, Z. Phys. Chem. (Frankfurt am Main), 73, 29 (1970).

110. J. L. Ord and F. C. Ho, J. Electrochem. Soc., 118, 46 (1971).

111. K. J. Vetter and J. W. Schultze, J. Electroanal. Chem., 34, 131, 141 (1972).

112. D. V. Kokoulina, Y. I. Krasovitskaya, and V. I. Krishtalik, Elektrokhimiya, 7, 1154, 1218 (1971).

113. I. V. Shimonis, A. A. Rakov, and V. I. Veselovskii, ibid., 6, 163 (1970).

114. G. F. Volodin and Y. M. Tiurin, ibid., 7, 233 (1971).

115. S. Shibata and M. P. Sumino, Electrochim. Acta, 16, 1089 (1971).

116. S. D. James, J. Electrochem. Soc., 116, 1681 (1969).

117. A. A. Yakovleva and V. I. Veselovskii, Elektrokhimiya, 7, 683 (1971).

118. V. S. Bagotskii, Y. B. Vassiliev, and I. I. Pyshnograeva, Electrochim. Acta, 16, 2141 (1971).

119a. D. C. Johnson, D. T. Napp, and S. Bruckenstein, ibid., 15, 1493 (1970).

119b. D. F. Untereker and S. Bruckenstein, J. Electrochem. Soc., 121, 360 (1974).

120. M. Costa, F. Chao, and A. Tadjeddine, C. R. Acad. Sci., Paris, Ser. C, 272, 821 (1971).

121. M. R. Tarasevitch and V. A. Bogdanovskaya, Elektrokhimiya, 7, 1519, 1684 (1971).

122. M. R. Tarasevitch and K. A. Radinskaya, ibid., 6, 812 (1970).

123. M. Bonnemay, E. Momot and G. Bronöel, C. R. Acad. Sci., Paris, Ser. C, 270, 2018 (1970); 271, 334 (1970).

124. M. R. Tarasevitch and V. S. Vilinskaya, Elektrokhimiya, 7, 710 (1971).

125. J. W. Schultze, Ber. Bunsenges. Phys. Chem., 73, 483 (1969).

126. M. R. Tarasevitch and V. A. Bogdanovskaya, Elektrokhimiya, 7, 1072 (1971).

127. V. I. Bagotzkii, V. I. Lukyanicheva, A. V. Yuzhanina, and B. V. Lenstner, ibid., 7, 1287 (1971).

128. A. A. Yakovleva and V. I. Veselovskii, ibid., 6, 967 (1970).

129. V. A. Oleinikova, M. V. Smirnov, and L. A. Tsiovkina, Chem. Abstr., 74, 133985 (1971); Elektrokhimiya, 7, 1218 (1965).

130. J. M. Schegel and R. Bauer, J. Chem. Soc., D, 1971, 483.

131. J. C. Stemmelin, L. A. Suárez Infanzón, and J. Brenet, C. R. Acad. Sci., Paris, Ser. C, 265, 141 (1967).

132. D. Inman and J. Braustein, J. Chem. Soc., D, 1966, 148.

133. E. Plumat, M. Boffe, F. Toussaint, and J. M. Joinville, Silicates Ind., 35, 185 (1970).

134. A. Ghosh and Th. B. King, Trans. Met. Soc. AIME, 245, 145 (1969).

135. R. L. Every and R. L. Grimsley, J. Electroanal. Chem., 9, 167 (1965).

136. A. Y. Sotnikov and N. M. Simkin, Elektrokhimiya, 7, 1002 (1971).

137. Y. K. Delimarskii, O. G. Zambilski, and O. A. Omelchenko, Zh. Prikl. Khim., 39, 1435 (1966); 41, 741 (1968).

138. T. Notoya, T. Ishikawa, and R. Midorikawa, Chem. Abstr., 69, 15308e (1969); 73, 51572e (1970).

139. R. Littlewood and J. Argent, Electrochim. Acta, 4, 155, 114 (1961).

140. J. Hladik, Y. Pointud, M. C. Bellisent, and G. Morand, ibid., 15, 405 (1970).

141. J. Hladik, Y. Pointud, M. C. Bellisent, and G. Morand, C. R. Acad. Sci., Paris, Ser. C, 269, 955 (1969).

142. S. Asakura and T. Mukaibo, Electrochim. Acta, 13, 881 (1968).

143. J. Hladik, J. Coustas, and G. Morand, J. Chim. Phys., 64, 1041 (1967).

144. A. de Haan and H. Van der Poorten, C. R. Acad. Sci., Paris, Ser. C, 261, 5462 (1965).

145. Y. M. Kolotyrkin, A. N. Chemodanov, V. E. Kosmatii, and M. A. Dembrovskii, Elektrokhimiya, 4, 1466 (1968).

146. J. A. Bittles and E. L. Littauer, Corros. Sci., 10, 29 (1970).

147. Y. M. Kolotyrkin and A. N. Chemodanov, Ann. Univ. Ferrara Sez., 5, 49 (1970).

148. Y. M. Kolotyrkin and A. N. Chemodanov, Elektrokhimiya, 5, 578 (1969).

149. N. Hackerman, E. S. Snavely, and L. D. Field, Corros. Sci., 7, 39 (1967).

150. A. N. Chemodanov, Y. M. Kolotyrkin, and M. A. Dembrovskii, Elektrokhimiya, 6, 460 (1970).

151. Y. M. Kolotyrkin, V. A. Gilman, and B. S. Kolomiets, Zashch. Metal, 5, 660 (1969).

152. Y. M. Kolotyrkin, A. N. Chemodanov, et al., ibid., 6, 678 (1970).

153. R. Yuchnievicz and P. Hayfield, Chem. Abstr., 72, 17766p (1970).

154. A. I. Koshev, E. P. Grigoreva, Y. D. Kudryatsev, and P. D. Samchenko, Tr. Novocherk. Politekh. Inst., 197, 79 (1969).

155. A. D. Styrkas and D. N. Efimov, Zh. Prikl. Khim., 43, 177 (1970).

156. P. R. Rowland, Nature, 171, 931 (1953).

157. P. H. Brouillet, Met. Corros., 30, 243 (1955).

158. R. Seegmiller and J. K. Gore, U.S. Patent 3,419,435 (December 31, 1968); Chem. Abstr., 70, 63493d (1969).

159. E. M. Savitskii, V. P. Polyakova, and N. B. Gorinz, Russian Patent 244839 (May 28, 1969); Chem. Abstr., 72, P27767d (1970).

160. C. T. J. Ahlers and R. N. Balluffi, J. Appl. Phys., 38, 910 (1967).

161. J. Toušek, Collect. Czech. Chem. Commun., 32, 2348 (1967).

162. A. M. Feltham and M. Spiro, Chem. Rev., 71, 177 (1971).

163. V. S. Bagotzkii, D. A. Khazova, and Y. B. Vasiliev, Elektrokhimiya, 6, 1367 (1970).

164. F. I. Kukoz and L. A. Kukoz, Zh. Prikl. Khim., 39, 705 (1966).

165. V. B. Obukhova and B. S. Krasikov, ibid., 44, 1025 (1971).

166. M. V. Pokrovskaya and B. S. Krasikov, Vestn. Leningrad. Uni. Ser. Fiz. Khim., 4, 152 (1968).

167. M. V. Pokrovskaya and B. S. Krasikov, ibid., 22, 118 (1968).

168. G. Hänsel, Metalloberfläche, 21, 238 (1967).

169a. F. M. Reid, Trans. Inst. Metal Finish., 48, 115 (1970).

169b. E. L. MacNamara, J. Electrochem. Soc., 109, 61 (1962).

170a. H. Benninghoff, Galvanotechnik, 60, 293 (1969).

170b. J. O'M. Bockris and S. Srinivasan, Fuel Cells: Their Electrochemistry, McGraw-Hill, New York, 1969.

L. G. Austin, Fuel Cells, NASA SP-120, Office of Technology Utilization, Washington, D.C., 1967.

G. Y. Young (ed.), Fuel Cells, Reinhold, New York, 1960, 1963.

W. Mitchell (ed.), Fuel Cells, Academic, New York, 1963.

171. M. Pourbaix, Atlas d'Equilibres Electrochimiques á 25°C, Gauthier Villars, Paris, 1963, p. 373.

172. R. N. Goldberg and L. G. Hepler, Chem. Rev., 68, 229 (1968).

173. V. I. Kravtsov and G. M. Petrova, Zh. Neorg. Khim., 9, 1010 (1964).

174. J. A. Page, Talanta, 9, 365 (1962).

175. V. I. Kravtsov, N. V. Titova, and G. P. Tsayun, Elektrokhimiya, 6, 573 (1970).

176. H. A. Laitinen and J. A. Plambeck, J. Amer. Chem. Soc., 87, 1202 (1965).

177. N. K. Pshenitsyn, N. A. Ezerskaya, and M. B. Bardin, Zh. Anal. Khim., 14, 446 (1959).

178. J. A. Page and G. Van Loon, Can. J. Chem., 44, 515 (1966).

179. V. I. Kravtsov and G. M. Petrova, Dokl. Akad. Nauk SSSR, 154, 433 (1964).

180. S. P. Gubin and V. S. Khandkarova, J. Organometal. Chem., 12, 523 (1968).

181. W. Mehl and F. Lohman, Electrochim. Acta, 13, 1459 (1968).

182. D. A. Petrii and Nguen Van Tue, Elektrokhimiya, 6, 408 (1970).

183. D. A. Petrii and Nguen Van Tue, ibid., 5, 494 (1969).

184. B. S. Krasikov and V. S. Krivonos, Zh. Prikl. Khim., 39, 1332 (1966).

185. V. S. Bagotzkii, S. S. Sedova, and B. Vasiliev, Elektrokhimiya, 4, 1113 (1968).

186. A. D. Semenova, T. G. Fedorova, and G. P. Khomchenko, Vestn. Mosk. Univ., Ser. II, 20, 47 (1965).

187. J. P. Hoare, J. Electroanal. Chem., 18, 251 (1968).

188. N. V. Pospelova, A. A. Rakov, and V. I. Veselovskii, Elektrokhimiya, 5, 797 (1969).

189. A. M. Skundin, V. Sh. Palanker, and V. S. Bagotzkii, ibid., 3, 240 (1967).

190. A. Nidola, Corsi. Semin. Chim., 12, 79 (1968).

191. R. L. Andrews, C. B. Kenahan, and D. Schlain, U.S. Bur. Mines Rep. Invest., 7023 (1967).

192. J. Llopis, J. M. Gamboa, and V. Menéndez, Collect. Czech. Chem. Commun., 36, 528 (1971).

193. K. I. Kryshchenko and A. D. Styrkas, Khim. Prom. (Moscow), 47, 363 (1971).

194. L. Y. Kadaner, T. N. Zagrebelneya, and Y. P. Vladimirova, Zh. Prikl. Khim., 40, 1378 (1967).

195. Y. M. Tiurin, E. P. Kovsman, and E. A. Karavaeva, ibid., 38, 1818 (1965).

196. M. Pourbaix, Atlas d'Equilibres Electrochimiques á 25°C, Gauthier Villars, Paris, 1963, p. 364.

197. W. M. Latimer, Oxidation Potentials, Prentice-Hall, New York, 1952.

198. W. P. Griffith, Quart. Rev. (London), 19, 254 (1965).

199. D. A. Buckingham, F. D. Dwyer, and A. M. Sargeson, Inorg. Chem., 5, 1243 (1966).

200. W. J. Hamer, M. S. Malmberg, and B. Rubin, J. Electrochem. Soc., 112, 750 (1965).

201. J. Heyrovsky, W. R. Crowell, and D. W. Engelkemair, J. Amer. Chem. Soc., 63, 2888 (1941).

202. L. Meites, ibid., 79, 4631 (1957).

203. R. E. Cover and L. Meites, ibid., 83, 4706 (1961).

204. R. E. Cover and J. G. Connery, Anal. Chem., 40, 87 (1968).

205. S. Lal and G. D. Christian, Rev. Polarogr. (Japan), 16, 109 (1970).

206. D. T. Sawyer and D. S. Tinti, Inorg. Chem., 2, 796 (1963).

207. A. M. Bond, G. A. Heath, and R. L. Martin, J. Electrochem. Soc., 117, 1362 (1970).

208. R. Buvet, J. Perichon, and S. Palous, Bull. Soc. Chim. Fr., 1963, 982.

209. M. B. Bardin and V. P. Goncharenko, Zh. Neorg. Khim., 15, 490 (1970).

210. J. C. Bavay, G. Novogrocki, and G. Tridot, Bull. Soc. Chim. Fr., 1967, 2030.

211. M. B. Bardin and V. P. Goncharenko, Zh. Neorg. Khim., 15, 490 (1970).

212. R. Kalvoda, J. Electroanal. Chem., 24, 53 (1970).

213. Y. Usatenko, L. F. Shvydka, A. I. Tolubara, and F. M. Tulyupa, Izv. Vyssh. Ucheb. Zaved., Khim. Khim. Tekh., 12, 1654 (1969).

214. C. M. Elson, J. Guless, I. J. Ilzkovitch, and J. A. Page, J. Chem. Soc., D, 1970, 875.

215. A. A. Sutyagina, I. N. Golianitskaya, and F. D. Vovchenko, Elektrokhimiya, 6, 908 (1972).

216. A. A. Sutyagina, I. N. Golianitskaya, and F. D. Vovchenko, Vestn. Mosk. Univ. Khim., 12, 176 (1971).

217. G. P. Khomchenko, N. G. Ul'ko, and G. D. Vovchenko, Elektrokhimiya, 1, 659 (1965).

218. G. P. Khomchenko, N. G. Ul'Ko, and G. D. Vovchenko, ibid., 3, 335 (1967).

219. J. F. Alder and B. Fleet, J. Electroanal. Chem., 30, 427 (1971).

220. V. P. Polyakova, E. M. Savitskii, and N. B. Gorina, Russian Patent 268112 (April 2, 1970); Otkrytya Izobiet. Prom. Obraztsy, Tovarnye Znaki, 47, 122 (1970).

221. J. Llopis and M. Vázquez, An. Real Soc. Espan. Fis. Quim., Ser. B, 63, 273 (1967).

222. J. M. Notley and J. Barr, German Offen. 2,052,458 (April 29, 1971).

223. L. Greenspan, Plating, 59, 137 (1972).

224. A. J. Appleby, J. Electrochem. Soc., 117, 1610 (1970).

225. D. H. Templeton, G. W. Watt, and C. S. Garner, J. Amer. Chem. Soc., 65, 1608 (1943).

226. R. M. Izatt, D. Eatough, and J. J. Christensen, J. Chem. Soc., 1967, 1301.

227. A. B. Fasman, G. G. Kulyukov, and D. V. Sokolskii, Zh. Neorg. Khim., 10, 1338 (1965).

228. J. P. Hoare, J. Electrochem. Soc., 111, 611 (1964).

229. H. B. Wellman, J. Amer. Chem. Soc., 52, 985 (1930).

230. A. A. Brinberg and A. S. Shamsiev, Zh. Obshch. Khim., 12, 55 (1942).

231. M. Pourbaix, Atlas d'Equilibres Electrochimiques á 25°C, Gauthier Villars, Paris, 1963, p. 359.

232. W. E. Bell, R. E. Inyard, and M. Tagami, J. Phys. Chem., 70, 3735 (1966).

233. N. G. Schmal and E. Minzl, Z. Phys. Chem. (Frankfurt), 47, 142 (1965).

234. J. S. Warner, J. Electrochem. Soc., 114, 68 (1967).

235. R. Kammel, T. Takei, H. Winterhager, and H. Yamamoto, Metalloberflaeche, 22, 107 (1968).

236. F. Müller, Z. Elektrochem., 34, 744 (1928).

237. K. J. Vetter and N. Berndt, ibid., 62, 378 (1958).

238. E. P. Kovsman, Y. M. Tiurin, and E. A. Karavaeva, Zh. Prikl, Khim., 37, 217 (1964).

239. F. Müller and A. Riefkohl, Z. Electrochem., 36, 181 (1930).

240. T. Heumann and R. Schürmann, Ber. Bunsenges. Phys. Chem., 67, 601 (1963).

241. F. Jirsa, Z. Phys. Chem., 113, 241 (1924).

242. Y. M. Kolotyrkin, G. N. Zotov, and Y. N. Bune, Elektrokhimiya, 6, 857 (1970).

243. R. K. Astakhova and B. S. Krasikov, Vestn. Leningrad. Univ., Ser. Fiz. Khim., 4, 116 (1969).

244. F. Streintz, Wied. Ann., 33, 471 (1888).

245. E. Rothé, Ann. Chim. Phys., 1, 336 (1904).

246. C. J. Gillespie and F. P. Hall, J. Amer. Chem. Soc., 48, 1207 (1926).

247. V. I. Kravtsov and M. I. Zelenskii, Vestn. Leningrad. Univ., Ser. Fiz. Khim., 21, 128 (1966); Elektrokhimiya, 2, 1138 (1966); ibid., 5, 247 (1969).

248. J. A. V. Butler and G. Drever, Trans. Faraday Soc., 32, 427 (1936).

249. S. E. S. El Wakkad and A. M. Sham El Din, J. Chem. Soc., 1954, 3094.

250. A. Hickling and G. G. Vrjosek, Trans. Faraday Soc., 57, 123 (1961).

251. T. R. Blackburn and J. J. Lingane, J. Electroanal. Chem., 5, 216 (1963).

252. Y. A. Mazitov, K. I. Rozental, and V. I. Veselovskii, Zh. Fiz. Khim., 38, 151 (1964).

253. J. Llopis, An. Real Soc. Espan. Fis. Quim., Ser. B, 61, 355 (1965).

254. T. Heumann and R. Schürmann, Z. Phys. Chem. (Leipzig), 226, 193 (1964).

255. J. E. Philpott, Platinum Metals Rev., 4, 12 (1960).

256. H. L. Grube, Metalloberflaeche, 5, 61 (1953).

257. V. V. Ostroumov, Zh. Fiz. Khim., 31, 402 (1957).

258. V. V. Ostroumov, Zh. Prikl. Khim., 31, 402 (1958).

259. V. V. Ostroumov, ibid., 31, 77 (1958).

260. G. C. Lambros, Met. Finish., 41, 73 (1943).

261. G. W. Watt and J. A. Cunningham, J. Electrochem. Soc., 110, 716 (1963).

262. L. Siaurukaite, S. I. Khotyanovich, and J. Matulis, Liet. TSR Mokslu Akad. Darb., Ser. B, 1970, 39.

263. F. H. Reid, Metall. Rev., 8, 167 (1963).

264. W. Keitel and H. E. Zschiegner, Trans. Electrochem. Soc., 59, 273 (1931).

265. I. F. Kusevitch and N. T. Kudytsev, Zashch. Metal, 6, 465 (1970).

266. C. Duval, in Nouveau Traite de Chimie Minerale, Vol. 19 (P. Pascal, ed.), Masson, Paris, 1956, p. 586.

267. M. Delepine, Bull. Soc. Chim. Fr., 6, 1471 (1939).

268. K. E. Johnson and H. A. Laitinen, J. Electrochem. Soc., 110, 314 (1963).

269. C. H. Liu, J. Phys. Chem., 66, 164 (1962).

270. H. A. Laitinen and B. B. Bathia, J. Electrochem. Soc., 107, 705 (1960).

271. B. Scrosati, Anal. Chem., 38, 1588 (1966).

272. H. B. Gray and C. B. Ballhausen, J. Amer. Chem. Soc., 85, 260 (1963).

273. H. Basch and H. B. Gray, Inorg. Chem., 6, 365 (1967).

274. F. A. Cotton and C. B. Harris, ibid., 6, 369 (1967).

275. S. S. Zumdahl and R. S. Drago, J. Amer. Chem. Soc., 90, 6669 (1968).

276. C. H. Lai and A. T. Hubbard, Inorg. Chem., 11, 2081 (1972).

277. W. P. Griffith, J. Lewis, and G. Wilkinson, J. Chem. Soc., 1961, 775.

278. P. G. Zambonin and J. Jordan, Anal. Lett., 1, 1 (1967).

279. Y. M. Maximov, V. E. Kazarinov, and O. A. Petrii, Electrokhimiya, 8, 1254 (1972).

280. T. M. Grishina, G. P. Khomchenko, and G. D. Vovchenko, Vestn. Mosk. Univ., Ser. II, 18, 48 (1963).

281. T. N. Stoyanovskaya, G. P. Khomchenko, and G. D. Vovchenko, ibid., 19, 403 (1964).

282. A. A. Sutyagina, I. V. Fadeeva, and F. V. Golyanitskaya, Zh. Fiz. Khim., 45, 1782 (1971).

283. O. A. Petrii and A. V. Safonov, Elektrokhimiya, 9, 125 (1973).

284. A. N. Frumkin, O. A. Petrii, and B. V. Damaskin, ibid., 6, 617 (1970).

285. A. N. Frumkin, J. Electrochem. Soc., 107, 461 (1960).

286. A. N. Frumkin and O. A. Petrii, Electrochim. Acta, 15, 391 (1970).

287. D. R. Flinn, M. Rosen, and S. Shuldiner, Collect. Czech. Chem. Commun., 36, 454 (1971).

288. N. A. Balashova and N. T. Gorokhova, Elektrokhimiya, 8, 760 (1972).

289. A. N. Frumkin, Zh. N. Malysheva, O. A. Petrii, and V. E. Kazarinov, ibid., 8, 599 (1972).

Chapter VI-8

RUTHENIUM

J. F. LLOPIS* and I. M. TORDESILLAS

Department of Surface Chemistry and Electrochemistry
Institute of Physical Chemistry, C.S.I.C.
Madrid, Spain

Ruthenium compounds show a whole gamut of oxidation states ranging from 0 (metal carbonyls) to VIII, and because of the tendency of its ions to form polynuclear complexes they appear also with apparent fractional valencies (3.5, 4.2, 4.5), although their more usual states

*Deceased.

The References for Chapters VI-8 and VI-9 appear on pages 322-328. A General Bibliography appears on pages 328 and 329.

are Ru(III) and Ru(IV). In aqueous solutions, Ru ions form complexes mainly in the valency range II to V with coordination number 6 and, to a lesser extent, 4.

1. STANDARD AND FORMAL POTENTIALS

1.1. AQUEOUS SOLUTION

As standard state of the element, the Ru solid metal is chosen. The standard potentials E° reported in Table 1.1.1 have been obtained by polarographic and potentiometric methods or evaluated from thermodynamic calculations.

In the latter case, Latimer's [1] values were calculated from well-known values of the formation enthalpies of $RuCl_3$(c) and RuO_2(s), and estimated values for their formation entropies. Recent data on the entropy and sublimation heat of Ru metal, the formation enthalpies and entropies of RuO_2(s) and RuO_4 (g), and thermal data for RuF_5(c) and RuS_2 (c) are reported, and a new calculation of the formation enthalpies for $RuCl_3$(c), $RuBr_3$(c), and RuI_3(c) have been made [2]. The free enthalpies calculated from these most recent data are higher than those given by Latimer which have been those employed by Pourbaix and Zoubov [3] to obtain the potential-pH diagram for Ru metal which is discussed later.

Polarographic and potentiometric data have been compiled with values derived from equilibrium constants and activity coefficients, and some potentials have been measured by comparison with other redox systems of known standard potentials. This method was used by Latimer and other authors to plot potential diagrams for Ru ions in aqueous solutions of HCl [4], $HClO_4$ [5, 6], F_3C-COOH [5], and NaOH [7, 8].

The Ru^{3+}(aq) ion is now well established as $[Ru(H_2O)_6]^{3+}$; $Ru(H_2O)_6{}^{2+}$ can be obtained when the tetrafluorate or p-toluenesulfonate is the anion. The standard potential of the system Ru^{3+}/Ru^{2+} was measured by Buckley and Mercer [9] using a glass electrode in solutions derived from p-toluenesulfonic acid at different ionic strengths and extrapolated to zero value.

Latimer [1] estimated E° = 0.45 V for the Ru^{2+}/Ru couple. If this value is combined with the above-reported potential for the system Ru^{3+}/Ru^{2+}, Ru^{2+} appears to be unstable with respect to disproportionation, which is in contradiction with the experimental results reported by Buckley and Mercer [9]. On this basis, it appears that a more appropriate potential for the system Ru^{2+}/Ru is E° = 0.2 V, but this leads to an E° of 2.1 V for the $RuO_2(c)/Ru^{3+}$, which seems too high when compared with the value E° = 1.6 calculated by Latimer.

Reversible cell [8, 10] and polarographic measurements [7] have led to the same potential for the $RuO_4{}^-/RuO_4{}^{2-}$ system. Equilibrium measurements in alkaline solutions have allowed the study of the $RuO_4/RuO_4{}^-$ half-reaction. In both cases the potentials are in good agreement with those estimated by Latimer.

TABLE 1.1.1. Standard and Formal Potentials in Aqueous Solutions

Half-reaction	Standard or formal potential (V)	Conditions	Refs.
1. $Ru^{2+} + 2e = Ru$	0.455	Thermodynamic data	[1], [4]
2. $RuCl_3 + 3e = Ru + 3Cl^-$	0.68	Thermodynamic data	[1]
3. $RuCl_5^{2-} + 3e = Ru + 5Cl^-$	0.60	1 M concentration	[11]
4. $Ru^{2+} + e = Ru^+$	0.03 ?	0.1 N HCl $C_{Ru^+} = C_{Ru^{2+}} = 0.01$ M	[14]
5. $Ru^{3+} + e = Ru^{2+}$	0.2487	Potential measurements (glass electrodes; I = 0) (extrapolated)	[9]
6. $RuCl_5^{2-} + e = Ru^{2+} + 5Cl^-$	0.3	Thermodynamic data	[1], [4]
7. $[Ru(NH_3)_6]^{3+} + e = [Ru(NH_3)_6]^{2+}$	0.24	Potential measurements	[15]
	0.214	Equilibrium constant	[15]
8. $[Ru(CN)_6]^{3-} + e = [Ru(CN)_6]^{4-}$	0.86 ± 0.03	$E_h^{\circ\prime}$. Polarographic and potentiometric data	[16], [17]
9. $[Ru(CN)_6]^{3-} + e = [Ru(CN)_6]^{4-}$	0.89	Polarographic data	[18]
10. Ru(IV) + e = Ru(III)	0.83	$E_h^{\circ\prime}$. Potentiometric data. HCl solution	[19], [20]
11. $Ru(OH)_2Cl^+ + Cl^- + 2H^+ + e = RuCl_2^- + 2H_2O$	0.96	$E_h^{\circ\prime}$. Potential extrapolation. $C_{HCl} = 0$	[21]
12. $Ru(OH)_2Br^+ + Br^- + 2H^+ + e = RuBr^- + 2H_2O$	0.82	$E_h^{\circ\prime}$. Potential extrapolation. $C_{HBr} = 0$	[21]
13. $RuO_4 + 8H^+ + 4e = Ru(IV) + 4H_2O$	1.40	$E_h^{\circ\prime}$. $HClO_4$, 1-6 N	[22]
14. $RuO_4^- + e = RuO_4^{2-}$	0.59	E_B°. Potential measurement. Na_2CO_3. NaOH. Polarographic data, $NaClO_4$	[7], [8]
15. $RuO_4^- + e = RuO_4^{2-}$	0.593	Equilibrium constant	[1], [11], [17]
16. $RuO_4 + 2Cl^- + 2e = [RuO_2Cl_4]^{2-}$	1.31	E_h°. Potential titration. 1 M HCl. Ce(IV)	[23]
17. $RuO_4 + e = RuO_4^-$	0.99	E_h°. Equilibrium constant	[10]
18. $H_2RuO_5 + e = RuO_4^- + H_2O$	1.00	Polarographic data	[7]

1.2. EQUILIBRIUM DATA

From the potential values collected mainly by Latimer [1] and Parsons [11], Pourbaix and Zoubov have constructed the potential-pH diagrams for the Ru-H_2O and the Ru-H_2O-Cl^- systems.

The diagram corresponding to the system Ru-H_2O was constructed with the following equations:

1. $H_2RuO_5 = HRuO_5^- + H^+$ $\quad \log \frac{H_2RuO_5}{HRuO_5^-} = 11.22 - pH$ (1.2.1)

2. Eqs. 14 and 15, Table 1.1.1

3. Eq. 18, Table 1.1.1

5. $RuO_4^- + 8H^+ + 5e = Ru^{2+} + 4H_2O$ $\quad E = 1.368 - 0.0945\ pH + 0.0118 \log \frac{RuO_4^-}{Ru^{2+}}$ (1.2.2)

6. $RuO_4^{2-} + 8H^+ + 4e = Ru^{2+} + 4H_2O$ $\quad E = 1.563 - 0.1182\ pH + 0.0148 \log \frac{RuO_4^{2-}}{Ru^{2+}}$ (1.2.3)

7. $HRuO_5^- + H^+ + e = RuO_4^- + H_2O$ $\quad E = 1.660 - 0.0591\ pH + 0.0591 \log \frac{HRuO_5^-}{RuO_4^-}$ (1.2.4)

8. $Ru(OH)_3 + 3H^+ + 3e = Ru + H_2O$ $\quad E = 0.738 - 0.0591\ pH$ (1.2.5)

9. $RuO_2 + H_2O + H^+ + e = Ru(OH)_3$ $\quad E = 0.937 - 0.0591\ pH$ (1.2.6)

10. $RuO_4 + 4H^+ + 4e = RuO_2 + 2H_2O$ $\quad E = 1.387 - 0.0591\ pH$ (1.2.7)

12. $RuO_4 + H_2O = HRuO_5^- + H^+$ $\quad \log HRuO_5^- = -12.10 + pH$ (1.2.8)

13. Eq. 1, Table 1.1.1

14. $RuO_4 + e = RuO_4^-$ $\quad E = 0.950 - 0.0591 \log RuO_4^-$ (1.2.9)

16. $Ru(OH)_3 + 3H^+ + e = Ru^{2+} + 3H_2O$ $\quad E = 1.304 - 0.1773\ pH - 0.0591 \log Ru^{2+}$ (1.2.10)

17. $RuO_4^{2-} + 5H^+ + 3e = Ru(OH)_3$ $\quad E = 1.649 - 0.985\ pH + 0.0197 \log RuO_4^{2-}$ (1.2.11)

18. $RuO_2 + 4H^+ + 2e = Ru^{2+} + 2H_2O$ $\quad E = 1.120 - 0.182\ pH - 0.0295 \log Ru^{2+}$ (1.2.12)

19. $RuO_4^{2-} + 4H^+ + 2e = RuO_2$ $\quad E = 2.005 - 0.1182\ pH + 0.0295 \log RuO_4^{2-}$ (1.2.13)

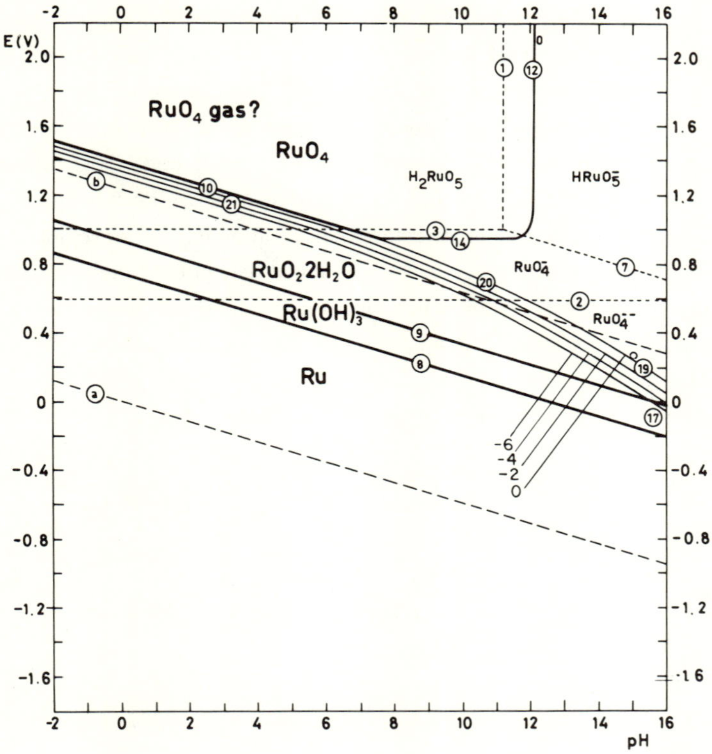

FIG. 1.2.1. Potential-pH diagram of the ruthenium-water system at 25°C. (Reprinted from Ref. 13, p. 8, by permission of Cebelcor.)

20. $RuO_4^- + 4H^+ + 3e = RuO_2 + 2H_2O$ $E = 1.533 - 0.0788\ pH + 0.0197 \log RuO_4^-$ (1.2.14)

21. $H_2RuO_5 + 4H^+ + 4e = RuO_2 + 3H_2O$ $E = 1.400 - 0.059\ pH + 0.0148 \log H_2RuO_5$ (1.2.15)

The diagram of Fig. 1.2.1 shows that Ru is very stable in acidic aqueous solutions in the absence of complexing agents but is attacked by oxidizing alkaline solutions; it appears to be less noble than Rh, Pd, Ir, and Pt. Ruthenium metal passivates at the anode in noncomplexing media.

Ru(IV) is the more stable oxidation state in the presence of oxygen and noncomplexing solutions, except at high pH where the more stable form is Ru(VI). All ruthenium ions tend to take up the IV or VI oxidation state by spontaneous oxidations or reductions as they meet the above-mentioned conditions. In very basic solutions the more stable ions are the orange ruthenates RuO_4^{2-}, which at pH $\simeq$ 12 change by dismutation to RuO_2, a solid black compound, and green RuO_4^- ions. The latter are unstable and at pH 12 liberate O_2 from water while they are reduced to RuO_4^{2-} ions; at pH $\simeq$ 7.5 this decomposition may give RuO_4 or unstable H_2RuO_5. In aqueous alkali the volatile yellow tetroxide RuO_4 forms perruthenates $HRuO_5^-$ which decompose to yield perruthenates RuO_4^- and ruthenates RuO_4^{2-}.

Eichner [12] has calculated the potential-pH diagram of Ru metal at high pH values and has also reached the conclusion that the VII and VIII oxidation states are unstable throughout this region. These states can only be obtained at the anode if the system OH^-/O_2 reacts very slowly. Thus anodic and cathodic potentiostatic polarization curves were obtained with Pt and vitreous carbon electrodes at pH 14, and it was found that in the latter case the limiting potential for oxidation increased from 0.6 V up to 1.0 V. This allowed the oxidation step of RuO_4^- and RuO_4 to be obtained, and with the RuO_4 formed the reduction step of this tetroxide was obtained.

The equations relative to the system Ru-Cl^--H_2O are [13]:

22. $RuCl_5^{2-} + e = Ru^{2+} + 5Cl^-$ $$E = 0.293 + 0.295\, p_{Cl^-} + 0.0591 \log \frac{RuCl_5^{2-}}{Ru^{2+}} \quad (1.2.16)$$

23. $RuCl_5OH^{2-} + H^+ + e = RuCl_5^{2-} + H_2O$ $$E = 1.200 - 0.0591\, pH + 0.0591 \log \frac{RuCl_5OH^{2-}}{RuCl_5^{2-}} \quad (1.2.17)$$

24. $H_2RuO_5 + 7H^+ + 5Cl^- + 4e = RuCl_5OH^{2-} + 4H_2O$ $$E = 1.587 - 0.1034\, pH - 0.0738\, pCl + 0.0148 \log \frac{H_2RuO_5}{RuCl_5OH^{2-}} \quad (1.2.18)$$

25. $RuO_4^- + 7H^+ + 5Cl^- + 3e = RuCl_5OH^{2-} + 3H_2O$ $$E = 1.783 - 0.1379\, pH - 0.0985\, pCl + 0.0197 \log \frac{RuO_4^-}{RuCl_5OH^{2-}} \quad (1.2.19)$$

26. $RuO_2 + 5Cl^- + 3H^+ = RuCl_5OH^{2-} + H_2O$ $$\text{Log } RuCl_5OH^{2-} = 12.65 - 5\, pCl - 3\, pH \quad (1.2.20)$$

27. $RuO_2 + 5Cl^- + 4H^+ + e = RuCl_5^{2-} + 2H_2O$ $$E = 1.948 - 0.2364\, pH - 0.2955\, pCl - 0.0591 \log RuCl_5^{2-} \quad (1.2.21)$$

The diagram for the Ru-Cl^--H_2O system has been plotted at a Cl^- free-ion concentration of 0.355 g/l (potential values are unwarranted) (Fig. 1.2.2). According to this diagram, Ru metal corrodes in HCl solutions and, in the presence of Cl_2 and/or Br_2, forms complex ions. The more stable oxidation state in these solutions is Ru(III), and only at high potentials is Ru(IV) produced. This is verified by anodic dissolution of Ru in HCl solutions, where Ru(III) is formed at potentials lower than 0.93 V and Ru(IV) at higher potentials.

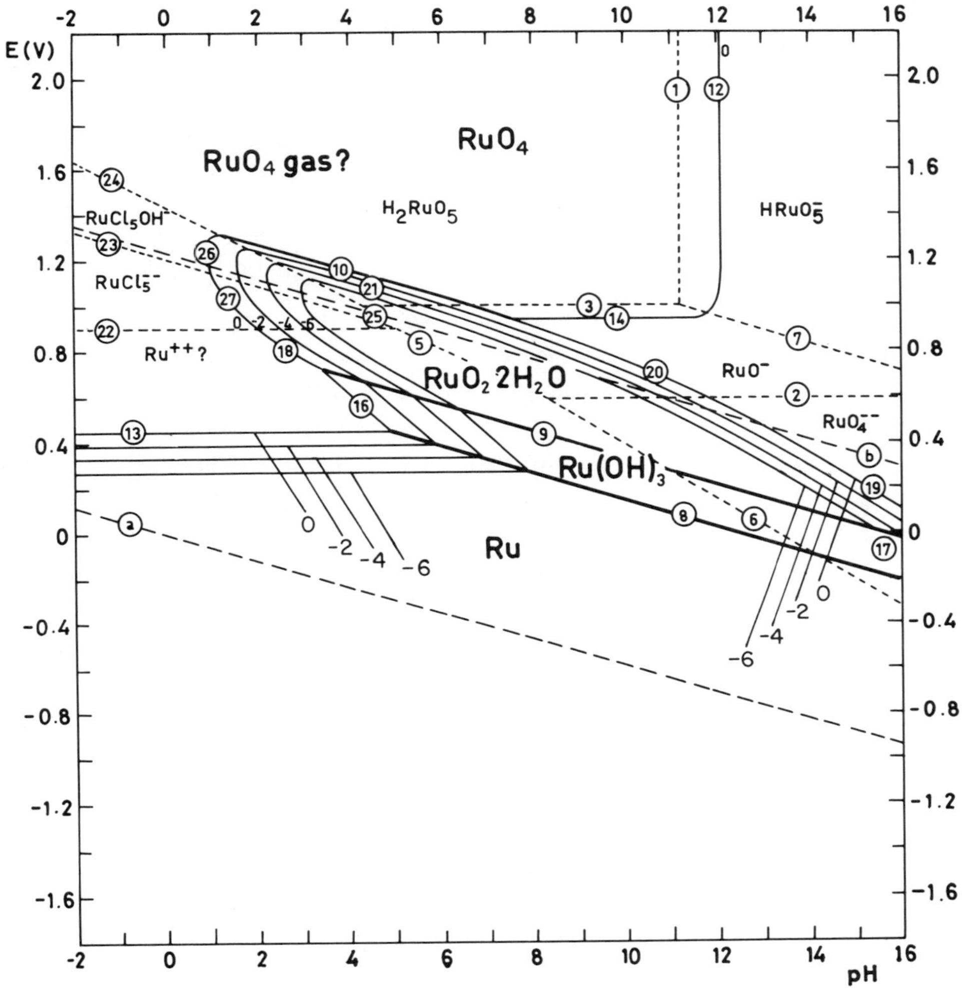

FIG. 1.2.2. Potential-pH diagram of the ruthenium-water system at 25°C in chloride solutions (355 mg Cl^-/1). (Reprinted from Ref. 13, p. 9, by permission of Cebelcor.)

2. VOLTAMMETRIC CHARACTERISTICS

2.1. POLAROGRAPHIC CHARACTERISTICS

2.1.1. Aqueous Solutions

The polarographic behavior of Ru has been studied in order to establish the potential of some redox systems of complex compounds or for microanalytical purposes. The polarography of Ru ions is hindered by the easy reduction of some Ru compounds by Hg. Furthermore, Evans et al. [25] proved that no polarographic reduction is possible on Hg electrodes for Ru complexes with ligands such as CN^-, SCN^-, NH_3, and pyridine [26].

On the other hand, Ru(III) chloride is not reduced at Hg electrodes because it decreases the overpotential of hydrogen ion discharge causing catalytic evolution of H_2. This phenomenon has been observed for all the platinum metals at concentrations of about 10^{-4} M, giving rise to two or three secondary waves, depending on the nature and concentration of the catalyst, at potentials lower than that of the hydrogen discharge (-1.2 V vs SCE in 1 N HCl solutions) [27-29].

A main wave appears at -1.0 V for all metals of the group, and Ru, Rh, and Ir show two further secondary waves which are only indicated by inflections in the polarograms at -0.8 and 0.6 V, respectively. In general, the catalytic effect, represented by the height of the first wave, is proportional to concentration (in the range of low values, from 10^{-6} to 10^{-5} M) and increases in the order of chlorocomplexes of Pd, Pt, Ir, Rh, and Ru. No data exist on Os.

In 0.01 M H_2SO_4 + 0.1 M K_2SO_4 solutions, polarographic reduction of Ru(III) sulfate yields a catalytic wave at -1.0 V (vs SCE) with a maximum that can be suppressed by addition of gelatin. The catalytic effect appears only in the concentration range from 5×10^{-10} to 2×10^{-7} M where it is proportional to Ru(III) concentration. The decrease in hydrogen overpotential amounts to 0.278 V.

This catalytic wave is also observed in mixed Ru(III) and Ru(IV) solutions in 0.9 M $HClO_4$ with a $E_{1/2}$ of -1.12 V (SCE) [30]. At higher values of pH, the $E_{1/2}$ shifts to more negative values, increasing to -1.85 V at pH 9. Limiting diffusion currents are proportional to Ru concentration in the range 10^{-5} to 3.7×10^{-6} M. A similar catalytic wave has been noticed in polarograms of Ru(IV) in 0.1 M NH_3 + 0.1 M ammonium tartrate solutions, with a $E_{1/2}$ of -1.577 (SCE) [31].

The polarographic reduction of Ru nitrosyl complexes reveals a very interesting behavior [32]. The chloronitrosyl complex exhibits a well-defined wave, the $E_{1/2}$-pH relation of which leads to the following reduction reaction:

$$RuNOCl_3 + 6H^+ + 5e \rightarrow RuCl_3 + NH_4^+ + H_2O$$

In contrast, the nitrosylnitrate complexes show irregular polarograms which are interpreted as corresponding to a gradual reduction of complexes of the type $RuNO(NO_3)_3$, $RuNO(NO_3)_3(OH)_2$, and $RuNO(OH)_3$. The potential range for the polarographic reduction of these complexes is larger than that for nitrosylchloro complexes. It has been concluded that the NO-Ru bonding energy is larger in the NO_3^- than in the Cl^- complexes.

Table 2.1.1 summarized most of the results concerning the polarographic behavior of Ru compounds. It also gives a number of supporting electrolytes used in analytical determinations of this metal.

TABLE 2.1.1. Polarographic Behavior of Ru Compounds

Substance	Product	Conditions	$E_{1/2}$ (V vs SCE)	$I = \frac{id}{cm^{2/3}t^{1/6}}$	$E_{3/4} - E_{1/4}$ (mV)	Remarks	Refs.
Ru(III)	Ru(II)	1 M $HClO_4$	-0.09	2.92	rev	NCE	[33]
Ru(III)	Ru(II)	1 M KNO_3 + 1 M HNO_3	-0.77			Hg_2SO_4/Hg	[34]
Ru(IV)	Ru(III)	1 M $HClO_4$	+0.65	0.94	rev	NCE	[33]
Ru(IV)	Ru(III)	1 M KNO_3 + 1 M HNO_3	-0.35			Hg_2SO_4/Hg	[34]
$RuCl_3$	$RuCl_2$ + Ru	Acetate buffer, pH 5		0.8	irr		[25]
$[Ru^{III}(NH_3)_5Cl]^{2+}$	$[Ru^{II}(NH_3)_5Cl]^+$	0.099 M K_2SO_4 + 0.1 M H_2SO_4	-0.27	2.13	sl irr	Triton X-100	[35]
$Ru^{III}(NH_3)_5X$	$Ru^{II}(NH_3)_5X$		-0.25	1.17	irr	Triton X-100	[35]
$RuNOCl_3$	$RuCl_3$, NH_4^+	0.1 M KCl + 0.25 M HCl, 20°C	-0.27		irr		[32]
$RuNO(NO_3)_2^{3-}$		2 M HNO_3	-0.22		irr		[32]
$Ru(C_5H_5)_2$	$[Ru(C_5H_5)_2]^+$	0.1 M $NaClO_4$ + 0.01 M $HClO_4$ + 90% C_2H_5OH	(+0.26)		rev		[36]

(continued)

TABLE 2.1.1 (continued)

Substance	Product	Conditions	$E_{1/2}$ (V vs SCE)	$I = \frac{i_d}{cm^{2/3}t^{1/6}}$	$E_{3/4} - E_{1/4}$ (mV)	Remarks	Refs.
$[Ru(C_5H_5)_2]^+$	$Ru(C_5H_5)_2$	0.1 M $NaClO_4$ + 0.01 M $HClO_4$ + 90% C_2H_5OH	+0.22		rev		[37], [38]
$[Ru(C_2O_4)_3]^{3-}$	$[Ru(C_2O_4)_3]^{4-}$	5×10^{-3} to 1.4×10^{-1} M $H_2C_2O_4$	-0.81		rev	Hg_2SO_4/Hg	[39]
$Ru[(C_2O_4)_2(H_2O)_2]^-$	$[Ru(C_2O_4)_2(H_2O)_2]^{2-}$	5×10^{-3} to 1.4×10^{-1} M $H_2C_2O_4$	-0.81		rev	Hg_2SO_4/Hg	[39]
$Ru[(C_2O_4)_2(H_2O)_2]^{2-}$	$[Ru(C_2O_4)_2(H_2O)_2]^-$	5×10^{-3} to 1.4×10^{-1} M $H_2C_2O_4$	-0.14		irr	Hg_2SO_4/Hg	[39]
$[Ru(C_2O_4)_4]^-$	$[Ru(C_2O_4)_4]^{2-}$	0.5 M $H_2C_2O_4$	-0.440		rev	0.006% Timol	[38], [40]
$[Ru(tart)_x]^{2-}$	$[Ru(tart)_x]^{3-}$	0.2 M H_2tart + 0.1 M NaOH	-0.510		rev		[37], [38]
$[Ru(citr)_x]^{3-}$	$[Ru(citr)_3]^{4-}$	0.15 M H_3citr + NaOH	-0.530		rev		[37], [38]
$[Ru(glucon)_x]^{y-}$	$Ru(glucon)_x^{(y+1)-}$	0.1 M sodium gluconate, pH 1	-0.67	1.18	rev		[41]
$[Ru(glucon)_x]^{y-}$	$Ru(glucon)_x^{(y+1)-}$	0.2 M sodium gluconate, pH 1	-0.67	1.17	rev		[42]

2.2. VOLTAMMETRIC DATA

Voltammetric techniques with stationary and rotating Pt and Au electrodes have been applied to a study of redox equilibria between low valence Ru complexes in acidic solutions, as well as to equilibria between Ru ions in a high oxidation state in very alkaline media. Generally, the half-wave potentials of these systems are too positive to allow the use of the DME.

Reduction of $[RuCl_5OH]^{2-}$ with stationary Pt electrodes in HCl solutions shows very low limiting diffusion currents with no dependence on the acid concentration, while the $E_{1/2}$ shifts to more negative values at low pH. These results are corroborated by using rotating electrodes. The irreversibility and low rate of the process suggest the formation of a labile intermediate $[Ru_2OCl_{10}]^{4-}$ [43]. In 6 N HCl solutions, $[RuCl_6]^{2-}$ is reduced to $[RuCl_5H_2O]^{2-}$ in a reversible wave [44, 45]. Voltammetry of cyanide complexes allowed the determination of the standard potential of the system $[Ru(CN)_6]^{4-}/[Ru(CN)_6]^{3-}$ as $E_h^\circ = 0.89$ (NHE) from the $E_{1/2}$ of polarization curves obtained with stationary Pt electrodes [16].

A comprehensive collection of voltammetric data concerning Ru compounds is given in Table 2.2.1.

TABLE 2.2.1. Voltammetric Data of Ru Compounds

Substance	Product	Conditions	Electrode	Technique	Potential vs SCE	Rev	Remarks	Refs.
Ru(II)	Ru(III)	Toluolsulfonic acid	Au	stat	0.1			[9]
Ru(III)	Ru(IV)	1 M H_2SO_4	Au	stat	1.0			[46]
Ru(III)	Ru(IV)	1 to 6 M $HClO_4$			1.3-1.4			[47]
Ru(III)	Ru(VI)	1 M H_2SO_4	Au	stat	1.4			[46]
Ru(III)	Ru(VIII)	1 M H_2SO_4	Au, Pt, Pd	stat	1.5			[46], [48]
Ru(VI)	Ru(VII)	0.1 M KOH + $NaClO_4$	Pt	stat				[7], [8]
Ru(VIII)	Ru(III)	1 N H_2SO_4	Pt	stat	-0.7			[46]

(continued)

TABLE 2.2.1 (continued)

Substance	Product	Conditions	Electrode	Technique	Potential vs SCE	Rev	Remarks	Refs.
Ru(VIII)	Ru(III)	$HClO_4$						[49]
Ru(VIII)	Ru(IV)	9 M $HClO_4$	Pt		1.11		NHE	[47]
Ru(VIII)	Ru(VI)	1 N H_2SO_4	Pt-Au		-0.2			[47]
Ru(VIII)	Ru(VII)	$NaClO_4$	Pt					[7]
$Ru(OH)_3$	Ru + $3OH^-$	4 N NaOH	Pt	E swp	-0.66			[12]
RuO_2	$Ru(OH)_3$	4 N NaOH	Pt	E swp	-0.39			[12]
RuO_4^{2-}	RuO_2 + $4OH^-$	4 N NaOH	Pt	E swp	-0.15			[12]
RuO_4^{2-}	RuO_4^-	KOH, pH 10.6	Pt	stat	(+0.598)	rev	NCE	[7]
RuO_4^-	RuO_4^{2-}	KOH, pH 10.6	Pt	stat	+0.606	rev	NCE	[7]
RuO_4^-	RuO_4	KOH, pH 10.6	Pt		+1.012	rev	NCE	[7]
RuO_4^-	RuO_4	KOH, pH 12 to 10	Vitreous carbon		+0.75			[12]
RuO_4	RuO_4^-	KOH, pH 10.6	Pt	stat	1.024	rev	NCE	[7]
$[RuOHCl_5]^{2-}$	$[RuCl_5H_2O]^{2-}$	2 N HCl + 1 N NaCl, Ru 10^{-3} to 10^{-5} M	Pt	stat	0.415-0.435	rev		[43], [50]
$[RuOHCl_5]^{2-}$	$[RuCl_5H_2O]^{2-}$	2 N HCl + 1 N NaCl, Ru 10^{-3} to 10^{-5} M	Pt	rot	E 0.415	rev		[43]

(continued)

TABLE 2.2.1 (continued)

Substance	Product	Conditions	Electrode	Technique	Potential vs SCE	Rev	Remarks	Refs.
$K_4[Ru_2OCl_{10}]$	$K_2[RuH_2OCl_5]$	2 M HCl + 1 M NaCl	Pt	stat	+0.415	irr		[43]
$[RuCl_5H_2O]^{2-}$	$[RuCl_6]^{2-}$	2 N HCl + 1 N NaCl	Pt	stat	E 0.8-1.0			[43], [45]
$[RuCl_6]^{2-}$	$[Ru(H_2O)Cl_5]^{2-}$	6 M HCl	Pt	rot	+0.72	rev		[44], [45]
					+0.9			[51]
$[Ru(CN)_6]^{4-}$	$[Ru(CN)_6]^{3-}$	0.2 M KCl	Pt	stat, rot	+0.72	irr		[16]
$Ru(CN)_6{}^{4-}$	$[Ru(CN)_6]^{3-}$	0.1 M KCl	Pt	stat, rot	+0.76	rev		[16]
$[Ru^{II}(NH_3)_5N_2]^{2+}$	$[Ru^{III}(NH_3)_5N_2]^{3+}$		Pt	rot microelectrode	+0.72			[35]
$[(Ru^{II}(NH_3)_5)_2N_2]^{4+}$	$[(Ru^{III}(NH_3)_5)_2N_2]^{6+}$		Pt	rot microelectrode	+0.78			[35]

3. KINETIC PARAMETERS AND DOUBLE-LAYER PROPERTIES

3.1. KINETIC PARAMETERS

No data on kinetic parameters of electrochemical reactions of Ru or its compounds have been found.

3.2. DOUBLE-LAYER PROPERTIES

The curves $(\partial E/\partial \mu_{H^+})_Q$ vs E_r obtained by Frumkin et al. [52] for Ru in 0.01 N HCl + 1 N HCl solutions (where E is the electrode potential vs a constant reference electrode and E_r is the potential referred to the NHE in the same solution) proved that even at this Cl^- concentration there is a continuous transition from the hydrogen adsorption region to that of oxygen adsorption. In the same solution the relationship between Γ_{H^+} (surface concentration of H^+) and E_r revealed a notable adsorption of H^+ on Ru electrodes at $E_r = 0$, which shows the zero charge potential ($E_{p.z.c.}$) to be in the negative region of potentials. Only at $E_r \simeq +0.61$ is $\Gamma_{H^+} = 0$. This potential can be considered as the p.z.c. for oxidized Ru.

In general, the values of $E_{p.z.c.}$ experimentally determined are considered unreliable and are better calculated from the work function of the metal Φ through the following equations:

$$E_{p.z.c.} = -4.78 + \Phi \quad \text{Frumkin}$$

$$E_{p.z.c.} = -4.21 + 0.88\Phi \quad \text{Antropov}$$

$$E_{p.z.c.} = -4.25 + 0.86\Phi \quad \text{Vasenin}$$

In Table 3.2.1 a number of values found in this way are given.

TABLE 3.2.1

Φ, 25°C (eV)	$E_{p.z.c.}$ (V vs NHE)	Formula	Refs.
4.52	-0.26	Frumkin	[53]
	-0.23	Antropov	[54]
	-0.36	Vasenin	[55]
4.53	-0.25	Frumkin	[53]
	-0.24	Antropov	[54]
	-0.37	Vasenin	[55]

4. ELECTROCHEMICAL STUDIES

4.1. CATHODIC BEHAVIOR OF Ru METAL

The hydrogen overpotential at Ru electrodes is very low [3], about 0.25 to 0.43 V in the current density range of 0.03 to 0.25 mA/cm^2 in 1 N H_2SO_4 solutions. The normal equilibrium potential for the Ru/H_2 electrode in 1 M H_2SO_4 solutions at pH = 0 and p_{H_2} = 1 atm is equal to zero as is also observed for Au, Pt, Rh, and Ir electrodes [58]. On the other hand, Ru possesses the property of absorbing large quantities of H_2 (about 1,500 times its volume) [3], and that allows use of the metal as a hydrogen electrode without a continuous bubbling of the gas over it. For the case of Ru deposited on Pt, the absorption capacity depends on the nature of the electrolyte.

The reduction potential of oxygen at this metal also depends on the electrolyte used. In general, as noted for the other noble metals, the theoretical and stable potentials of this system are never attained. The slope of the Tafel equation for oxygen reduction at Ru in 1 N H_2SO_4 and 0.1 N NaOH solutions is -80 and -51 mV, respectively, the exchange current being in both cases 10^{-12} A/cm^2. A study of this process with a ring-disk electrode in basic media indicated that the surface oxides of the electrode play an active role in the reduction. Thus the Ru oxide formed at potentials lower than 1.0 V enhances the decomposition of H_2O_2 produced as an intermediate, while those formed at higher potentials inhibit this process [59, 60].

Bright Ru behaves as an inert electrode with respect to a redox process as long as the metal ions do not participate in the electrochemical reactions proceeding at the electrode. This assumes the potential corresponding to the redox system, i.e., Ru can be used as the indicator electrode for measuring redox potentials.

4.2. ANODIC BEHAVIOR OF Ru: CORROSION

According to the potential-pH diagram [3, 13], Ru metal passivates in the absence of complexing agents by anodic polarization. In $HClO_4$ solutions, oxidation of the metal surface starts at 0.55 V, a RuO_2 layer is formed at 1.03 V, and RuO_4 is produced at potentials of about 1.4 V. These potentials shift to values of 0.25, 0.7, and 1.25 V, respectively, as the temperature is raised to 50°C [61, 62]. Ru is oxidized at 0.8 V in 0.1 N H_2SO_4 solutions, and at higher potentials ($\simeq$1.4 V) the simultaneous evolution of O_2 and formation of RuO_4 with a 25 to 30% current efficiency occurs [63, 64].

Ru dissolves in basic media with the production of oxy-compounds depending upon experimental conditions; for example, RuO_4^{2-} is formed at pH 14 [65]. Large amounts of RuO_4^{2-} along with smaller amounts of RuO_4^- and RuO_4 are produced in NaOH solutions under controlled conditions [64, 66]. The variation with time of Ru dissolved at different current densities shows that prior to dissolution an induction period of up to 2 or 3 hours is required during which the electrode surface saturates with oxygen or an oxide layer of $RuO_2.H_2O$ is formed. Dissolution rates do not increase linearly with the current density [67].

Charging curves for Ru electrodeposited on Pt reveals that in 0.1 N KOH solutions Ru oxidizes more easily than Pt, the bonding energy of Ru—O being higher than that of Pt—O [63].

Complexing agents increase the corrosion of Ru, particularly Cl^- ions [63, 66]. Ru dissolves into concentrated solutions of HCl as Ru(III) with 100% yields at potentials from 0.8 to 0.95 V (SCE). At higher potentials Ru(IV) is formed with very low current efficiency [62].

Radiochemical techniques have allowed the study of the electrochemical corrosion of this metal in the passivation region at high potentials. Absorption spectra and potentiometric analysis showed that no $RuCl_6^{2-}$ or $RuCl_6^{3-}$ species are formed under these conditions, but chloro-oxy, chloro-hydroxo, or even polynuclear complexes are probably produced [68].

Superimposition of ac to dc causes Ru to corrode easily at frequencies lower than 4 Hz. By operating with a square wave ac of 1 Hz, Ru dissolves into the $RuCl_6^{3-}$ and $RuCl_6^{2-}$ forms at current densities at which the Cl_2 evolution potential or the passive state is not attained during the positive half-cycle, but hydrogen discharge takes place during the negative half-cycle of the wave [62].

Earlier work on charging curves at Ru in 0.01 N H_2SO_4 solutions seemed to indicate an initial oxidation of the metal at 0.8 V and further RuO_4 formation at 1.4 V [69-71]. Later investigations showed that oxidation begins at a potential as low as 0.55 V with subsequent RuO_4 production at the mentioned potential [61]. Passivation of Ru is supposed to occur by formation of an anodic film several molecules thick, with the oxidation region overlapping that of hydrogen adsorption. This was substantiated experimentally in 4 N HCl solutions by the potential sweep method which showed some current peaks at potentials that seem to correspond either to reduction of ruthenium oxide or Cl_2 adsorbed on the electrode surface [62].

4.3. ELECTROCHEMICAL BEHAVIOR OF Ru COMPOUNDS

As indicated above, Ru occurs in several oxidation states and even with fractional valences due to the formation of polynuclear complexes, the more usual valences being Ru(III) and Ru(IV) in acid media and Ru(VI) in basic solutions. The volatile tetroxide RuO_4, formed by oxidation at high potentials, is also important.

Ru(I) compounds are unstable and reduce H_2O (its existence is debated [14, 72]). Aqueous solutions of Ru(II) are blue and barely stable since this species is easily oxidized to Ru(III) and Ru(IV). The Ru(II)/Ru(III) system has been studied as ruthenocene complexes with Hg electrodes [36, 73], as well as in p-toluene sulphonic acid solutions with DME [9] and as cyanocomplexes in 2 M KCl with Pt microelectrodes [16]. The Ru(II) coordination number is usually 6, but some species with a value of 4 are known.

The stability of Ru(III) compounds with coordination number 6 is rather appreciable. A whole series of Ru(III) complexes with H_2O and Cl^- ligands, from $[Ru(H_2O)_6]^{3+}$ to $[RuCl_6]^{3-}$, is known. By anodic oxidation in 1 M H_2SO_4, Ru(III) yields a yellow product of complex composition in which the apparent valence of Ru is 3.65 [74]. By chemical oxidation and depending on the oxidizing agent employed, several compounds of different valences are formed. In general, anodic oxidation products of Ru(III) in 1 M H_2SO_4 depend on the potential applied so that at values between 1.0 and 1.07 V (SCE) and using Au electrodes, a stable red Ru(IV) compound is formed [46]; at higher potential, 1.4 V, a colloidal black precipitate of Ru_2O_5 appears which changes to the very unstable compound $[RuO_2(SO_4)_2]^{2-}$ [46]. This decomposes in an open circuit according to

$$2Ru(VI) \rightarrow Ru(IV) + Ru(VIII)$$

At a potential of 1.5 V (SCE), $Ru_2(SO_4)_3$ is oxidized to RuO_4 through the successive formation of Ru(IV), (V), and (VI). The same reactions are observed with Pt and Pd anodes [46, 48]. In 1 to 6 M $HClO_4$ solutions a similar oxidation of Ru(III) up to Ru(VIII) at potentials of 1.3 to 1.4 V (NHE) may take place [6, 47].

In the presence of Cl^- ions, oxidation of Ru(III) to Ru(IV) goes through reactions in which the Cl^- ions take an active part [19]. For instance, oxidation of $[RuCl_5H_2O]^{2-}$ in 2 N HCl + 1 N NaCl solutions at 0.8 to 1.0 V with Pt electrodes produces a dark compound whose composition, determined by absorption spectra and polarographic measurements, is $[RuCl_6]^{2-}$ and not $[RuCl_5OH]^{2-}$ [50, 51].

Ru(IV) is a stable oxidation state that can form complexes with coordination number 6 and, in the case of chloro compounds, also take up H_2O and OH^- as ligands. This oxidation state may also form polynuclear complexes with Ru(III) [75].

On the other hand, Ru(IV) compounds are easily reduced or oxidized by electrolysis. In this way, red Ru(IV) is reduced to Ru(III) in 1 M H_2SO_4 solutions, going reversibly through the formation of a polynuclear complex of apparent oxidation state 3.5 [24] and other yellow intermediates with an apparent oxidation state of 3.25 [74]. Polarography of 0.03 M Ru(IV) solution in 1 M H_2SO_4 gives rise to a greenish brown Ru(III) compound [42]. It seems that a direct change of Ru(IV) to Ru(III) is self-catalyzed by hydrogen. Oxidation of that Ru(IV) solution with Au electrodes at 1.07 V slowly gives rise to a dark colloidal solution, and at 1.2 V (SCE) a precipitate, possibly of a Ru(V) peroxide, appears [46].

In 1 M $HClO_4$ at low temperature (from 0 to -5°C), Ru(IV) is reduced with a Pt electrode to Ru(III) in only 80% yield. For complete reduction the potential must be decreased from 0.9 to 0.3 or 0.4 V (NHE) [6, 47].

Reduction of the dark red Ru(IV) in 0.1 to 1 M $HClO_4$ solutions proceeds through three potential steps. It is argued that this phenomenon may be due to the existence of two different hydrolyzed Ru(IV) species, the more hydrolyzed one being more easily reduced to Ru(III). Colorimetric measurements show that the first two steps lead to Ru(III) formation and the third one corresponds to the Ru(III) → Ru(II) reduction [33, 77].

In $HClO_4$-$NaClO_4$ solutions the three steps are ascribed to the reactions Ru(4) → Ru(3.5) → Ru(3) → Ru(2) [6, 76]. On the other hand, Ru(IV) in 6 M $HClO_4$ slowly but spontaneously reaches an equilibrium state with an apparent oxidation number of 4.2. In $HClO_4$ media at concentrations from 1 to 9 M and temperatures from 0 to 25°C, the complete oxidation of Ru(IV) is not attained, hence oxidation of H_2O at the Pt electrode, catalyzed by Ru, is supposed to occur.

The polarography of $Ru(NO_3)_4$ has been referred to elsewhere (see Table 2.1.1) [34].

Reduction of Ru(IV) in Cl^- solutions presents very interesting features. At a high Cl^- concentration, Ru(IV) is reduced to the metal at a DME in HCl + KCl solutions through an intermediate, probably Ru(II). The half-wave potential depends on temperature and Ru(IV) and HCl concentrations [78]. The species $RuCl_6^{2-}$ in 4 N HCl solutions adjusted to pH values from 1.5 to 2.5 and at -0.47 V (SCE) is reduced at a Hg cathode to the blue $RuCl_4^{2-}$ with a 100% current efficiency [79, 80]. The same reaction product is obtained with solutions of commercial ruthenium chloride [indeed, a mixture of Ru(IV) and Ru(III) in 2 N HCl] [19]. If a 6 N HCl solution is used, the reduction of $RuCl_6^{2-}$ proceeds reversibly at a $E_{1/2}$ of 0.72 to 0.82 V (SCE) with the formation of $RuCl_5H_2O^{2-}$ [45]. Brown $RuCl_5OH^{2-}$ is reduced with Pt electrodes in 2 N HCl + 1 N NaCl solutions at a potential of 0.3 to 0.5 V (SCE) to $RuCl_5H_2O^{2-}$ [43].

A study of the Ru(IV)/Ru(III) system as a function of H^+ and Cl^- concentrations shows that the redox potential is most markedly affected by variation of the Cl^- concentration [21, 81]. On the other hand, potentiometric and spectrophotometric investigations with such different agents as $TiCl_3$, ascorbic acid, or hydroquinone proved that reduction of $RuCl_5OH^{2-}$ to Ru(III) concentrations of HCl lower than 2 N proceeds gradually through formation of intermediates with fractional oxidation number [75]. The results lead to the conclusion that a binuclear complex of Ru(IV) and Ru(III) in a 1:1 ratio with the formula Cl_5Ru^{IV}-OH... $Ru^{III}Cl_5^{4-}$ is formed. The maximum stability of this compound is attained in the HCl concentration range of 0.2 to 0.4 N; it decomposes in 2 N HCl. Thus Ru(IV) behavior in HCl solutions is similar to that in other media ($HClO_4$, H_2SO_4, etc.); intermediates with fractional oxidation numbers are formed in all of them. In the latter case, because of the weak complexing properties of ClO_4^- and SO_4^{2-}, the binuclear species assume a colloidal form in their solutions, while Cl^- ligands are strongly bonded and the above-mentioned intermediate displays a marked ionic character.

Ru(VI) is stable in highly alkaline solutions. In $NaClO_4$ solutions it can be reduced at a DME to Ru(IV) [7]. Oxidation of Ru(VI) with Pt electrodes in 0.1 N KOH + $NaClO_4$ solutions occurs at half-wave potentials that are independent of pH [7, 8]. Similarly, RuO_4^- is reduced to RuO_4^{2-} at Pt electrodes in $NaClO_4$ solutions [7]. In very basic media RuO_4^- transforms easily to RuO_4^{2-} with O_2 evolution [7, 8].

RuO_4 is reduced with either Pt or Au electrodes in 1 M H_2SO_4 solutions at -0.7 V (SCE). Reduction takes place stepwise, with color changes, to Ru metal. The steps are as follows: orange Ru(VII) → green Ru(VI) → black colloidal Ru(V) → light red Ru(IV) → yellow Ru(III) → green [Ru(III) + Ru(II)] (?) → colorless Ru(III) → Ru(0) [46].

In the case of NO complexes with NO_3^-, $C_2O_4^{2-}$, CH_3—COO^-, and Cl^- ligands, the potentiometric oxidation curves exhibit potential steps that lead all the compounds to the same final product (RuO_4). They also show that the oxidation state of Ru in the NO complex with NO_3^-, $C_2O_4^{2-}$, and CH_3—COO^- ligands is Ru(II), while Ru(III) is the state for the NO—Cl complex [32].

5. APPLIED ELECTROCHEMISTRY

5.1. ELECTRODEPOSITION

As the equilibrium diagram shows, Ru precipitates readily in metallic form by chemical reduction. Cathodic deposition is only possible in solutions of Ru ions with a valence lower than Ru(IV) [48]. The accurate deposition potential has not been determined [83].

Deposition of Ru from chloro-complex solutions shows low current efficiencies because the potential of hydrogen discharge is readily attained even at low current densities [84]. According to polarographic measurements, a cathodic potential of -1.8 V (SCE) is required to obtain the complete deposition of the metal [85]. The electrochemical equivalent of Ru is 0.9485 g/A-hr [86].

A summary of experimental conditions, solutions, and nature of Ru deposits is represented in Tables 5.1.1 and 5.1.2.

In order to improve the quality and thickness of Ru deposits, several complexing agents have been tried: amine sulfonic acid, etc. According to Reid, the nitrosyl-sulfamate solution was the most satisfactory electrolyte for Ru electroplating.

The conclusion was reached that electrolysis by ac instead of dc should favor the quality of Ru plating [93]; in fact, a lighter color, a smooth surface, and a more uniform thickness of deposit was obtained. The method consisted of changing the polarity of current periodically, and the cathodic phase should be no longer than 40 sec while the anodic phase should have a period long enough to consume 5 to 50% the quantity of electricity spent during the cathodic phase. In practice, prior experiments are necessary to determine both periods.

The mechanism of Ru deposition on Au, Pt, Pd, and Pb electrodes has been studied by means of radioactive Ru in $Ru_2(SO_4)_3$-H_2SO_4 solutions at temperatures higher than 75°C since the critical temperature and potential are 65°C and 0.25 V (NHE), respectively [89].

TABLE 5.1.1. Electroplating of Ru: Aqueous Solutions

Electrolyte	t (°C)	Cathode	Potential vs s.c.s.	Current density (mA/cm^2)	Deposit appearance	Coatings thickness (cm)	% Efficiency	Remarks	Refs.
Ru(III)-Ru(IV) acid solutions	25				Black, no adherent				[90]
$RuCl_3$ + fuming HNO_3	45	Pt		2-3	Grey, adherent				[63]
	45	Pt		10	Bright				[63]
Ru(III), 5 g/l; H_2SO_4, 40 g/l	70			80			1		[88]
	93	Pt	-0.02		Very bright, good adherence			8 mg/cm^2	[89]
	93	Au	-0.4						[89]
Ru(III), 4 g/l; NH_2SO_3H		Ag, other metals		1.6		0.00025			[91]
Ru(as sulfamate), 5 g/l NH_2SO_3H, 10-50 g/l	50-70			20-80	Good, free from cracks	0.0005		Unstable solution, evolution of RuO_4	[88], [92]
RuO_4 + concn HCl	45	Pt		2-5	Grey, mat				[63]
	45	Pt		10	Bright				[63]
$RuNOCl_3$, 5×10^{-5} to 5×10^{-3} M; 0.3 N HNO_3					Dark		9		
$RuNOCl_3$, 5×10^{-5} to 5×10^{-3} M; 0.3 M HCl or 0.3 M H_2SO_4	25	Ni, Cu	-0.40	10-30	Brightness of silver		100		[90], [93]

	25	Au, Pt	-0.60	10-30	Good adherence	100		
$RuNOCl_3$, 5×10^{-3} <u>M</u>; 0.5 <u>M</u> HCl		Cu		2-5	Good appearance, adherent			[94]
Nitrosyl-sulfamate solution of Ru, 5 g/l Ru	70			40	Bright, free from cracks		Stable solution. Small evolution of RuO_4	[88]
$Na_2[(RuNO_2)_4OH] + NH_2SO_3H$					Very good		Less stable solution	[88]
Soluble anode Ru, 0.1 <u>N</u> KOH	45	Pt		2-10	Dark, small adherence			[63]

TABLE 5.1.2. Electroplating of Ru: Fused Salts

Bath composition	t (°C)	Cathode	Current density (mA/cm^2)	Deposit appearance	Coatings thickness (cm)	% Efficiency	Remarks	Refs.
Mole %, 50.6 LiCl + 22.7 NaCl + 26.7 KCl + Ru		Au		Insufficient				[95]
Weight %, 53 NaCN + 47 KCN + Ru	560	Pt, other metals	5-18	Bright, good coherence	5×10^{-3}	11-45	Alternating current	[96]
		Pt, other metals		Bright, smooth	6.25×10^{-3}	30	Cathodic period, 60 sec, 10 mA/cm^2. Anodic period 7.5 sec, 3 mA/cm^2	

5.2. ELECTROPOLISHING

Electropolishing of Ru or Ru alloys with metals of the Pt group is carried out with ac in HCl solutions saturated with 20% NaCl or in KCN solutions. The current density applied at room temperature may yary from 0.02 to 0.2 A/cm^2 in Cl^- solutions and from 0.02 to 0.5 A/cm^2 in cyanide media, the electrolysis time lasting from 10 to 20 min [95].

Chapter VI-9

RHODIUM

J. F. LLOPIS* and I. M. TORDESILLAS

Department of Surface Chemistry and Electrochemistry
Institute of Physical Chemistry, C.S.I.C.
Madrid, Spain

---

*Deceased.

The References for Chapters VI-8 and VI-9 appear on pages 322-328. A General Bibliography appears on pages 328 and 329.

Rhodium metal shows in its compounds a variety of oxidation states that spans from -I to +VI although the Rh(III) state appears to be the most stable one. The lower oxidation states (+I, 0, -I) correspond to carbonyl complexes while the IV and VI species only occur in alkaline media. Along with the other metals of the group, Rh tends to form complex ions of coordination number 6 with aquo- and halo-ligands.

1. STANDARD AND FORMAL POTENTIALS

1.1. AQUEOUS SOLUTIONS

The Rh solid metal has been chosen as the standard state of the element. Thermodynamic data on Rh compounds necessary to evaluate the standard potential E° of its electrochemical systems have been recently reviewed by Goldberg and Hepler [2]. The authors have used the entropy values listed by Kelley and King [99]. The free enthalpy of Rh_2O_3(c) was calculated from experimental data on the heat of sublimation of Rh metal [100-103], vapor pressures of the oxide [104], and the heat of reduction of the oxide by hydrogen [105]. An estimation of the free enthalpy of Rh_2O_3 hydration was obtained from solubility [106] and hydrolysis [106, 107] data on the Rh oxide/hydroxide couple and comparison with similar $M_2O_3/M(OH)_3$ systems. Hence the free energy of formation of $Rh(OH)_3$(c) and finally the standard potential of the fundamental reaction $Rh^{3+} + 3e = Rh(c)$, E° = 0.7 V were derived [2].

The thermodynamic data of the $RhCl_6^{3-}$ complex were deduced from its stability constant [108], and the heat of reduction of the complex by means of Co(c) [109].

For solid halides the values were obtained from equilibrium studies on $RhCl_3$ decomposition at high temperatures which, on the other hand, have shown that RhCl and $RhCl_2$ do not exist as solid compounds [110]. Similar work on the thermal decomposition of the potassium and barium salts of $RhCl_6^{3-}$ has been performed [111, 112].

With regard to direct potential measurements, Grube et al. [113, 114] and others [115] have published basic results on the Rh(III)/Rh(IV), Rh(IV)/Rh(VI), and Rh(III)/Rh(VI) systems. Recently, Amosse, Ruband, and Barbier [116] have obtained the value E° = 0.758 ± 0.002 V for the system Rh/Rh^{3+} in an aqueous solution of $NaClO_4$ at pH 5 and extrapolated to zero ionic strength. This potential is in agreement with the value E° = 0.7 V deduced from thermodynamic calculations.

In his electrochemical work on the oxygen electrode, Hoare [117] has measured the potential of a Rh electrode covered by a film of adsorbed oxygen in aqueous solution. Depending upon experimental conditions, two types of films have been found that the author identified as the surface oxides RhO and Rh_2O_3. The free energy for the latter compound, calculated from its standard potential, appears to be notably less negative than for Rh_2O_3 (c). If this result is correct, the surface oxide is less stable than the crystalline one.

The standard potential for cyanide complexes was estimated by Willis [118] from polarographic data (Table 1.1.1).

From the free energy values collected by Bremen [119], calculated by Latimer [1] or evaluated by Zoubov and Pourbaix [3, 120], the latter authors have constructed the potential-pH diagram for the system $Rh\text{-}H_2O$ (Fig. 1.1.1).

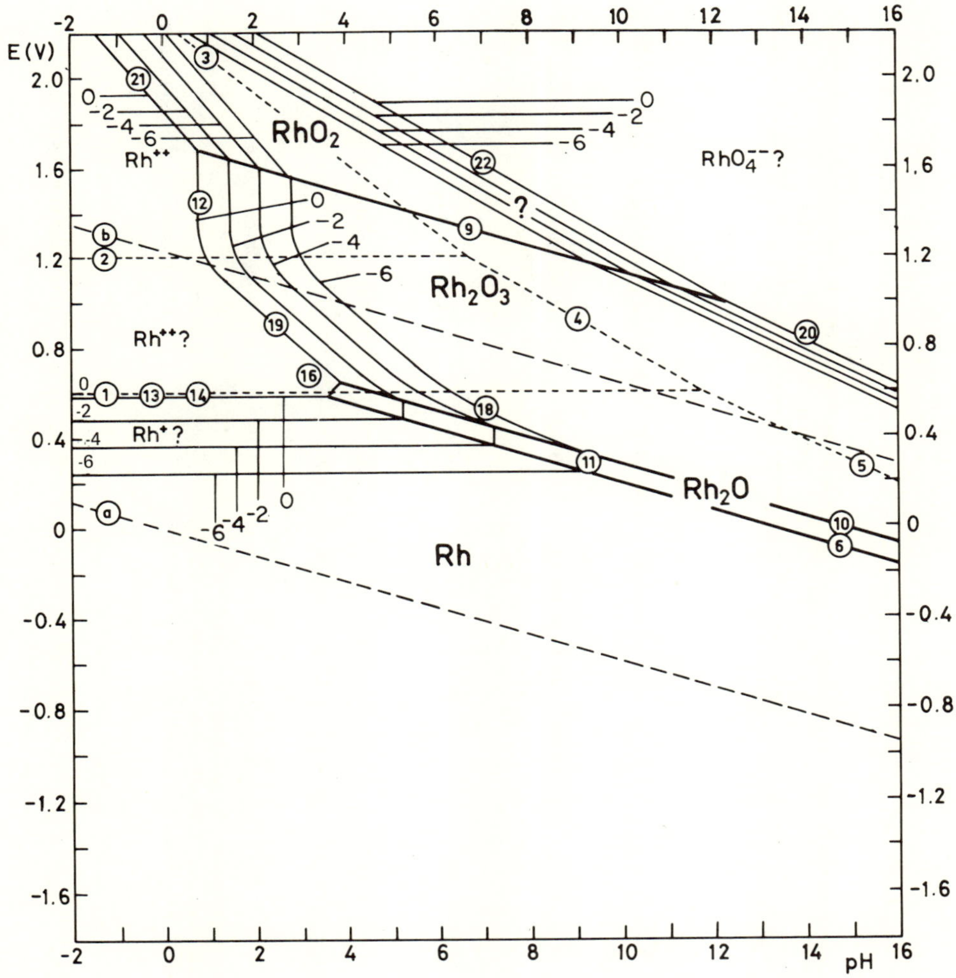

FIG. 1.1.1. Potential-pH diagram of the system rhodium-water at 25°C. (Reprinted from Ref. 120, p. 5, by permission of Cebelcor.)

TABLE 1.1.1. Standard and Formal Potentials in Aqueous Solutions

Half-reaction	Standard and formal potential (V)	Conditions	Refs.
1. $Rh^{+} + e = Rh$	≃0.6	Estimated value	[1]
2. $Rh^{2+} + 2e = Rh$	≃0.6	Approximated, calculated value	[1], [122]
3. $Rh(sup) + 2H^{+}(aq) + 2e = Rh(c) + H_2O(liq)$	+0.81	Electrochemical measurements of surface oxides	[117]
4. $Rh^{3+}(aq) + 3e = Rh(c)$	+0.7	Thermodynamic data and solubility products	[2]
5. $Rh^{3+}(aq) + 3e = Rh(c)$	+0.758	Extrapolated at zero ionic strength. $NaClO_4$, pH = 5	[116]
6. $Rh(OH)_3(c) + 3e = Rh(c) + 3OH^{-}(aq)$	0.0	Calculated value	[2]
7. $Rh_2O_3(sup) + 6H^{+}(aq) + 6e = 2Rh(c) + 3H_2O(liq)$	+0.88	Electrochemical measurements of surface oxides	[117]
8. $RhO_4{}^{2-}(aq) + 4H_2O(liq) + 6e = Rh(c) + 8OH^{-}(aq)$	0.0	Approximated value	[2]
9. $RhO_4{}^{2-}(aq) + 8H^{+}(aq) + 6e = Rh(c) + 4H_2O(liq)$	+1.1	Approximated value	[2]

10. $2RhO_2 + H_2O + 2e = Rh_2O_3 + 2OH^-$	>0.9	E° alkaline solutions in presence of Cl_2	[1]
11. $RhO^{2+}(aq) + 2H^+(aq) + e = Rh^{3+}(aq) + H_2O\ (liq)$	+1.4	HNO_3, $HClO_4$	[113]
12. $RhO^{2-}(aq) + 4H_2O(liq) + 3e = Rh(OH)_3(c) + 5OH^-(aq)$	-0.1	Calculated value	[2]
13. $RhO_4^{2-}(aq) + 8H^+(aq) + 3e = Rh^{3+}(aq) + 4H_2O(liq)$	1.5	Calculated value	[2]
14. $RhO_4^{2-} + 2H_2O + 2e = RhO_2 + 4OH^-$	>0.9	E° alkaline solutions in presence of Cl_2	[1]
15. $RhO^{2-}(aq) + 6H^+(aq) + 2e = RhO^{2+}(aq) + 3H_2O(liq)$	+1.46		[122]
16. $RhCl_6^{3-}(aq) + 3e = Rh(c) + 6Cl^-(aq)$	+0.5	Calculated value	[108]
	+0.44		[122]
17. $RhCl_6^{2-}(aq) + e = RhCl_6^{3-}(aq)$	+1.2	Estimated value	[1]
18. $RhO_2 + 6Cl^- + 4H^+ + e = RhCl_6^{3-} + 2H_2O$	<+1.4	HCl	[1]
19. $RhO_4^{2-} + 6Cl^- + 8H^+ + 3e = RhCl_6^{3-} + 4H_2O$	<-1.4	HCl	[1]
20. $Rh(CN)_6^{3-}(aq) + e = Rh(CN)_6^{4-}(aq)$	0.9	From polarographic data. Estimated value	[118]

The construction of Fig. 1.1.1 was based on:

Homogeneous reactions

	Electrochemical reactions
1. $Rh^{2+} + e^- = Rh^+$	$E = 0.600 + 0.0591 \log\frac{(Rh^{2+})}{(Rh^+)}$
2. $Rh^{3+} + e^- = Rh^{2+}$	$E = 1.198 + 0.0591 \log\frac{(Rh^{3+})}{(Rh^{2+})}$
3. $RhO_4^{2-} + 8H^+ + 3e^- = Rh^{3+} + 4H_2O$	$E = 2.261 - 0.1576\ pH + 0.0197 \log\frac{(RhO_4^{2-})}{(Rh^{3+})}$
4. $RhO_4^{2-} + 8H^+ + 4e^- = Rh^{2+} + 4H_2O$	$E = 1.995 - 0.1182\ pH + 0.0148 \log\frac{(RhO_4^{2-})}{(Rh^{2+})}$
5. $RhO_4^{2-} + 8H^+ + 5e^- = Rh^+ + 4H_2O$	$E = 1.717 - 0.0946\ pH + 0.0118 \log\frac{(RhO_4^{2-})}{(Rh^+)}$

Heterogeneous reactions with two solid compounds

	Electrochemical reactions
6. $Rh_2O + 2H^+ + 2e^- = 2Rh + H_2O$	$E = 0.796 - 0.0591\ pH$
7. $2RhO + 2H^+ + 2e^- = Rh_2O + H_2O$	$E = 0.882 - 0.0591\ pH$
8. $Rh_2O_3 + 2H^+ + 2e^- = 2RhO + H_2O$	$E = 0.871 - 0.0591\ pH$
9. $2RhO_2 + 2H^+ + 2e^- = Rh_2O_3 + H_2O$	$E = 1.730 - 0.0591\ pH$
10. $Rh_2O_3 + 4H^+ + 4e^- = Rh_2O + 2H_2O$	$E = 0.877 - 0.0591\ pH$

Heterogeneous reactions with one solid compound

	Chemical reactions
11. $2Rh^+ + H_2O = Rh_2O + 2H^+$	$\log(Rh^+) = 3.31 - pH$
12. $2Rh^{3+} + 3H_2O = Rh_2O_3 + 6H^+$	$\log(Rh^{3+}) = 2.56 - 3\ pH$

	Electrochemical reactions
13. $Rh^+ + e^- = Rh$	$E = 0.600 + 0.0591 \log(Rh^+)$

14.	$Rh^{2+} + 2e^- = Rh$	$E = 0.600 + 0.0295 \log(Rh^{2+})$
15.	$Rh^{3+} + 3e^- = Rh$	$E = 0.799 + 0.0197 \log(Rh^{3+})$
16.	$2Rh^{2+} + H_2O + 2e^- = Rh_2O + 2H^+$	$E = 0.396 + 0.0591\ pH + 0.0591 \log(Rh^{2+})$
17.	$2Rh^{3+} + H_2O + 4e^- = Rh_2O + 2H^+$	$E = 0.801 + 0.0295\ pH + 0.0295 \log(Rh^{3+})$
18.	$Rh_2O_3 + 6H^+ + 4e^- = 2Rh^+ + 3H_2O$	$E = 0.975 - 0.0886\ pH - 0.0295 \log(Rh^+)$
19.	$Rh_2O_3 + 6H^+ + 2e^- = 2Rh^{2+} + 3H_2O$	$E = 1.349 - 0.1773\ pH - 0.0591 \log(Rh^{2+})$
20.	$2RhO_4^{2-} + OH^+ + 6e^- = Rh2O_3 + 5H_2O$	$E = 2.211 - 0.0985\ pH + 0.0197 \log (RhO_4^{2-})$
21.	$RhO_2 + 4H^+ + e^- = Rh^{3+} + 2H_2O$	$E = 1.881 - 0.2364\ pH - 0.0591 \log(Rh^{3+})$
22.	$RhO_4^{2-} + 4H^+ + 2e^- = RhO_2 + 2H_2O$	$E = 2.452 - 0.1182\ pH - 0.0295 \log(RhO_4^{2-})$

Some of the chemical species used in the reactions represented in Fig. 1.1.1 have not been experimentally identified or isolated, and others exist mainly in complexed form, such as the Rh^{3+} ion. As is well known, Rh, like the other Pt metals, very readily forms complex ions. Inspection of Fig. 1.1.1 shows that Rh is a very stable metal since its stability domain is above that of water in aqueous solutions at any pH in the absence of complexing agents.

Oxidizing alkaline or acidic solutions and strongly corrosive media such as aqua regia do not attack the compact metal and, generally speaking, Rh is more resistant than Pt to chemical action [121]. On the other hand, the stability of Rh is related to the physical structure of the metal so that finely divided Rh (black Rh) can be dissolved in hot, concentrated sulfuric acid or aqua regia [120].

1.2. MOLTEN SALTS

Laitinen et al. [123-127] have measured the standard potential of the Rh(0)/Rh(III) systems in fused alkali chloride and sulfate eutectics. The standard potentials for Rh/Rh(III) given in Table 1.2.1 were extrapolated to the unit concentration of the species in the melt on the molarity M, molality m, and mole fraction x scales. Thus there are three potential scales: E°_M, E°_m, and E°_x. As indicated in the table, the reference electrodes were (1.0 M)/Pt(0) and Ag(I) 1 m/Ag(0).

TABLE 1.2.1. Electrode Potentials in Fused Salts of the Rh(0)/Rh(III) System

Fused salts	E_M° (V)	E_m° (V)	E_x° (V)	Reference electrode	T (°C)	Refs.
1. $Li_2SO_4 + K_2SO_4$ (80:20 M%)		+0.387 ± 0.003		Ag/1 m Ag^+	625	[125-127]
2. Li_2SO_4-Na_2SO_4-K_2SO_4 (78:8.5:13.5 M%)	+0.407	+0.370	+0.267	Ag/1 M Ag^+	575	[126],[127]
3. LiCl-KCl (59:41 M%)	-0.196	-0.201	-0.231	Pt/1 M Pt^{2+}	450	[123],[124]

2. VOLTAMMETRIC CHARACTERISTICS

2.1. POLAROGRAPHIC CHARACTERISTICS

The polarography of the compounds $RhCl_3$, Na_3RhCl_6 and $Rh(NH_3)_5Cl_2$ in complexing and noncomplexing solutions has been widely investigated. The difficulties met within the polarographic work arise from the following properties of the Pt metal group: spontaneous oxidation of the electrode by the metal complex, production of catalytic hydrogen waves that in many cases prevent the use of acidic solutions [27, 28], and high reduction potentials very near to those of the supporting electrolyte that make the determination of limiting currents confusing [129]. On the other hand, Rh metal is able to form different complex ions with mixed ligands. The numerical relation of these depends on experimental conditions (concentrations, temperature, aging of the solution, etc.) [25]. This poses problems of identification and gives rise to contradictory data. Therefore, the spectrophotometric method has been of great help in establishing and identifying the chemical species being studied by polarography [128].

Generally, the Rh(III) complexes, e.g., the halo compounds, yield a polarographic wave with two or more steps, for which the total limiting current is proportional to the Rh(III) concentration [25, 108, 128] (Table 2.1.1). The electron exchange number of the total process as calculated by Willis [118] from Tafel lines and by comparison with polarograms of Co complexes appeared to be 1 or 2. Later coulometric studies and electrolysis at controlled potential gave a value of 3, which seems to be more suitable for the reduction of Rh(III) [128]. However, in the cases of fluoride solution [130] and pyridine and γ-picoline complexes [131], Rh(I) and Rh(II) appear as relatively stable intermediate states in the reduction process. Kravtsov and Shereshevskaya [132] assume that the first wave of the reduction polarogram of Rh(III) chloro complex corresponds to the step Rh(III) → Rh(I). The polarographic results suggest that except for $RhBr_6^{3-}$ [133], all the Rh complexes behave irreversibly.

From an analytical point of view, the best results in the determination of Rh have been produced by pyridine solutions [129] and in presence of Ir by EDTA [134], thiourea, and trilon B solutions [51, 136].

The study of Rh(III) solutions by oscillographic polarography was undertaken by Beran and Dolezal [135, 137, 139] and Douglas and Magee [138]. The former authors found ethylenediamine sulfate solutions a satisfactory media for determination of Rh in the presence of Pt and Au, but the experimental work was easier in HCl solution. This medium is useful for analysis even in the presence of large amounts of Pt, Pd, and Au. On the other hand, Douglas and Magee showed that $Rh(SCN)_6^{3-}$ solutions produce a well-defined wave at -0.6 V which allows the determination of Rh concentration down to 5×10^{-5} M.

The reduction of $Rh(H_2O)_6^{3+}$ in 0.13 M ($HClO_4$ + $NaClO_4$) solutions at a DME was studied by Van Loon [128]. The polarograms exhibit two waves, the first of which was irreversible and diffusion controlled with $E_{1/2}$ = -0.38 V. The second wave showed a marked maximum that did not change notably by addition to solution of maximum suppressors such as gelatin or methyl red. This wave is not controlled by a diffusion process since the limiting current did not depend on the mercury height and had a negative temperature coefficient.

The polarography of halo complexes of Rh(III) was studied by Cozzi and Pantani [108, 133, 140] as a function of complex concentration and the pH of the solution. The authors tried to obtain stabilized solutions of these complexes, and the work was completed with a spectrophotometric investigation of the chemical species involved. In every case the polarograms showed two steps of different height and required the addition of gelatin to the solution in order to obtain well-defined waves. The limiting current of the first step was lower than that of the second step and decreased at increasing concentrations of Rh(III). This current increases, under the same experimental conditions, following the series $Cl^- < Br^- < I^-$, so that for the iodo complex the first step becomes higher than the second one, especially in the absence of surface-active substances. The total limiting current of both waves is diffusion controlled. According to the authors, the first wave would correspond to the reduction of Hg_2^{2+} formed by spontaneous oxidation of the DME by the Rh(III) complex. The second wave would be related to direct decomposition of the halo complex at the surface of the mercury drop which would consist mainly of Rh amalgam. Coulometric determinations yielded an exchange of three electrons for the reduction process. The values of the Tafel slope and the temperature coefficient of the limiting current and half-wave potential show that the reduction of the chloro and iodo complexes proceeds irreversibly while that of the bromo complex is reversible.

In every case the half-wave potential becomes more negative as the concentration of halide increases. This fact suggests the mediation in the process of several types of complex with a different number of halo ligands.

TABLE 2.1.1. Polarographic Characteristics

Substance	Product	Conditions	$E_{1/2}$ (V vs SCE)	I	$E_{3/4} - E_{1/2}$	Remarks	Refs.
$RhCl_3$		0.5 M $HClO_4$	-0.39			n = 2	[118]
Rh(III)	Rh(0)	1 M $HClO_4$	+0.035	5.65		Stabilized soln	[108]
Rh(III)	Rh(0)	1 M $NaClO_4$	-0.018	5.2		Stabilized soln	[108]
$[Rh(H_2O)_6]^{3+}$	Rh(0)	0.13 N ($HClO_4$ + $KClO_4$), pH 1.5-2.7	-0.38		irr	2nd wave not diffusion cond.	[128]
Rh(III)	Rh(0)	0.5 M K_2SO_4	+0.001	5.14	irr	Not stabilized soln	[108]
Rh(III)	Rh(0)	1 M KNO_3	-0.012	4.63	irr	Stabilized soln	[108]
$(RhCl_6)^{3-}$	Rh(0)	6 M HCl	-0.28		irr	Two waves	[108]
$(RhCl_6)^{3-}$	Rh(0)	6 M HCl	-0.26		irr	compd wave	[128]
$RhCl_3(H_2O)_3$		0.1 M HCl + NaCl, pH 1.3-3.8	-0.07		irr	compd wave	[128]
Rh(III)	Rh(0)	1 M KCl	-0.158	5.10	irr	Stabilized soln	[108]
Rh(III)	Rh(0)	4 M (HBr + NaBr) + 0.03% gelatin, pH 1	-0.366	4.43	rev	$E_{1/2}$ depending on pH and Br concn	[133]

$Rh(CN)_6^{3-}$	$Rh(CN)_6^{4-}$	1 M KCN	-1.47	irr	n = 1	[118]
$Rh(SCN)_6^{3-}$	Rh^{1+} ? pol $Rh^{3+} \rightleftharpoons Rh^{2+} — Rh^0$?	0.9 M KSCN	-0.39	Partially irr	n = 2	[118]
Rh(III)	Rh(0)	1 M KCNS, 0.01% gelatin	-0.487	74 (irr)	Stabilized soln at 100°C	[144]
$[Rh(NH_3)_5Cl]^{2+}$	$Rh(NH_3)_5Cl$	0.5 M K_2SO_4	-0.96	rev	n = 2	[118]
$[Rh(NH_3)_5Cl]^{2+}$	$Rh(NH_3)_5Cl$	1 M $NaNO_3$	-0.96	rev	n = 2	[118]
$[Rh(NH_3)_5Cl]^{2+}$	$Rh(NH_3)_5Cl$	1 M NH_4Cl	-0.93	rev	n = 2	[118]
$[Rh(C_2O_4)_2Cl]^{3-}$			-0.43	irr		[142]
$[Rh(C_2O_4)_3]^{3-}$		0.5 M $H_2C_2O_4$	-0.83	irr	n = 2	[142]
$RhCl_3$	Rh(II)	M KCl + M pyridine	-0.414	irr	n = 1	[118]
$RhCl_3$	Rh(II)	M KBr + M pyridine	-0.411	irr	n = 1	[118]
Rh(III)	Rh(I)	KCl + pyridine	-0.42	irr		[131]
Rh(III)	Rh(I)	0.1 M KCl, 0.1 M HCl + 0.2 M γ-picoline	-0.47	irr		[131]

The research by Van Loon and Page [128] on the reduction of Rh(III) in 0.1 M (HCl + NaCl) solutions with a DME also yielded polarograms with two waves. The first wave is composite and its total limiting current depends on the Rh concentration; the second wave is very similar to that for the Rh-aquo complex and shows a maximum that is not controlled by a diffusion process.

Recently, Kravtsov and Shereshevskaya [132] investigated solutions of Rh(III) in 4 M ($HCl + HClO_4$) and reached the conclusion that at low Rh concentrations (3×10^{-5} M) the first wave corresponds to the reversible reduction process Rh(III) + 2e $\rightleftarrows$ Rh(I), i.e., to ($RhCl_2^-$). The second wave, which is irreversible and diffusion controlled, is ascribed by the authors to reduction of the $RhCl_4^-$ ion. This complex is supposed to be in equilibrium with the species $RhCl_5^{2-}$ and $RhCl_6^{3-}$ which are the predominant species in the solution.

Polarographic and spectrophotometric studies on the formation of Rh complexes with oxalate, citrate, and tartrate ions from chloro compounds disclose that they are built up gradually. It appears that the intermediate complex with mixed chloride and organic anion ligands is reduced at more negative potentials than those of the corresponding chloro complex, but that the species completely free of chloride ligands could not be reduced at the DME [141, 142].

Rhodium, like the other metals of this group, shows a strong catalytic influence on the hydrogen discharge at Hg. This effect is more intense than that of Pt. Thus the polarogram for solutions of 10^{-6} N Rh in 0.1 N HCl exhibits a catalytic wave with a potential shift of 0.2 V. At concentrations of Rh higher than 5×10^{-5} N, a second wave arises at potentials from 0.7 to 0.8 V that is also related to the catalytic formation of hydrogen. The height of this wave increases with increasing concentration of Rh [129].

2.2. VOLTAMMETRIC CHARACTERISTICS

Voltammetric work with Pt rotating microelectrodes on the Rh chloro complex has led to the conclusion that a 0.1 M Na_2SO_4 solution is the best supporting electrolyte. In this solution, cationic aquo-chloro complexes are formed that migrate easily to the cathode. The half-wave potential falls in the 1.0 to 1.1 V region against a Pt anode, and the limiting diffusion current is proportional to concentrations in the 1×10^{-4} to 1×10^{-3} M range. It was found that three electrons are exchanged in the electrochemical process.

3. KINETIC PARAMETERS AND DOUBLE-LAYER PROPERTIES

3.1. KINETIC PARAMETERS

Few studies have been devoted to the kinetics of electrochemical processes with Rh compounds and, in consequence, data on kinetic parameters are very scarce.

Kravtsov and Shereshevskaya [132, 145] investigated the kinetic characteristics of Rh-chloro-complex reduction at the DME and a rotating disk of Rh. Polarograms for Rh solutions of 10^{-5} M concentration display two reduction waves, the first of which is reversible with a 30 mV slope and corresponds to the equilibrium

$$[RhCl_5H_2O]^{2-} + 3H_2O + 2e \rightleftharpoons [RhCl_2(H_2O)_4]^- + 3Cl^-$$

that is, to formation of a Rh(I) complex.

The second wave shows the slope dE/(d log i) = 82 mV, which is related to an irreversible process with the exchange of three electrons. At concentrations higher than 5×10^{-4} M, only a single wave is obtained, and it coincides with the second one described above. The authors explain the results as due to the slow electrochemical reduction of the complex $[RhCl_4(H_2O)_2]^-$ to Rh metal; this complex is in equilibrium with the $[RhCl_5H_2O]^{2-}$ and $[RhCl_6]^{3-}$ ions which are predominant in the solution.

Similar results were obtained for Rh deposition on rotating disk electrodes of this metal. Thus a first-order reaction with respect to Rh concentration was observed together with a dE/(d log i) = 88 mV slope. The presence of 4-chloro ligands in the complex discharged at the electrode was confirmed. However, the reaction rate (per cm^2 of apparent surface) of the chloro complex at the DME is an order of magnitude higher than that at the metal electrode at the same electrode potential.

3.2. POTENTIAL OF ZERO CHARGE

Several potentials of zero charge (p.z.c.) for Rh measured by different methods and in various solutions are given in Table 3.2.1.

TABLE 3.2.1. Potential of Zero Charge for Rh (NHE)

p.z.c.	Conditions	Method	Refs.
-0.03	Na_2SO_4	Radiotracers	[146]
+0.18 ± 0.025	10^{-3} to 10^{-2} M Na_2SO_4 or Cs_2SO_4, pH 2.5	Radiotracers radicals Adsorption	[147]
+0.28	0.02 N H_2SO_4	Capacity	[148]
0.0[a]	1 N KCl + 0.01 N HCl	Charging curves	[149]
-0.012			[150]

[a]Potential at zero total charge of the electrode.

Petrii et al. [149, 151] have studied the variation of the p.z.c. for the metals of the Pt group with the solution pH, and they found that the potentials become more negative with increasing pH values. In the case of Rh, this variation amounts to -45 mV in 1 N Na_2SO_4 + H_2SO_4 (pH 1.67 to 3.56) solutions. This increment is higher than that observed for Pt under the same experimental conditions.

In the case of solid electrodes, the metal/solution interphase potential depends not only on the surface charge but also on the amount of hydrogen adsorbed and the specific adsorption of ions.

Thus these authors [52, 149] have developed from a thermodynamic view a series of equations that relate the potential variation of an electrode that adsorbs hydrogen with the activity of H^+ in the solution under isoelectric conditions and the dependence on H^+ adsorption Γ_{H^+} on this potential E_r. Comparison of theoretical and experimental results for black Rh in 1 N Na_2SO_4 + 10^{-2} N H_2SO_4 and 1 N KCl + 10^{-2} N HCl solutions indicates that at low potentials the Rh electrode behaves as a reversible hydrogen electrode. In acidic chloride solutions the adsorption of atomic hydrogen, A_H, becomes zero at potentials $\geq$ 0.4 V. On the contrary, in alkaline chloride or acidic sulfate solutions there is no potential region where $A_H = 0$, i.e., there is an overlapping of the hydrogen and oxygen adsorption region.

The relationship between the electrode potential E_r and the surface density of adsorbed H^+, Γ_{H^+}, deduced by Frumkin et al., explains fairly well the experimental results in acidic solutions, but in alkaline solutions the agreement is only observed at potentials below 0.3 V.

The p.z.c. for Rh in 1 N Na_2SO_4 + 0.01 N H_2SO_4 and 1 N KCl + 1 N HCl solutions deduced from these relationships is about 0.10 and 0.0 V, respectively. The value found in sulfate solution coincides with that obtained by radiotracer methods. In alkaline media the Rh electrode seems to show no real potential with zero charge [149, 150].

3.3. DOUBLE-LAYER CAPACITANCE

The capacitance of a Rh electrode with fresh and aged surface as a function of overpotential and ac frequency has been measured by Breiter et al. in 8 N H_2SO_4 solutions using the impedance bridge technique [152, 153]. The double-layer capacitance of a bright Rh electrode previously activated by ac was measured by anodic charging curves from +117 mV in 3 N $HClO_4$ solutions and yielded the value 30 $\mu F/cm^2$ [154]. On the theoretical side, Frumkin and Petrii [52] have calculated the differential capacitance curves for black electrodes in 1 N Na_2SO_4 + 0.01 N H_2SO_4 solutions; in a way similar to that of Pt, the Rh capacitance shows a marked maximum at the hydrogen adsorption potential.

Finally, Hoare [155] has measured the capacitance of the double layer on bright Rh in 2 N Na_2SO_4 solutions saturated with N_2 by the galvanostatic "single-pulse" method using small Rh beads as the electrodes. A capacitance value between 16 and 18 $\mu F/cm^2$ was obtained when the Rh surface was completely free of oxygen.

3.4. HYDROGEN AND OXYGEN ADSORPTION

The surface properties of the Rh electrode and, particularly, hydrogen and oxygen adsorption have generally been studied by anodic and cathodic charging curves, differential capacitance measurements of the double layer, and triangular potential sweeps.

As in the case of Pt, the Rh electrode exhibits three more or less well-defined potential regions, according to experimental conditions, which correspond to hydrogen adsorption, the double-layer region, and oxygen adsorption. Hydrogen adsorption on Rh electrodes with fresh and aged surfaces was investigated by the impedance method in H_2SO_4 solutions [152, 153]. The results revealed a different behavior of the electrode according to the surface state, such that the extent of hydrogen adsorption on a fresh Rh surface was lower than that observed on aged electrodes.

Böld and Breiter [156] applied a linear potential sweep analysis [157] to bright Rh electrodes in 2.3 M H_2SO_4, 5 N $NaClO_4$ + 10^{-3} to 10^{-1} N $HClO_4$, and 0.1 N NaOH solutions. The E-i curves obtained are very similar to those observed for Pt. Oxygen adsorption spans over a wide potential range, from 0.55 to 1.3 V, but oxygen reduction is strongly inhibited so that the reduction maximum shifts to more negative values the higher the anodic potential reached. In the cathodic region, hydrogen adsorption and desorption show a single peak, in contrast with that on Pt where two maxima are noted. On increasing the pH of solution the ionization potential of hydrogen shifts to more positive values and both an extension of the hydrogen adsorption and retrenchment of the double-layer region is observed. In alkaline solutions the region of hydrogen adsorption overlaps that of reduction of adsorbed oxygen.

In the case of Rh it is always found that $Q_{anod} > Q_{cat}$, a fact that is attributed to the simultaneous adsorption and evolution of oxygen. On the other hand, no distinct boundary between the oxygen reduction and hydrogen adsorption regions can be noticed if the potential sweep is faster than 1 V/sec.

Following this work, Breiter [158] also determined the differential heat of hydrogen adsorption as a function of the surface coverage θ by this technique. The results disclose the existence of only strongly bonded hydrogen. In the region of the maximum the coverage, $\theta = 0.53$, is almost independent of temperature in the range -10 to 80°C. At $\theta < 0.1$ the extent of hydrogen adsorption is proportional to the square root of gas pressure in the system H_2/2.3 M H_2SO_4/Rh. The free energy and entropy of adsorption were calculated from these results, and the author concluded that hydrogen is adsorbed in atomic form, the adsorption complying with the Langmuir isotherm.

Adsorption measurements with black Rh, Ru, and Pt electrodes in 1 N NaCl + 0.01 M NaOH solutions over the potential range -0.1 to 0.7 V, joined to capacitance studies [159], lead to the conclusion that in the double-layer region the platinum surface is free of hydrogen and Rh shows a low hydrogen adsorption. By contrast, a large coverage is observed on the Ru surface. The hydrogen-metal bond exhibits a dipole character.

The charging curve technique employed by Tyurin [160] with black Rh and Rh deposited on Pt electrodes discloses the three potential regions described above in acidic media, but in alkaline media only the oxygen and hydrogen adsorption regions occur. At increasing pH the energy of the Rh—H bond increases and so does the extent of the hydrogen region. In their studies on the variation of the charging curves on black Rh with solution pH, these authors [161] found an empirical relationship, $A_H = f_1\ (pa^{-0.52})$, between the quantity of atomic hydrogen adsorbed, A_H, the equilibrium pressure of H_2, p, and the activity of the H_3O^+ ion, a, f_1 being the relation function. From this relationship $A_H = f_2(\mu_H - 0.26\mu_{H^+})$ is obtained. By taking into account the thermodynamic equations deduced by Frumkin et al. [52], the differential coefficient $(\partial A_H)/(\partial \Gamma_{H^+}) = -2.8$ was calculated. The low value of this coefficient with respect to that for Pt may be ascribed to the positive charging of the Rh surface in the hydrogen region or to formation of Rh—H dipoles with their negative ends pointing constantly toward the solution.

Later papers by Tyurin and Kossaya [162] report on the adsorption of hydrogen and oxygen by Rh electrodes in buffer solutions. They found that from pH 7 the extent of hydrogen and oxygen adsorption increases with the solution pH and also that the suitability of the equation $(\partial A_H/\partial \ln a)_p = -2(\partial A_H/\partial \ln p)_a$ was satisfactory over the pH range from 0.3 to 10.

Research on oxygen adsorption in alkaline and acidic solutions and N_2 atmosphere by charging curves [154, 160, 163] showed the formation of a film of adsorbed oxygen at potentials below that of oxygen evolution. The film is electronically conducting and about a monolayer thick. The formation of a Rh_2O_3 layer has also been suggested [154, 164], but it seems that anodic polarization allows oxygen to be dissolved into the outer layer of the metal surface with the production of a Rh—O alloy [165, 166]. The potential sweep method disclosed that oxygen dissolution in Rh occurs at potentials higher than those for Pt and also involves a stronger bonding of oxygen to metal, as proved by a more negative reduction potential [156, 160, 164]. This rather supports the hypothesis of the formation of a Rh—O alloy layer which would be more difficult to reduce than a layer of adsorbed oxygen.

Work on adsorption from solutions saturated with O_2 [166] has provided evidence for the existence of two types of Rh—O bonding: oxygen from the solution weakly adsorbed to the Rh surface and oxygen strongly adsorbed by anodic polarization.

The anodic oxidation of Pt, Ir, and Rh in acidic, neutral, and alkaline solutions was investigated by Shibata and Mitsui [167] through measurements of the double-layer capacitance at charging potentials. The authors noticed the surface formation of oxygen compounds through the reaction sequence: M → M—OH → M—O → M—O_2H where the bonding strength varies from simple adsorption to a chemical bond according to the metal and the pH of the solution.

Shibata [168] also used the charging curve method to study Rh oxidation under different experimental conditions. The anodic curves showed three regions with approximately the same quantity of current in each. The author ascribed these to: 1) adsorption of OH, 2) formation of RhO, and 3) formation of Rh_2O_3. In the cathodic curves only two potential regions appear, in which the charge ratio of the first to the second region is two. Application of the same technique to black Rh electrodes in 0.1 N H_2SO_4, HCl, and NaOH solutions by Grishina et al. [169] yielded charging curves that varied markedly with the electrolyte. The authors observed that the M—H bonding increased in strength in accordance with the relation HCl < H_2SO_4 < KOH. Similar results were obtained for the M—O bond by applying triangular and trapezoidal potential sweeps to Rh electrodes in 1 N H_2SO_4 and 0.1 N KOH solutions [170, 197], the oxygen bonding being stronger in alkaline than in acidic solutions. Thus the experimental results described again lead to the conclusion that oxygen is linked to rhodium by two types of bond, the first of which would be met at potentials lower than 1.0 V and would be of a chemisorptive nature [171].

Tarasevich et al. [172] have determined the extent of hydrogen and oxygen adsorption on finely divided Rh by comparison of results produced by charging curves with those from B.E.T. area measurements. They found that the surface fraction of Rh covered by hydrogen, and also that of Pt, at zero potential is unity. Bright Rh and Pt electrodes produced results equivalent to those mentioned before.

3.5. IONIC ADSORPTION

Cd^{2+} and Br^- adsorption on Pt electrodes covered by electrodeposited Rh was studied by Balashova et al. [173] as a function of potential and the nature of the electrolyte at potentials near to that of the zero charge. The structure of the double layer on Rh in solutions of 10^{-2} N H_2SO_4 + 10^{-2} N $CdSO_4$, 10^{-3} N H_2SO_4 + 10^{-2} N NaBr, and 10^{-2} N NaOH + 10^{-2} N NaBr was found to be similar to that on Pt under the same solutions. The work of Petrii and Topolev [174] on Rh in solutions of HCl, HBr, and 0.1 N H_2SO_4 disclosed that anion adsorption on this metal in the hydrogen and double-layer regions is higher than that on Pt, Br^-, and Cl^- ions being adsorbed at zero potential vs the reversible hydrogen electrode. This can be explained by the more negative value of the p.z.c. on Rh compared to that of Pt.

Lastly, an overpotential increase for oxygen adsorption in the presence of Cl^- ions has been reported for Rh electrodes in 3 M H_2SO_4 solutions [175].

3.6. ADSORPTION OF CO

The adsorption of CO on Rh electrodes has been particularly investigated with respect to its influence on hydrogen adsorption and electrocatalytic properties of the metal [176, 177].

Gilman [178] has studied the adsorption of CO and hydrogen on Rh by the potentiodynamic multipulse method, the amount of CO adsorbed being measured by anodic potential sweeps. It was concluded that in the potential range -0.2 to +0.5 V, the adsorption of CO proceeds rapidly up to formation of a monolayer and is controlled by the diffusion process.

The hydrogen adsorption isotherm and the variation of the hydrogen saturated coverage with CO adsorption seem to indicate that an adsorbed CO occupies one hydrogen adsorption site. In the regions where hydrogen saturation is not reached, the hydrogen equilibrium coverage decreases linearly with the increase in the extent of CO adsorption.

4. ELECTROCHEMICAL STUDIES

4.1. ANODIC BEHAVIOR OF Rh: CORROSION

Rh behaves anodically as a very noble metal that dissolves by electrolysis only in highly concentrated complexing media and at high overpotential. Thus, by means of prolonged electrolysis in concentrated H_2SO_4 solutions, Rh can be dissolved at the anode to form a sulfate complex [179]. In 6 N HCl the metal dissolves at high positive potentials and low current densities, yielding a yellow Rh(III) complex [180]. An increase in temperature and HCl concentration raises the rate of the process; as a rule, however, superimposition of a square or sine-wave alternating current is the most effective way to increase corrosion rates [168]. This method has been applied by Krasikov and Milenin [181] to Rh in concentrated H_2SO_4 solutions by means of an asymmetric alternating current of 0.1 Hz. During the anodic phase of the cycle a Rh oxide is formed which dissolves as the polarity of the current is changed. Oxide film formation proceeds through an electrochemical process while the dissolution process seems to occur through a combination of chemical and electrochemical mechanisms. Spectrophotometric, polarographic, and ion exchange methods disclose the dissolution product of Rh in concentrated H_2SO_4 to be a Rh sulfate similar to that obtained in the $Rh(OH)_3$-H_2SO_4 system [179].

The influence of alternating current on the dissolution of Rh in HCl was studied by Llopis, Tordesillas, and Muñiz [180]. Superimposition of a square or sine-wave alternating current produces an intense corrosion of the electrode and leads to the formation of a cherry-red solution of a Rh complex. Spectrophotometry shows the complex to be $(RhCl_6)^{3-}$. As noticed for the other metals of this group, no corrosion is observed at a constant current density of direct current, unless a threshold current density, i_a, of alternating current is reached. This threshold value is higher as the frequency of the alternating current is increased. The oscillogram of a superimposed square wave current of 10 Hz shows that corrosion of the metal is produced as soon as the potential of the electrode reaches the value of hydrogen evolution during the negative half-cycle of the alternating current.

The anodic dissolution of metals of the Pt group, Rh included, in organic solutions of methanol and ethylene glycol has been studied by Kovsman, Tyurin, and Karavaeva [182], who found that corrosion rates increase in the series Pt, Rh, Ir, and Pd, and that they are independent of temperature.

4.2. OXIDATION AND PASSIVATION

Among the possible Rh oxides, only Rh_2O_3 seems to be stable [183] since the RhO and Rh_2O referred to by some authors [105] are probably mixtures of Rh_2O_3 and Rh [117, 183], RhO_2 has only been obtained in hydrated form. On the other hand, according to Hoare [165, 170], and as indicated above for oxygen adsorption, the anodic oxidation of Rh produces an alloy system Rh—O in which oxygen is strongly bonded to Rh. This system is probably also produced by heating the metal in air. The anodic oxidation of Rh starts at potentials lower than that of Pt, and according to Butler and Drever [184] it consists in a chemisorption of oxygen on some points of the surface followed by adsorption of "peroxidic" oxygen. This oxidation can proceed even at potentials not excessively positive. The results obtained by Llopis and Vázquez [154] through cathodic charging curves in $HClO_4$ seem to confirm that interpretation. The previous oxidation of the electrode at potentials between 0.5 and 1.2 V (SCE) gives rise to a cathodic charging curve with a single step due to reduction of an oxidation product, but at oxidation potentials higher than 1.2 V (SCE) the cathodic curve shows two steps. This suggests the formation of two different surface oxides that the authors assume to be Rh_2O_3 and RhO_2. By using anodic and cathodic charging curves in HCl solutions, these authors [154] drew the conclusion that at potentials more positive than 0.7 V (SCE) the Rh surface is covered with oxygen even at high Cl^- concentrations. This oxidation determines the passivation of the metal and this, in turn, inhibits the discharge of Cl^- ions, therefore increasing the overpotential of Cl_2 evolution. On the other hand, a rise in concentration of Cl^- increases the overpotential of oxygen adsorption on Rh. This causes the competition between the Cl^- discharge and oxygen adsorption processes to depend on HCl concentration [185].

Rh is used as the anode for oxidation processes at high potentials such as persulfate formation and organic oxidations. The electrocatalytic properties of Rh for the oxidation reactions taking place on the electrode surface are probably related to its adsorption capacity. Thus the anodic processes proceeding on Rh at high H_2SO_4 concentration (12 to 18 N) and low temperature were studied by Pospelova et al. [186] with bright Rh electrodes and Rh electrodes oxidized at high anodic polarization. Comparison of the results obtained by coulometric and radiotracer methods has led to a determination of a relation between the amount of the HSO_4 species strongly adsorbed on the electrode as a function of the surface state and the current efficiency of $H_2S_2O_8$ formation [187].

4.3. CATHODIC BEHAVIOR

The prospects of Rh metal as a hydrogen electrode have been studied by Gortalova et al. [188] who measured the cathodic polarization of a 99.9% Rh wire and Rh deposited on a Pt electrode in acidic (H_2SO_4) and alkaline (NaOH) solutions at several concentrations. The pH values determined with a Rh-hydrogen electrode differ by less than 0.2 pH units from those obtained with the Pt-H_2 electrode.

The overpotential of H_2 on Rh in 2 N H_2SO_4 solutions was studied by Hoare [189] who found a Tafel slope of 0.118.

4.4. ELECTROCHEMICAL BEHAVIOR OF Rh COMPOUNDS

Grube and Gu [113] have investigated the anodic oxidation of Rh(III) in H_2SO_4 and $HClO_4$ solutions. Under these conditions Rh(VI) is formed, probably via Rh(IV), which disproportionates into Rh(III) and Rh(VI). These conclusions were confirmed by Grube and Mayer by potentiometric titrations with Fe(II) and iodometry of the high oxidation states of Rh in acidic solutions [190]. Rh ions with high oxidation states are strong oxidizing agents and, therefore, are not to be expected in the electrolysis of HCl solutions.

Dwyer and Schafer [115] have studied the Rh(III)/Rh(IV) system in H_2SO_4 obtained by anodic oxidation of Rh(III) solutions in this medium. Rh(IV) appears in the form of the non-complex RhO^{2+} ion at high acid concentration. Hydrate RhO_2 is precipitated by hydrolysis in acid solutions.

5. ELECTRODEPOSITION

Rhodium is the platinum group metal which has been most used practically in electrodeposited form in both the decorative field (jewelry) and in technical applications. According to Laister [191], the ratio of Rh used in plating in 1954 with respect to the other metals of the group was 100:1. Therefore, Rh plating has been broadly studied and has even been the subject of Ministry of Aviation Process Specifications (D.T.D. May 1961) [92].

Lately, wider attention has been paid to developing new plating processes with the aim of obtaining thicker coatings for technical uses and industrial applications (electric contacts).

The problems met with in electroplating with the platinum group metals arise mainly from their great capacity to form different complex ions. This requires a strict determination and control of the experimental conditions for the preparation and maintenance of the electrolytic bath. These so-called inert metals are characterized by high deposit overpotentials and fine grain coatings with low cathodic efficiencies. Those experimental conditions depend on the metal and the type of complex ion of the solution [92, 191-193].

Rh plates possess an extremely high hardness and high internal stress.

The Rh plating baths consist of acidic electrolytes. Those usually applied are: 1) phosphates solutions (thin coatings, generally used in jewelry); 2) sulfate and phosphate mixtures (plate thickness up to 2.5×10^{-5} cm, decorative field); and 3) sulfate electrolytes (deposit thickness over 2.5×10^{-4} cm, industrial applications).

Typical conditions of sulfate and phosphate-sulfate electrolytes are [92]:

Rhodium as sulfate or phosphate concentrate	2 g/l
Sulfuric acid	20 ml/l
Temperature	35-40°C
Current density	1-4 A/dm^2

Although other plating baths are used to obtain thick deposits, the acid sulfate electrolyte remains the most advantageous one from the economic point of view because of its cathodic efficiency. A summary of the operational conditions for this electrolyte was gathered by Reid [92] and is given in Table 5.1.

The data show that the cathodic efficiency of Rh deposition decreases at very acidic solutions. However, a certain degree of acidity of the bath is necessary to prevent hydrolysis of the metal salt. Thus a high acidity of 100 ml/l is required for a Rh content of 10 g/l so that the tendency of the deposit to pitting striation and cracking is reduced. Current densities should not be higher than 2 A/dm^2 since above this value a notable decrease in current efficiencies is observed. Stirring of the solution seems very convenient and can be carried through by several means: mechanical stirring, electrolyte flowing, bubbling of compressed air, moving of the electrodes, etc. According to Laister [191], stirring of the solution increases the deposit roughness required to obtain coating thicknesses higher than 3×10^{-4} cm. Deposition at 50°C seems to be satisfactory since it gives higher current efficiency and produces a minimum internal stress of the plate.

The method and procedure for the preparation of the Rh electroplating bath have shown no influence on the plating process except for the deposit hardness [199].

A loss of brightness of the deposit has been observed at thicknesses higher than 2.5×10^{-5} cm, a fact related to the simultaneous discharge of hydrogen. Hence the effect of some additives to the solution that increase the overvoltage of hydrogen evolution (such as reducible metallic salts, organic oxy-acids, and unsaturated acids) has been studied [200]. Another procedure has involved the addition to the electrolyte of chemicals that increase the oxygen overvoltage at the anode; in this way peroxy compounds are formed that depolarize the hydrogen discharge at the cathode. The Rh coatings usually obtained from sulfate solutions containing 10 g/l of Rh are bright up to thicknesses of 2.5×10^{-4} cm; with thicker deposits the surface begins to darken and becomes grey at 1.2×10^{-3} cm. On the other hand, the internal stress of the deposit reaches values from 3×10^{-3} to 6.5×10^{-3} kg/cm^2, and this leads to the development of cracks amenable to detection by microscopy or electrographic testing.

The cracking depends on the internal stress of the deposit and such other parameters as the nature of the substrate and the degree of plate adherence to the substrate. Therefore, the latter should be a ductile metal and its surface slightly attacked before proceeding to Rh electroplating [92].

TABLE 5.1. Formulations of Rhodium Sulfate Electrolytes

Range of deposit thickness (cm)	Rhodium content (g/l)	Sulfuric acid content (ml/l)	t (°C)	Current density (A/dm^2)	Cathode efficiency (%)	Agitation	Refs.
"Heavy" (0.0025)	10	100	50	0.5	90		[194]
General purpose	4	?	35	1.0		None	[191]
0.0005	10	25	50	2	90	Vigorous	[195]
Up to 0.00025	3.7	17	40-50	1.5	60-70	Yes	[196]
0.00025 to 0.00125	5.3	15	40-50	1.3	60-70	Yes	[196]
0.00025	10	"Lowest possible"	70-80[a]	0.5-1.0	60-70	Vigorous	[197]
0.00025 to 0.00050	10	30	50	1.5	85-98	Yes	[198]
0.0025 to 0.0050	20	50	50	1.2		Yes	
0.00025 to 0.00225	10	100	50	1.2		Yes	
0.0000375 to 0.0000625	2	20	30-50	1.2			
0.00025 to 0.00050	4	20	30-50	1.2			
0.001 to 0.002	10	50	30-50	1.2			

[a]Bright deposits, 0.001 to 0.002 cm thick.

The influence of metallic and organic impurities on Rh electrodeposition has also been recorded by Reid.

TABLE 5.2. Effect of Impurities in Bright-Rhodium-Plating Solutions (Phosphate-Sulfate)[a]

Impurity	Concentration (g/l)	Reflectivity of deposit, % sterling silver	Limiting thickness of bright deposit (in. $\times 10^{-6}$)[b]
Nil		81	10
Nickel	0.1	83	10
	0.2	87	10
Nickel + lead	0.2, 0.005	87	45
Silver	0.1		Severe dark streaks
	0.01	86	42
Copper	0.1	83	125 (73)
Iron(II)	0.1	85	10
Iron(III)	0.1	86	10
Iron(II) + lead	0.1, 0.005	85	275
Iron(III) + lead	0.1, 0.005	84	205
Tin(II)	0.1		Very low efficiency
	0.01	87	10
Tin(II) + lead	0.01, 0.005	84	47
Lead	0.005 to 0.01	85	35
Bismuth	0.1		Severe dark streaks
	0.025	86	40
Zinc	0.1	70	36 (60)
	0.01	83	35 (79)
Beryllium	0.1	86	10
Beryllium + lead	0.1, 0.005	86	98

[a]Electrolyte composition:

Rhodium (as phosphate solution)	2 g/l
Sulfuric acid	10 ml/l
Current density	4 A/dm^2
Room temperature	

[b]Figure in parentheses is reflectivity of thickest deposit when this differed significantly from the value in the third column.

As mentioned above, several electrolytes other than sulfates and phosphates have been tested for Rh electroplating. Thus Piontelli [201] has reported on sulfamate baths that produce very bright coatings with little tendency to cracking but low cathodic efficiency. Another electrolyte reported as a satisfactory bath is a 1 M $(NH_4)_3Rh(NO_2)_6$ + 6 M NH_2SO_3H solution [92]. Several electroplating patents have been registered that indicate the use of perchlorate and perchloric acid [202], mixtures of sulfates and chlorides [203], etc.

In order to obtain deposits with low internal stress, a necessary requirement for the technical applications of Rh coatings, new plating processes have been developed which include the use of special additives. Among the electrolytes proposed is one based on the addition of large amounts of magnesium sulfate or sulfamate to a rhodium sulfate solution [204]. A typical composition of this bath is:

Rhodium	5 g/l
Sulfuric acid	10 ml/l
Magnesium sulfamate	30 g/l
Magnesium sulfate	10 g/l

Reid has studied another type of electrolyte which is characterized by the addition of selenic acid to a sulfate solution [205]. The composition is:

Rhodium (as sulfate)	10 g/l
Selenic acid	0.4 g/l
Sulfuric acid	50 ml/l

The deposits produced from this type of electrolyte are bright and show a protective value on metal substrates against liquid corrodants.

REFERENCES

1. W. M. Latimer, The Oxidation States of the Elements and Their Potentials in Aqueous Solutions, 2nd ed., Prentice-Hall, Englewood Cliffs, New Jersey, 1952.
2. R. N. Goldberg and L. G. Hepler, Chem. Rev., 68, 229 (1968).
3. M. Pourbaix, N. de Zoubov, and J. Van Muylder, Atlas d'Equilibres Electrochimiques a 25°C, Gauthier-Villars, Paris, 1963.
4. W. M. Latimer and H. J. Hildebrand, Reference Book of Organic Chemistry, 3rd ed., New York, 1951/58, p. 450.
5. H. H. Cady, UCRI-3757 (1957) 1/81, 70, 78, N.S.A. 11 (1957), No. 8825.
6. D. K. Atwood and T. de Vries, J. Amer. Chem. Soc., 84, 2659-2661 (1962).
7. M. D. Silverman and H. A. Levy, ibid., 76, 3319-3321 (1954).
8. R. E. Connick and C. R. Hurley, ibid., 74, 5012-5015 (1952).

9. R. R. Buckley and E. E. Mercer, J. Phys. Chem., 70, 3103-3106 (1966).

10. R. E. Connick and C. R. Hurley, ibid., 61, 1018 (1957).

11. R. Parsons, Handbook of Electrochemical Constants, London, 1959, p. 70.

12. P. Eichner, Bull. Soc. Chim. France, No. 6, (1967), p. 2051.

13. N. de Zoubov and M. Pourbaix, Rapports Techniques, No. 58, Cebelcor, Brussels, 1958.

14. G. Grube and H. Nann, Z. Elektrochem., 45, 874-880 (1939).

15. J. F. Endicott and H. Taube, Inorg. Chem., 4, 437-445 (1965).

16. D. D. de Ford and A. V. Davidson, J. Amer. Chem. Soc., 73, 1469-1474 (1951).

17. G. Charlot, D. Bezier, and J. Courtot, Selected Constants, Oxidation-Reduction Potentials, New York, 1958, p. 27.

18. W. P. Griffith, Quart. Rev. (London), 16, 188-207 (1962).

19. G. Grube and G. Fromm, Z. Elektrochem., 47, 208-211 (1941); 46, 661-667 (1940).

20. R. Charonnat, in Nouveau Traite de Chimie Minerale, Vol. 19 (P. Pascal, ed.), Paris, 1958, p. 49.

21. J. R. Backhouse and F. P. Dwyer, J. Proc. Roy. Soc. N. S. W., 83, 138-145 (1949).

22. P. Wahner and J. C. Hindman, J. Amer. Chem. Soc., 72, 3911-3918 (1951).

23. J. L. Woodhead and J. R. Fletcher, AERE-R-4123 (1962) 1-30, 15, N.S.A. 17 (1962), No. 14198.

24. L. N. Lazarev and Ya. S. Khvorostin, Zh. Neorg. Khim., 13(9), 2512-2517 (1968).

25. R. B. Simpson, R. L. Evans, and H. A. Saroff, J. Amer. Chem. Soc., 77, 1438 (1955).

26. J. B. Willis, ibid., 67, 547 (1945).

27. P. Herasymenko and I. Slendyk, Collect. Czech. Chem. Commun., 5, 479 (1933).

28. I. Slendyk and P. Herasymenko, Z. Phys. Chem., A162, 223 (1932).

29. M. A. El Guebely, J. Chim. Phys., 53, 753 (1956).

30. M. V. Susic, Bull. Inst. Nucl. Sci. "Boris Kidrich" (Belgrade), 8, 53 (1958).

31. S. Baumgarten, R. E. Cover, H. Hofssass, S. Karp, Ph. B. Pinches, and L. Meites, Anal. Chim. Acta, 20, 397 (1959).

32. E. N. Jenkins, Atomic Energy Research Establishment R-3491 (1960).

33. L. W. Niedrach and A. D. Taevebaugh, J. Amer. Chem. Soc., 73, 2835 (1951).

34. G. J. Alkire, H. W. 13481-IOF-2 (1949), 1-36, 7-11.

35. J. I. Itzkovich and A. J. Page, Can. J. Chem., 46, 2743-2747 (1968).

36. J. A. Page and G. Wilkinson, J. Amer. Chem. Soc., 74, 6149 (1952).

37. M. Branica and S. Mesarić, Croat. Chem. Acta, 30, 89 (1958).

38. S. Mesarić and M. Branica, Proceedings of the 2nd United Nations International Conference on Peaceful Uses Atomic Energy, Geneva, 1958, Vol. 28, 1959, p. 81.

39. D. A. Wagnerowá, Collect. Czech. Chem. Commun., 27, 1130 (1962).

40. S. Mesarić and M. Branica, Croat. Chem. Acta, 30, 81 (1958).

41. D. T. Sawyer, R. J. Kula, and A. L. Budd, J. Electroanal. Chem., 4, 242 (1962).

42. D. T. Sawyer, R. S. Georg, and J. B. Bagger, J. Amer. Chem. Soc., 81, 5893 (1959).

43. N. K. Pshenitsyn and N. A. Ezerskaya, Zh. Neorg. Khim., 5, 1068 (1960).

44. N. K. Pshenitsyn and N. A. Ezerskaya, Zh. Anal. Khim., 16, 196 (1961).

45. N. K. Pshenitsyn and N. A. Ezerskaya, Zh. Neorg. Khim., 6, 613 (1961).

46. M. A. Guebely and M. Haissinsky, J. Chim. Phys., 51, 290-295 (1954).

47. P. Wehner and J. C. Hindman, J. Amer. Chem. Soc., 72, 3911-3918 (1950).

48. M. A. El Guebely, Anal. Chim. Acta, 15, 580-584 (1956).

49. P. Wehner and J. C. Hindman, ANL-4378 (1949); H. Cady, NCRL-3757, 1-81 (1957).

50. N. K. Pshenitsyn and N. A. Ezerskaya, Teoriya i Prakt. Prolyarogr. Analiza, Akad. Nauk, Muldavsk. SSSR, 1962, p. 145.

51. N. K. Pshenitsyn and N. A. Ezerskaya, Vops. Analiza Blagorodn. Metal Akad. Nauk SSSR, 1963, 76-80.

52. A. N. Frumkin, O. A. Petrii, A. Kossaya, and V. Topolev, J. Electroanal. Chem., 16, 175-191 (1968).

53. A. I. Levin, Teoreticheskoie osnovi Elektrokhim, Sverdlovsk, 1959, p. 87.

54. L. I. Antropov, Usp. Khim., 25, 1043 (1956).

55. R. M. Vasenin, Zh. Fiz. Khim., 28, 1672 (1954).

56. H. B. Michaelson, J. Appl. Phys., 21, 536 (1950).

57. N. E. Khomuntov, Zh. Fiz. Khim., 39, 532-533 (1965).

58. J. P. Hoar, G. W. Castellan, and S. Schuldinger, J. Phys. Chem., 62, 1141 (1958).

59. D. S. Gnanamuthu and J. V. Petrocelli, J. Electrochem. Soc., 114, 1036-1041 (1967).

60. L. N. Nekrasov and E. J. Kruscheva, Elektrokhimiya, 3, 166-172 (1967).

61. J. Llopis and M. Vázquez, Electrochim. Acta, 11, 633-640 (1965).

62. J. Llopis, I. M. Tordesillas, and J. M. Alfayate, ibid., 11, 625-626 (1965).

63. T. N. Stoyanovskaya, G. P. Khomchenko, and G. D. Vovchenko, Vestn. Mosk. Univ., Ser. II, 17(5), 30-33 (1962).

64. O. A. Petrii and V. E. Kazarinov, Electrokhimiya, 1, 1389-1391 (1965).

65. J. Llopis, Anal. Real Soc. Espan. Fis. Quim., Ser. B, 61, 355-373 (1965).

66. Y. Kobayashi, H. Yamatera, and H. Okuno, Bull. Chem. Soc. Japan, 38, 1911-1915 (1965).

67. A. A. Goryunov and A. N. Ryabov, Zh. Neorg. Khim., 10, 2596-2601 (1965).

68. J. Llopis, J. M. Gamboa, and J. M. Alfayate, Electrochim. Acta, 12, 57-65 (1967).

69. T. N. Stoyanovskaya, G. P. Khomchenko, and G. D. Vovchenko, Vestn. Mosk. Univ., Ser. II, 18, 20 (1963).

70. G. P. Khomchenko, T. N. Stoyanovskaya, and G. D. Vovchenko, Zh. Fiz. Khim., 38, 434 (1964).

71. T. N. Stoyanovskaya, G. P. Khomchenko, A. I. Pletyushkina, and G. D. Vovchenko, Vestn. Mosk. Univ., Ser. II, 18, 6, 50 (1963).

72. R. Backhouse and F. P. Dwyer, J. Proc. Roy. Soc. N. S. W., 83, 146-149 (1949).

73. T. Kuwana, D. E. Bublitz, and G. Hoh, J. Amer. Chem. Soc., 82, 5811-5817 (1960).

74. F. Brito and D. Lewis, Proceedings of the 9th International Conference on Coordination Chemistry, St. Moritz Bad, 1966, p. 340.

75. N. K. Pshenitsyn and G. I. Ginzburg, Zh. Neorg. Khim., 2, 112-120 (1957).

76. D. K. Atwood, Dissertation Purdue University, Lafayette, Indiana, 1961.

77. V. M. Vdovenko, L. N. Lazarev, and Ya. S. Khvorostin, Sov. Radiochem., 7, 228-234 (1965).

78. F. Pantani, J. Less-Common Metals, 4, 116-123 (1962).

79. G. A. Rechnitz, Inorg. Chem., 1, 950-954 (1962).

80. G. A. Rechnitz and H. A. Catherin, ibid., 4, 112-114 (1965).

81. O. E. Zvyagintsev, N. I. Kolhin, A. N. Ryabov, T. D. Avtokratova, and A. A. Gorgunov, Khimiya Ruteniya, Nauka, Moscow, 1965.

82. O. E. Ziyagintsev and A. Kurbanov, Zh. Neorg. Khim., 3, 2662 (1958).

83. J. C. Griess, ORNL-795 (1950), 1-56; N.S.A. 11 (1957), No. 8283.

84. A. I. Levin and B. A. Pankratov, Zh. Prikl. Khim., 32, 1787-1793 (1959).

85. K. T. Faler and D. R. Anderson, IDO-14536 (1961), 1-27, 22; N.S.A. 15 (1961), No. 14334

86. B. E. Conway, Electrochemische Tabellen, Frankfurt, 1957, p. 333; Electrochemical Data, Elsevier, Amsterdam, 1952, p. 333.

87. O. E. Zvyagintsev and A. Kurbanov, Zh. Neorg. Khim., 6, 2216-2218 (1961).

88. F. H. Reid and J. C. Blake, Trans. Inst. Metal Finish., 38, 45 (1961).

89. M. Haissinsky and M. A. El Guebely, J. Chim. Phys., 53, 744-752 (1956).

90. J. C. Griess, J. Electrochem. Soc., 100, 429-433 (1953).

91. R. U. Volterra, U.S. Patent 2,600,175 (1952), Metals and Controls Corp.

92. F. H. Reid, Electrodeposition of Pt-Group Metals, Met. Rev., 8, 167-211 (1963).

93. M. H. Lietzke and J. C. Griess, J. Electrochem. Soc., 100, 434-441 (1953).

94. A. C. Littlejohn, Atomic Energy Research Establishment, C/r 1892, (1958), 4 pp.

95. R. H. Atkinson, Trans. Faraday Soc., 26, 490-496 (1930).

96. R. H. Roda, Plating, 49, 69-71 (1962).

97. G. W. Jernstedt, U.S. Patents 2,451,340 and 2,451,341 (1948), Westinghouse Electric Corp.

98. E. Raub and G. Buss, Z. Elektrochem., 46, 195 (1940).

99. K. K. Kelley and E. G. King, U.S. Bur. Mines Bull. 592, (1961).

100. T. P. J. H. Babeliowsky, Physica, 28, 1160 (1962).

101. L. H. Dreger and J. L. Margrave, J. Phys. Chem., 65, 2106 (1961).

102. R. F. Hampson and R. F. Walker, J. Res. Natl. Bur. Stand., 65A, 289 (1961).

103. M. B. Panish and L. Reif, J. Chem. Phys., 34, 1915 (1961).

104. M. G. Schmahl and E. Minzl, Z. Phys. Chem. (Frankfurt), 41, 78 (1964).

105. L. Wöhler and N. Jochum, Z. Phys. Chem. (Leipzig), A167, 169 (1933).

106. J. S. Forrester and G. H. Ayres, J. Phys. Chem., 63, 1979 (1959).

107. W. Plumb and G. M. Harris, Inorg. Chem., 3, 542 (1964).

108. D. Cossi and F. Pantani, J. Inorg. Nucl. Chem., 8, 385 (1958).

109. G. Gire, Ann. Chim., 4, 370 (1925).

110. W. E. Bell, M. Tagami, and R. E. Inyard, J. Phys. Chem., 70, 2084 (1966).

111. F. Puche, Ann. Chim., 9, 233 (1938).

112. G. Gire and F. Puche, C.R. Acad. Sci., Paris, 200, 670 (1935).

113. G. Grube and B. Gu, Z. Elektrochem., 43, 397 (1937).

114. G. Grube and H. Autenrieth, ibid., 44, 296 (1938).

115. F. P. Dwyer and H. N. Schafer, J. Proc. Roy. Soc. N. S. W., 82, 294 (1948).
116. J. Amose, M. Ruband, and M. J. Barbier, C. R. Acad. Sci., Paris, C, 273, 1708-1710 (1971).
117. J. P. Hoare, J. Electrochem. Soc., 111, 232 (1964).
118. J. B. Willis, J. Amer. Chem. Soc., 66, 1067 (1944).
119. L. Bremen, Chem. Rev., 52 (1953).
120. M. Pourbaix and J. Van Muylder, Rapports Techniques, No. 59, Cebelcor, 1958, p. 2.
121. E. Rhodas, Chim. Ind. (Milan), 75(6), (1956).
122. G. Milazzo, Electrochemistry, Theoretical Principles and Practical Applications, Elsevier, Amsterdam, 1963.
123. H. A. Laitinen and C. H. Liu, J. Amer. Chem. Soc., 80, 1013 (1958).
124. H. A. Laitinen and J. A. Plambeck, ibid., 87, 1202 (1965).
125. J. A. Plambeck. J. Chem. Eng. Data, 12, 77, 82 (1967).
126. C. H. Liu, J. Phys. Chem., 66, 164 (1962).
127. K. E. Johnson and H. A. Laitinen, J. Electrochem. Soc., 110, 314 (1963).
128. G. Van Loon and J. A. Page, Talanta, 12, 227-235 (1965).
129. G. Semerano and E. Gagliardo, Ric. Sci., 20, 1676-1684 (1950).
130. P. W. West, J. F. Dean, and E. J. Breda, Collect. Czech. Chem. Commun., 13, 1 (1948).
131. F. Pantani, J. Electroanal. Chem., 5, 40 (1963).
132. V. I. Kravtsov and I. I. Shereshevskaya, Elektrokhimiya, 7, 1677-1680 (1971).
133. D. Cozzi and F. Pantani, J. Electroanal. Chem., 2, 72 (1961).
134. N. A. Ezerskaya and V. N. Filiomonova, Zh. Anal. Khim., 17, 972-978 (1962).
135. P. Beran, Chem. Zvesti, 14, 735-742 (1960).
136. F. Pantani and P. G. Desideri, Talanta, 5, 69 (1960).
137. P. Beran and J. Dolezal, Collect. Czech. Chem. Commun., 23, 608 (1958).
138. W. H. Douglas and R. J. Magee, J. Electroanal. Chem., 5, 171 (1963).
139. P. Beran and J. Dolezal, Chem. Listy, 51, 1289 (1957).
140. F. Pantani, Ric. Sci., 33(II-A), 217 (1963).
141. F. Pantani, ibid., 34(II-A), 41 (1964).
142. D. M. Wagnerova, Collect. Czech. Chem. Commun., 26, 2076 (1961).
143. D. Cozzi and F. Pantani, J. Electroanal. Chem., 2, 230 (1961).
144. M. B. Bardin and V. I. Shapiro, Uch. Zap. Kishinev. Gos. Univ., 68, 64-70 (1963).
145. V. I. Kratsov, E. G. Taventarny, and N. I. Lyzlov, Elektrokhimiya, 8, 941 (1972).
146. T. P. Birintseva and B. N. Kabanov, Zh. Fiz. Khim., 37, 2600 (1963).
147. N. A. Balashova, N. T. Govokhova, and M. I. Kulezneva, Elektrokhimiya, 4, 871 (1968).
148. B. S. Krasikov, Zh. Prikl. Khim., 37, 2420 (1964).
149. A. N. Frumkin and O. A. Petrii, Electrochim. Acta, 15, 391 (1970).
150. O. A. Petrii, A. N. Frumkin, and Yu. Kotlob, Elektrokhimiya, 4, 871 (1968).
151. A. N. Frumkin and O. A. Petrii, ibid., 5, 476 (1969).

152. M. Breiter, K. Kammermaier, and C. A. Knorr, Z. Elektrochem., 60, 119-127 (1956).

153. M. Breiter, K. Kammermaier, and C. A. Knorr, ibid., 60, 452-462 (1952).

154. J. Llopis and M. Vázquez, Electrochim. Acta, 9, 1655 (1964).

155. J. P. Hoare, ibid., 13, 417 (1968).

156. W. Böld and M. Breiter, ibid., 5, 169 (1961).

157. F. G. Will and C. A. Knorr, Z. Elektrochem., 64, 270-275 (1960).

158. M. Breiter, Electrochim. Acta, 7, 25 (1962).

159. G. P. Khomchenko, A. F. Lunev, and K. N. Bogdanovskaya, Elektrokhimiya, 1, 1352-1355 (1965).

160. Yu. M. Tyurin, Dokl. Akad. Nauk SSSR, 126, 827 (1959).

161. Yu. M. Tyurin and A. M. Tsybulerskaya, ibid., 159, 1140 (1964).

162. Yu. M. Tyurin and A. M. Kossaya, Elektrokhimiya, 4(1), 103-107 (1968).

163. M. W. Breiter, C. A. Knorr, and R. Meggle, Z. Elektrochem., 59, 153 (1955).

164. E. I. Khruscheva, N. A. Shumilova, and M. R. Tarasevich, Sov. Electrochem., 2, 258-262 (1966).

165. J. P. Hoare, Electrochim. Acta, 11, 203 (1966).

166. J. P. Hoare, Adv. Electrochem. Electrochem. Eng., 6, 201 (1966).

167. Sh. Shibata and T. Mitsui, Yamagata Daigaku Kiyo, Shizen Kagaku, 6(4), 423-431 (1967).

168. S. Shibata, Bull. Chem. Soc. Japan, 38(8), 1330-1337 (1965).

169. T. M. Grishina, G. P. Khomchenko, and G. S. Vovchenko, Vest. Mosc. Univ., Ser. II, 17(2), 53-56 (1962).

170. J. P. Hoare, The Electrochemistry of Oxygen, Wiley-Interscience, New York, 1968.

171. R. Kh. Burshtein, M. R. Tarasevich, and A. Radyushkinak, Elektrokhimiya, 6, 1611 (1970).

172. M. R. Tarasevich, A. Radyushkinak, and R. Kh. Burshtein, Sov. Electrochem., 3, 399-403 (1967).

173. I. A. Balashova, I. T. Gorokhova, and Kh. A. Davydova, Elektrokhimiya, 6, 436 (1970).

174. V. V. Topolev and O. A. Petrii, ibid., 6, 1726 (1970).

175. V. I. Kravtsov and B. V. Simakov, Vestn. Leningrad. Univ., Ser. Fiz. Khim., 20(4), 103-105 (1965).

176. G. L. Padyukova, A. B. Fasman, and I. V. Khizhnyark, Elektrokhimiya, 4(2), 191-196 (1968).

177. I. I. Kartaradze and I. P. Sokolova, Zh. Fiz. Khim., 42, 1286 (1968).

178. S. Gilman, J. Phys. Chem., 71, 4339-4343 (1967).

179. S. I. Ginzburg, N. N. Chalisova, and B. S. Krasikov, Zh. Neorg. Khim., 11, 1384 (1966).

180. J. Llopis, I. M. Tordesillas, and M. Muñiz, Electrochim. Acta, 10, 1045 (1965).

181. B. S. Krasikov and A. M. Milenin, Korrosiya Metal i Issled. v Obl. Elektrokhimiya, Akademii Nauk SSSR, Moscow, 1965, p. 87.

182. E. P. Kovsman, Yu. M. Tyurin, and E. A. Karavaeva, Zh. Prikl. Khim., 37, 217 (1964).

183. N. V. Sidgwick, The Chemical Elements and Their Compounds, Oxford University Press, London, 1950, p. 1581.

184. J. A. V. Butler and G. Drever, Trans. Faraday Soc., 32, 427 (1936).

185. J. Llopis, Catal. Rev., 2(2), 161-220 (1968).

186. N. V. Pospelova, A. A. Rakov, and V. I. Veselovsky, Elektrokhimiya, 6, 57 (1970).

187. N. V. Pospelova, A. A. Rakov, and V. I. Veselovsky, ibid., 6, 722 (1970).

188. T. A. Gortalova, S. I. Berezina, and G. S. Hemskii, ibid., 4(12), 1473 (1968).

189. J. P. Hoare and S. Schuldiner, J. Chem. Phys., 25, 786 (1956).

190. G. Grube and K. H. Mayer, Z. Elektrochem., 43, 404 (1937).

191. E. H. Laister, Metal Ind., 85, 427, 469 (1954).

192. K. Schumpelt, in Electroplating, (W. Blum and W. R. Mayer, eds.), Wiley, New York (1953).

193. H. Benninghoff, Galvanotechnic, 57, 451-463 (1966).

194. K. Schumpelt, ibid., 37, 1052 (1950).

195. H. J. Wiesner, Proc. Amer. Electroplaters' Soc., 39, 79 (1952).

196. E. A. Parker, Plating, 42, 882 (1955).

197. J. Fischer and F. Leonhard, Metall, 10, 608 (1956).

198. H. J. Wiesner and H. A. Meers, Plating, 43, 347 (1956).

199. F. H. Reid, Trans. Inst. Metal Finish., 33, 105 (1955-1956).

200. German Patent 692,122 (1940).

201. R. Piontelli, Chim. Ind. (Milan), 22, 61 (1940).

202. British Patent 856,867 (1960).

203. J. C. Withers, W. A. D. D. Tech. Rep. No. 60-718 (1961).

204. U. S. Patents 2,895,889 and 2,895,890 (1959).

205. F. H. Reid, Trans. Inst. Metal Finish., 36, 74 (1958-1959).

GENERAL BIBLIOGRAPHY

G. Charlot, A. Collumeau, and M. J. C. Marchot, Selected Constants. Oxidation-Reduction Potentials of Inorganic Substances in Aqueous Solution, I.U.P.A.C., Butterworths, London, 1971.

Gmelins Handbuch. Anorganischem Chemie, 8th ed., System No. 63, Ruthenium, Verlag Chemie, Weinheim, 1970.

Gmelins Handbuch. Anorganischem Chemie, 8th ed., System No. 64, Rhodium, Verlag Chemie, Berlin, 1938.

R. N. Goldberg and L. G. Hepler, "Thermochemistry and Oxidation Potentials of the Platinum Group Metals and Their Compounds," Chem. Rev., 68, 229 (1968).

M. W. Latimer, Oxidation Potentials, 2nd ed., Prentice-Hall, Englewood Cliffs, New Jersey, 1952.

J. Llopis, "Corrosion of Platinum Metals and Chemisorption," Catal. Rev., 2(2), 161-220 (1968).

G. Milazzo, Electrochemistry. Theoretical Principles and Practical Applications, Amsterdam, Elsevier, 1963.

M. Pourbaix, N. de Zoubov, and J. Van Muylder, Atlas d'Equilibres Electrochimiques a 25°C, Gauthier-Villars, Paris, 1963.

F. H. Reid, "Electrodeposition of the Platinum Group Metals," Metall. Rev., 8(30), 1963.

O. E. Zvyagintsev, N. I. Kolhin, A. N. Ryabov, T. D. Avtokratova, and A. A. Goryunov, Khimiya Ruteniya, Inst. Obsch. Neorg. Khim. N. S. Kurnakov. Akad. Nauk. URSS. Izd. Nauka, Moscow, 1965.

SUBJECT INDEX

A

B

C

D

E

R

S